LEGUMES

Their Production, Improvement and Protection

LEGUMES

Their Production, Improvement and Protection

Edited by

M.PRAKASH & S.MURUGAN

Department of Agricultural Botany
Faculty of Agriculture, Annamalai University
Tamil Nadu, India

SSPH

2009

SATISH SERIAL PUBLISHING HOUSE

403, Express Tower, Commercial Complex, Azadpur, Delhi-110033 (India)
Phone : 011-27672852, Fax : 91-11-27672046
E-mail :info@satishserial.com, hkjain1975@yahoo.com
Website : www.satishserial.com

LEGUMES : Their Production, Improvement and Protection
(Proceedings of National Seminar on Advances in Legumes Research, March 6th and 7th, 2008, Faculty of Agriculture, Annamalai University, Tamil Nadu, India)

Published by :

SATISH SERIAL PUBLISHING HOUSE
403, Express Tower, Commercial Complex, Azadpur, Delhi-110033 (INDIA)
Phone : 011-27672852 Fax : 91-11-27672046
E-mail : info@satishserial.com, hkjain1975@yahoo.com

ISBN 81-89304-70-4

Typeset at: Typographiya, Delhi # 98 11 48 23 28
Printed at: Salasar Imaging Systems, Delhi-35

Preface

Legumes are a large, diverse family ranging from herbaceous annuals to woody perennials that, because of their capacity to fix nitrogen, are essential components in natural and managed terrestrial ecosystems. Legumes have been domesticated for the production of food, feed, forage, fiber, industrial and medicinal compounds, flowers, and other end uses. The proceedings covers a wide-ranging survey of all the major legumes from their evolution to their potential for further development and improvement as economically important food crops. The contents of the book have been divided into five themes viz., Crop Improvement, Crop Production, Physiological Bio-Chemical and Bio-Technological advances, Crop Protection and Policies and Programmes. It covers the changes that have occurred in the course of plant domestication leading to our present pulse crops and oilseed legume crops. It also includes reviews on the benefits to be gained from evaluation and improvement of grain legume genetic resources. This book will serve as a suitable study and reference material for postgraduate students and researchers.

Editors

Contents

Preface v

SECTION-I : CROP IMPROVEMENT

Chapter-1 : Scenario of Legumes in India 1
Muralidharan, V.

Chapter-2 : Crop Improvement Strategies in Pulses Improvement 9
Raveendran, T.S. and P. Jayamani

Chapter-3 : Transgenic Approach for Pulses Improvement........... 33
Nadarajan. N and P. Jayamani

Chapter-4 : Genetic Analysis for Physiological Traits in Mungbean (*Vigna radiata* L. Wilczek) 45
Sunil Kumar, B., Murugan, S. and M. Prakash

Chapter-5 : Enhancement of Gamma Rays Mutagenicity by Ethidium Bromide and Ethyl Methane – Sulphonate in Cowpea 51
Thirugnana Kumar, S., S. Thirumal Raj, G. Kanagaraj, R. Balamurugan and P.A. Velusamy

Chapter-6 : VRI (Gn) 6 – New High Yielding Spanish Bunch Groundnut 61
Viswanathan, P.L., P. Vindhiyavarman, G. Nallathambi, D. Kumaresan, A. Mothilal, S. Chidamabaram, K. N. Ragumoorthi, K. Sachithanantham, K. Subramanian, K. Manian, M. Balusamy and V. Muralidharan

Chapter-7 : Heterosis for Yield and Yield Components in Greengram (*Vigna radiata* (L.) Wilczek) 65
Saravanan, K., T. Sabesan and M. Prakash

Chapter-8 : Combining Ability and Heterosis for Yield and Yield Component Characters in Greengram (*Vigna radiata* (L.) Wilczek.) 71
Anitha Vasline, Y. and V. Suguna

Chapter-9 : Interspecific Hybridization in Vigna Species 77
Ramamoorthi, N., M. Pandiyan, SK.Ganesh and S. Jebaraj

Chapter-10: Genetic Analysis for Yield Attributes in Vegetable Cowpea (*Vigna unguiculata* (L.) Walp.) 89
Manivannan, R., Dr. K. Sekar and Dr. Mohan Andre Savery

Chapter-11 : Future of Pulse Breeding in the Post PPV & FRA, 2001 Era 103
Sabesan, T.

Chapter-12 : Wide Hybridization in *Vigna species* 109
Pandiyan. M , N.Ramamoorthi, S.K.Ganesh and S. Jebaraj

Chapter-13 : Association and Path Analysis for Earliness Breeding in Cow Pea (*Vigna unguicul*ata (L.) Walp). 121
Anandan, A. and R. Eswaran

Chapter-14 : Spectrum of Induced Chlorophyll Mutations in Cowpea (*Vigna unguiculata* L. Walp.) 125
Kanagaraj, G., S. Thirugnana Kumar, S. Thirumal Raj, R. Balamurugan and P.A. Velusamy

SECTION-II : CROP PRODUCTION

Chapter-15 : Pulse Production Technology 135
Ravichandran, M.

Chapter-16 : Legumes in Sustainable Cotton Farming 149
Sankaranarayanan, K., C.S. Praharaj and N.Gopalakrishnan

Chapter-17 : Effect of Grain Cowpea Inter Cropping in Castor 167
Thanunathan, K., K.Muthulingam, M.Thiruppathi and M.Ravichandran

Chapter-18 : The Response of Seed Enhancement Techniques to Mid Storage Correction in Groundnut (*Arachis hypogaea* (L.)) Cv Vri 2 Kernels 171
Sathiya Narayanan, G., M. Prakash, J. Gokulakrishnan and B. Sunil Kumar

Chapter-19 : Pattern of Pea Seed (*Pisum sativum (L.))* Deterioration 175
Paramasivam,V., S.Sundareswaran, R.Vaidyanathan and D.Ananda Nayaki

Chapter-20 : Effect of Integrated Nutrient Management on the Performance of *Dolichos Lablab Var. Typicus* in Lignite Mine Spoil 179
Karuppaiah, P. and K. Manivannan

Chapter-21 : Chemical Impregnation Treatment for Greengram (*Vigna radiata* (L.) Wilczek.) 191
Gnanasekaran, J., S. Padmavathi, G. Sathiyanarayanan and M. Prakash

Chapter-22 : Influence of Mother Crop Nutrition on Seed Yield and Quality of Blackgram (*Vigna ungiculata* (L.)) 195
Manonmani, V· and P. Srimathi

Chapter-23 : Studies on the Production Potentiality of Blackgram Varieties Under Rice Fallow Condition 201
Shobana, R., P. Thangavel and V. Imayavaramban

Chapter-24 : Effect of Different Levels of Phosphorus and Potassium on Growth, Flowering and Yield of Vegetable Cowpea Cv Tux. 944 205
Anuja, S., K.Ilavarasi and S. Archana

SECTION-III : PHYSIOLOGICAL BIO-CHEMICAL AND BIO-TECHNOLOGICAL ADVANCES

Chapter-25 : Studies on the Management of Differentially Aged Seed and their Progenies in Cowpea (*Vigna sinensis* (L.) Savi) Cv.Co 6 by Foliar Spray of Nutrients and Hormones 209
Sasikala, K.

Chapter-26 : Studies on the Effective Role of Biofertilizers on the Germination, Growth and Yield of Greengram (*Vigna radiata* (L.)) 217
Rajasekaran,S., P. Sundaramoorthy, K. Sankar Ganesh, K. Sumathi and I. Baskaran

Chapter-27 : Physiological Aspects of Grain Yield in Soybean 223
Vijayaraghavan, H., S. Vincent and K. Balakrishnan

Chapter-28 : Physiological Constraints in Productivity of Grain Legumes 235
Balakrishnan K., R. Balasubramanian and H.Vijayaraghavan

Chapter-29 : Studies on the Effect of 28 - Homobrassi-nolide on Biochemical Parameters and Yield of Groundnut Cv. Vri 2. 251
Prakash, M., S. Suganthi, J. Gokulakrishnan, B. Sunil Kumar and T. Sabesan

Chapter-30 : Groundnut Research on Drought Tolerance or Wue at NRCG: From Breeding to Farmer's Field 257
Nautiyal, P.C., M.S. Basu, R. K. Mathur, Rukam Singh and T. Radhakrishnan

Chapter-31: Field Screening of Greengram (*Vigna radiata* (l.) Wilczek) Genotypes to Drought Tolerance 289
Vincent, S., H. Vijayaragavan, and K. Balakrishnan

Chapter-32 : Screening of Guar Genotypes for Drought Tolerance Traits Under Simulated Moisture Stress Conditions ... 303
Chakraborty, A., P. Sudhakar, A. Siva Sankar, P. Latha and B.G. Singh

Chapter-33 : Effect of Bio-Regulators on the Yield of Greengram ... 309
Ananthi K., Mallika Vanagamudi, M. Gomathy, K. Krishna Surendar, S. Srimathi and T. Sivashankaridevi

Chapter-34 : A Step-wise Protocol for Standardization of *In Vitro* Pollen Germination Medium for Pigeonpea and Its Wild pecies 317
Jayaprakash, P.

SECTION-IV : CROP PROTECTION

Chapter-35 : Effect of Age of Plants and Standardisation of Inoculum Level of *Macrophomina phaseolina* Causing Root Rot of Blackgram 325
Rettinassababady. C. and C. Issac Sunil

Chapter-36 : Field Screening of Cowpea Accessions for Resistance Against Selected Insect Pests 331
Sajad Hussain Mir and V. Selvanarayanan

Chapter-37 : Study on Efficacy of Neem Based Products Against Pulse Beetle During Storage of Mung Bean (*Vigna radiata* (L.) Wilczek) Seeds 339
Ramanadane, T., K. Paramasivam and C. Rettinassababady

Chapter-38 : Management of Root Rot of *Vigna mungo* L. Using Biocontrol Agents Multiplied in Vermicompost 343
Saravanan, T., R. Ramamoorthy and K. Subbiah

SECTION-V : POLICIES AND PROGRAMMES

Chapter-39 : Legume Sector in India: An Analysis of Issues, Potentials and Strategies 351
Suresh Kumar, D., S. Gurunathan and M. Chandrasekaran

Chapter-40 : Recent Trends in Pulses in Tamil Nadu – An Overview 369
Subramanian, A and S. Murugan

Chapter-41 : Problems Faced by Women Agricultural Labourers in Groundnut Farming Operations 377
Bharathiraja, P. And Santha Govind

SECTION - I
CROP IMPROVEMENT

Chapter 1

Scenario of Legumes in India

MURALIDHARAN, V.
National Research Centre for Groundnut, Junagadh - 362 001 (Gujarat)

Indian agriculture most continuously evolve to remain ever responsive to manage the change and to meet the growing and diversified needs of different stakeholders in the entire production to consumption chain.

1. Pulses

India is the largest producer of pulses, accounting for about 25 percent of the global share. Being an inseparable ingredient in the diet of the vast majority of population and mainstay of sustainable crop production, pulses continue to be an important component of the rainfed agriculture since time immemorial. About a dozen of pulse crops, namely chickpea, pigeonpea, mungbean, urdbean, lentil, field pea, lathyrus, cowpea, common bean, moth bean, horse gram, and rice bean are cultivated on 22.47 million ha area under varied agro-ecological conditions.

1.1 Indian Scenario

Pulse production in the country has fluctuated widely between 11 and 15 million tonnes with no significant growth trend between 1991 and 2005. The latest estimates indicate that the present production of pulses is 13.11 million tonnes with productivity of 595 kg/ha. Stagnant growth in pulses production as compared to population growth rate of 1.44% has led to progressive decline in per capita availability of pulses (41.6 g in 1991 to 34 g in 2005). To meet the domestic demand, the country has to import pulses to the tune of 1.5-2.0 million tonnes. The projected pulse requirement for the year 2025 is 29.44 million tonnes which necessitates annual growth rate of 4.02% in production. To meet the projected requirement, the productivity needs to uplift at 1178 kg per ha and about 2.5 million ha additional area has to be brought under pulses.

The estimate for 2006-07 showed total pulse production of 14.23 million tonnes from 22.47 million ha area. The major pulse producing states are Madhya Pradesh, Mahrashtra, Rajasthan, Uttar Pradesh and Andhra Pradesh, which together account for 74% of production and 70% of the area. The state-wise and crop-wise trends in area, production and yield under different pulses are given in during 1991-2005, area under pulses fluctuated between 20.35 and 24.66 million ha while production and productivity varied between 11.08 and 14.94 million tonnes, and 533 and 637 kg/ha, respectively.

National pulses scenario (based on mean of the last five years 2001-05)

Crop	Area (m ha)	Production (mt)	Yield (kg/ha)	Major states
Chickpea	6.26	5.00	798	M.P., U.P., Rajasthan, Maharashtra, A.P
Pigeonpea	3.47	2.32	670	Maharashtra, U.P., M.P., Karnataka, A.P
Lentil	1.43	0.95	664	U.P., M.P, Bihar, W.B, Rajasthan
Dry peas	0.68	0.62	911	U.P., M.P., Rajasthan, Orissa, Bihar
Urdbean	3.31	1.43	432	A.P., Maharashtra, U.P., M.P., TN
Mungbean	3.19	1.18	371	Maharashtra, Rajasthan, A.P., Bihar, Karnataka
Total Pulses	21.75	12.78	588	M.P., U.P., Rajasthan, Maharashtra, A.P

The oilseed production, which used to be only 5 million tonnes in 1950s, has increased to 24 million tonnes in 1990s and since then hovering around 20 million tonnes. The country is experiencing deficit in edible oil availability and importing almost 40 to 50% of its requirement draining foreign exchange worth Rs.9500 to 10,000 crores.

1.2 Global Scenario

Pulses are the second most important group of crops after cereals. In 2005, the global pulses production was 61.72 million tonnes from an area of

73.2 million ha with an average yield of 843 kg/ha. Dry beans contributed about 31% to global pulses production followed by dry peas (18.3%), chickpea (14.9%), broad beans (7.5%), lentils (6.5%), cowpea (6%) and pigeonpa (5.3%). In the triennium (2003-05) average of 60.01 million tonnes production shows a positive annual growth of 0.72% per annum over 53.94 million tonnes recorded in 1991-93. Comparative data of the eighties reveal phenomenal annual growth of 2.85% mainly attributed to positive growth of 0.87% in area and 1.83% in productivity. Developing countries contribute about 74% to the global pulses production and the remaining comes from developed countries. India, China, Brazil, Canada and Myanmar are the major pulse producing countries with relative share of 25%, 10%, 5%, 5% and 4%, respectively. Countries recording annual production growth of more than 4% are Myanmar (11.48%), Canada (10.80%), Germany (8.27%), Sudan (8.08%), Spain (7.37%), Ethiopia (4.92%), China (4.67%) and Syria (4.12%). Crop-wise growth recorded in major countries is

Global scenario of pulses (based on mean of the last five years (2001-05)

Crop	Area (mha)	Production (mt)	*Yield* (kg/ha)	Major countries
Chickpea	10.34	8.02	776	India, Turkey, Pakistan, Iran, Mexico, Canada
Pigeonpea	4.49	3.13	697	India, Myanmar, Uganda, Malawi, Tanzania, Nepal
Lentil	3.87	3.42	885	India, Canada, Turkey, Australia, Nepal, Bangladesh
Dry peas	6.28	10.84	1727	Canada, France, Russia, India, Ukraine, Australia
Dry beans	26.18	18.45	704	Brazil, India, China, Myanmar, Mexico, USA
Total pulses	71.12	58.24	819	India, China, Canada, Brazil, Myanmar, Nigeria

Crop-wise importing and exporting countries for major pulse crops

Crop	Importer	Exporter
Dry bean	India (10), USA (7), Japan (6), UK (5), Mexico (5), Italy (4), Pakistan (4), Brazil (4), Cuba (3), Netherlands (3) Venezuela (3)	Myanmar (33), China (23), USA (10), Canada (8), Argentina (7), UK (2)

Contd...

Dry peas	India (33), Bangladesh (10), Belgium (8), Spain (8), Netherlands (4), China (4), Italy (4), Pakistan (3)	France (30), Canada (25), Australia (14), Ukraine (7), Russia (5), USA (3), Germany (3), Denmark (3), UK (3)
Lentil	Sri Lanka (10), Egypt (9), Pakistan (6), India (6), Colombia (6), Bangladesh (6), Algeria (6), Spain (6)	Canada (34), Australia (24), Turkey (12), USA (10), India (9), China (2).
Chickpea	India (25), Pakistan (22), Spain (7), Bangladesh (7), Algeria (4), Saudi Arabia (3), Italy (3)	Mexico (19), Iran (19), Canada (15), Turkey (14), Australia (13), Ethiopia (7), USA (3), Tanzania (3)
Pigeonpea	USA, UK, Kuwait, Singapore, Saudi Arabia, Malaysia	Myanmar (90), Tanzania (4), Canada (1), Mozambique (1)
Total Pulses	India (16.7), Spain (10.7), Egypt (4.9), Italy (4.8), Bangladesh (4.8), Belgium (4.6), Netherlands (3.0), Pakistan (3.0), USA (2.9), Cuba (2.5), UK (1.9), China (1.9)	Canada (26.2), Myanmar (10.2), China (8.9), France (8.7), Australia (8.3), USA (6.3), UK (4.7), Turkey (3.6), India (2.9), Argentina (2.1), Ukraine (2.1), Syria (1.3)

Figures in parentheses show percentage share of global import/export

2. Groundnut

Groundnut (*Arachis hypogaea* L.) is an important oilseed crop of India and has all the potential of becoming an important ancillary food crop too. It is one of the few crops that can be cultivated even on marginal lands under low input conditions. However, the average productivity of groundnut in India is quite low (1000kg/ha) compared to that of the USA (3000 kg/ha), China (2600 kg/ha), Argentina (2100 kg/ha) and Indonesia (1550 kg/ha). The vegetable oils form an essential ingredient of Indian cookery. In India, oilseeds are the second largest commodity after cereals in acreage, production and value

The demand of edible oils in the country has been increasing over years due to changing food habits, enhanced purchasing capacity and growing population. Thus, the urgency of increasing domestic production of edible oils in general and groundnut in particular is more than that of increasing food grain production. On the one hand, the genetic yield potential of groundnut yet remains to be fully exploited; on the other hand, considerable scope exists for enhancing tolerance of various biotic and abiotic stresses by employing either conventional breeding techniques or biotechnological tools. For expansion of area under groundnut, additional/new niches for groundnut in the eastern and the north-eastern regions can be developed by introducing groundnut to fit in the existing cropping systems. As groundnut is by and large cultivated under rain-fed conditions, improvement in yield can be brought about by enhancing drought tolerance as well as by adopting improved water management practices. Contamination of kernels by aflatoxins, which lowers the quality of the produce and many a time renders it unfit for export, can be minimized through exercising certain precautions while are harvesting and post-harvest handling of the produce.

2.1 Indian Scenario

India accounts for about 35% of the global area under this crop and about 30% the global production of groundnuts. Andhra Pradesh and Guajrat are the major groundnut producing states of India. Karnataka, Tamil Nadu and Maharashtra are the other major groundnut growing states. Rajasthan, Madhya Pradesh and Uttar Pradesh also contribute substantially. Groundnut is also grown in many other states.

Decadal annual compound growth rates (%) of groundnut in India (1950-2000)

Period	Area	Production	Productivity
1951-1960	17.21	21.60	3.72
1961-1970	4.77	0.17	-4.38
1971-1980	-1.07	1.41	2.51
1981-1990	5.86	14.58	8.22
1991-2000	-4.06	-1.98	6.64

3. Soya bean

World Oilseeds production was estimated about 387 million tonnes during the year 2005-06. Soybean is the most important oilseed crop in the world and constitutes about 42 and 56 per cent of area and production of total oilseeds (2003-04), respectively. India ranks fifth in the area and production in the world after USA, Brazil, Argentina and China. The contribution of India in world soybean area and production is about 7.8 and 4.2 per cent, respectively. In terms of productivity of soybean, India is at the level of only 53.5 per cent of the world average.

3.1 Indian Scenario

There has been unprecedented increase in area and production of soybean in India over a period of three and a half decades and no other oilseed crop has shown such tremendous increase. The area and production of soybean increased from 1.53 m ha and 0.89 mt in 1986-87 to 6.50 m ha and 7.85 mt in

2003-04 (accounting for 325% increase in area and 782% increase in production) respectively. However, the increase in productivity was rather low (108%). It has revolutionized rural economy and has uplifted socio-economic status of farmers in general and of Madhya Pradesh in particular.

World area, production and productivity of oilseeds and soybean

Year	*Oilseeds in world**			*Soybean in world***		
	Area (m/ha)	Production (mt)	Productivity (kg/ha)	Area (m ha)	Production (m t)	Productivity (kg/ha)
1997-98	177.31	286.79	1617	67.12	143.40	2137
1998-99	184.83	294.61	1593	70.98	160.09	2260
1999-00	188.55	304.33	1614	71.90	157.56	2190
2000-01	188.64	313.94	1664	74.10	161.23	2170
2001-02	188.58	324.90	1722	76.83	176.76	2300
2002-03	186.52	330.19	1770	78.83	180.91	2290
2003-04	199.42	335.22	1681	83.50	188.93	2260
2004-05	206.22	381.41	1849	91.44	204.26	2230

*USDA, Foreign Agricultural Service-www.fas.usda.gov/psdonline/psdgetreport
**Oilseeds Station: A statistical compendium 2002, DOR, Hyderbad and FAO

The commercial cultivation of soybean crop in India started in late sixties, thereafter, it made phenomenal growth, which has never been registered in any other crop in the country. The percent share of soybean in total oilseeds area and production increased from 0.19 to 27.62 and 0.14 to 28.77, respectively during 1970-2004.

Area, production and yield of soybean in India

Year	Area (m ha)	Production (m t)	Yield (kg/ha)	% share in oilseeds area	% share in oilseeds production
1970-71	0.032	0.014	426	00.19	00.14
1975-76	0.093	0.091	977	00.55	00.86
1980-81	0.610	0.440	728	03.46	04.69
1985-86	1.340	1.020	764	07.04	09.42
1990-91	2.560	2.600	1015	10.60	13.97
1995-96	5.030	5.090	1012	19.37	23.02
2000-01	6.420	5.270	822	27.61	28.64
2002-03	5.860	4.560	777	27.30	28.90
2003-04	6.500	7.860	1210	27.73	31.22
2004-05*	7.460	7.510	1007	27.62	28.77

Source: DOD, Hyderabad and ICAR Data Book 2006, (*Estimate)

In Indian scenario, Madhya Pradesh contributes about 67 and 56 per cent in total area and production of soybean and is called as 'Soya State'. Madhya Pradesh, Maharashtra and Rajasthan together contribute about 97 and 96 per cent to total area and production of soybean in the country, respectively.

At present about 80% of total soybean produced in the county goes to solvent extraction mills, 10% is retained seed and remaining 10% is utilized for different food uses such as soya milk, tofu, soya nuts, fortification of wheat flour, etc. Of the 80% going to solvent extraction plan, 18% is oil and 82% is de-oiled cake (DOC). Until last 5-6 years, entire DOC was being exported leading international market dependence. However, at present about 30% of DOC produced is utilized domestically for feed (about 80%) and food (about 20%). The policy to be conceived is to aim at de-linking of soybean farming and industry from foreign dependence. The progressive increase in domestic consumption of DOC provides silver lining in this direction.

4. Research Priorities for Achieving the Near Self-sufficiency

This requires a paradigm shift in research, technology generation and dissemination, and commercialization along with capacity building in frontier areas of research. The major R&D issues identified are low genetic yield potential, poor and unstable yield, huge post-harvest losses, inadequate adoption of improved technology and low profitability which need to be tackled through integration of conventional approaches with cutting edge technologies such as genomics, molecular marker assisted breeding, transgenics, molecular approaches for stress management, high input use efficiency, quality improvement, resource conservation technologies, value addition and food safety. Exploitation of heterosis and yield genes from wild relatives have also been identified as promising avenues for breaking yield plateaus.

REFERENCES

Vision 2025. Indian Institute of Pulses Research, ICAR, Kanpur, India, 2007, P:46. Ed. Dr. Masood Ali, Director.

Vision 2025. NRCG Perspective Plan-National Research Centre for Groundnut, ICAR, Junagadh, India. 2007. P: 50. Ed. Dr. M.S. Basu, Director.

Vision 2025. National Research Centre for Soya bean, ICAR, Indore, India, 2007. P: 46. Ed. Dr. G.S. Chauhan, Director.

□□□

Chapter 2

Crop Improvement Strategies in Pulses Improvement

RAVEENDRAN, T.S. AND P. JAYAMANI
Centre for Plant Breeding and Genetics, Tamil Nadu Agricultural University Coimbatore

Introduction

Pulses are not only the important food grain to supply protein which forms part of the vegetarian diet but also useful in many ways. Pulses are rich source of protein and possess 2-3 times more protein than that of many other cereals. Pulses enrich the soil fertility by fixing atmospheric nitrogen in the root nodules and improve the soil structure. They are ideal crops for mixed and intercropping and also serve as nutritious vegetable and fodder. The current productivity level of pulses is very low and could not meet the per capita requirement of 80 g / day pulses as recommended by the FAO/WHO. The current per capita availability of pulses is below 40 g at the National level.

In Tamil Nadu, the total area under pulses is around 5.90 (2004-05) with a production of 2.16 lakh tons. The average productivity of pulses is 367 Kg/ha which is far below the average productivity of the country as well that of the global productivity.

Area, production and productivity of pulse crops in Tamil Nadu

S.No.	Crop	Area (Lakh ha.)	Production Lakh tones	Productivity (Kg/ha)
1	Blackgram	2.26	0.83	367
2	Greengram	1.55	0.62	399
3	Horsegram	0.68	0.20	298
4	Redgram	0.43	0.29	667
5	Bengalgram	0.06	0.04	614
7	Other Pulses	0.21	0.04	204
	TOTAL	**5.90**	**2.16**	**367**

The requirement of protein in Indian diet has to be met through pulses especially for the vegetarians. The country will need 30.3 million tons of pulses by 2020 AD since the population will touch 1350 million by that time. But the pulses production remains stagnated for about 4 decades and it should be increased through many ways.

The following are the main reasons for low productivity of pulses

1. Around 78% of the area is under rainfed conditions
2. The soils where the pulse crops are grown are poor fertility
3. Unfavourable weather conditions like erratic and uncertain rainfall, low and high temperature and moisture stress at various crop growth stages
4. Poor moisture retention capacity
5. Soil salinity and alkalinity
6. Storage grain pests cause considerable losses
7. Farmers devote poor attention to the pulses cultivation
8. Mostly grown as mixed crop, intercrop, bund crop etc.
9. Highly susceptible to drought and water logged conditions
10. Susceptibility to pest and diseases
11. Poor storability and lack of storage facility
12. The crops react generously with the environmental fluctuations

Redgram

Breeding Objectives

Following objectives were envisaged in consideration with agro-climatic conditions in different regions, local preferences, distribution of diseases and pests, extent of irrigation resources and cropping systems:

- High yielding varieties possessing acceptable grain quality and stability in production which can be attained by incorporation of resistance/ tolerance to major insect pests (pod borer, podfly), diseases (wilt, sterility mosaic, Phytophthora stem blight and Alternaria blight) and abiotic stresses *viz.*, waterlogging, drought, acidity, salinity and extremes of temperatures.
- Short duration varieties or hybrids for multiple cropping
- Efficient and productive hybrids for regions having adequate resources and better management conditions
- Relatively photoperiod and thermo-insensitive varieties which can be grown in non-traditional seasons.
- Genotypes for adaptation to different cropping systems such as pure cropping, mixed or intercropping, alley cropping, multiple cropping and multiple harvest (ratoonability).
- Varieties for special uses such as fodder, agro-forestry, vegetables and lac insect rearing, and
- Varieties responsive to inputs.

Breeding Methods

Pulses are very easy crops for conventional breeding and all the methods of crop breeding have been practiced. The varieties developed through various methods have been listed out here.

Plant Introductions

It is the oldest and simplest method of meeting varietal needs of a given situation as well as an effective mean to overcome the narrow genetic diversity observed in a crop. Varieties cultivated in one country/ region are often introduced and acclimatized in the prevailing conditions and, if found suitable, released as variety. There are several varieties introduced in India from other countries. The variety Hy 3C was a selection from exotic line PI 2817-2. An exotic line Brazil 1-1 has been used as source of earliness in pigeonpea, which in combination with NP (WR) 15, NP 41 and NP 69 has led to the development of early maturing varieties Mukta, sharad and Pusa Ageti, respectively.

Pure Line Selection

Pure line selection from genetically heterogeneous population and land races has been an important breeding method in pigeonpea. The selection forces in this method of breeding act on the already existing genetic variability of highly heterogeneous populations for extracting varieties with uniform morphological and phenological traits like height, maturity, seed colour, etc. Frequent outbreak of diseases and insect pests in epidemic form also provides strong selection pressure to identify disease-free plants from heterogeneous landraces, leading to the development of disease resistant varieties through direct selection in this crop. A good number of varieties (Table 1) have been developed through pure line selection. Some of the popular varieties are T 7, UPAS 120, Bahar, BDN 2, C 11, No. 148, T 15-15, LRG 30, Hy 3C, Narendra Arhar 1 and Maruthi (ICPL 8863).

Recombination Breeding

Recombination breeding consisting of controlled crossing between the parents of choice followed by pedigree selection or its various modifications in the segregating generations for targeted trait (s) is the widely employed method for pigeonpea improvement. Breeders have shown great interest on various aspects of this method in the past, as efficiency of the breeding method is most critical for pigeonpea in view of high proportion of genotype x environment interaction. Indirect selection for yield *via* pods per plant and seed weight has most effectively been pursued as compared to direct selection for yield. This method has been useful in breeding for highly heritable traits such as resistance to diseases, seed size, seed colour, growth habit and seed number per pod, whereas bulk population and single seed descent methods appear to be better options for breeding high yielding varieties. Many varieties have been bred using this method (Table 1). Besides pedigree selection and bulk population methods, recurrent selection and population improvement methods have been suggested as ways to accumulate desirable genes and facilitate the breaking of linkages in pigeonpea.

Table 1: Improved varieties of pigeonpea developed through different methods of breeding

Method of breeding	Varieties developed
Pure line selection	T 7, T 17, UPAS 120, T 21, BDN 2, No 148, Hy 3 C, LRG 30, AL 15, C 11, Maruthi, T 15-15, SA 1, Bahar, TTB 7, Abhaya (ICPL 332), Birsa Arhar-1, Sarita (ICPL 85010), Narendra Arhar 1, Amar (KA 32-1), Malviya Vikalp (MA 3)
Recombination Breeding	WB 20, Manak, Pusa 84, TAT 10, Pragati (ICPL 87), Pusa 33, Jagriti (ICPL 151), JA 4, BSMR 176, GT 100 Vamban 1, Asha (ICPL 87119), AL 201, Pusa 9, BSMR 736, TS 3, Jawahar (KM 7), Durga (ICPL 84031), MAL 13
Mutation Breeding	Co 3, TT 5, TT 6, TAT 10, Co 5

Mutation Breeding

Mutation breeding has been employed to generate additional variability since mid fifties. Initial experiments were mainly aimed at observing morphological variations induced by mutagenesis. Experiments were expanded further to study effectiveness and efficiency of a wide array of physical and chemical mutagens. A wide array of economically useful mutants, both induced and spontaneous, such as resistant to key diseases, male sterility, short duration, various plant types, seed size and plant height have been reported. Some of them have been either released directly as variety (Table 1) or used as donor for improving the traits. TAT 10 is an extra short duration variety developed through inter-mutant crosses. Irradiation of T 21 seeds with gamma rays led to the development of bold seeded variety TT 6.

Heterosis Breeding

The variability available in the base population has been exhausted and the varieties, so far evolved could not make any headway in breaking the yield plateau which makes the breeder to go for non-conventional and specialized procedures. Exploitation of hybrid vigour or heterosis has successfully been attempted in a number of crops to break the yield ceiling. Among pulses, pigeonpea has the highest out-crossing of about 20-25% depending on several environmental factors. Identification of stable genetic male sterility coupled with its often cross-pollinated nature has been instrumental in opening up the avenue for exploitation of heterosis. Six hybrids using genetic male sterility were developed and released for cultivation in different agro-ecological zones. These hybrids are ICPH 8, PPH 4, CoPH 1, CoPH 2, AKPH 4101 and AKPH 2022. Although these hybrids had definite yield superiority over varieties, they did not witness a significant spread due to inherent difficulties of genetic male sterility system in seed production. The need to rogue out fertile segregants within female rows increases seed production cost. Imperfect elimination of fertile sibs has direct bearing on the quality of hybrid seeds. To overcome this problem, development of cytoplasmic male sterility system (CMS) has been explored through mission-mode project on development of CMS based hybrid in pigeonpea.

Two approaches namely, use of mutagens and interspecific hybridization were followed for development of a stable cytoplasmic genetic male sterility system in pigeonpea. Mutagens were ineffective in creating stable cytosterility whereas inter-specific hybridization with *C. scarabaeoides*, *C. sericeus*, *C. cajanifolius* and *C. volubilis* has met with some success. Two CMS lines, GT 288A and 67A were found 100% sterile over the years and locations. One CMS line developed at ICRISAT is also used in hybrid programme. Fertility restoration was observed in fertile derivatives of inter-specific cross, *C. scarabaeoides x C. cajan.* A large number of *C. cajan* accessions of early and medium durations were used to produce test crosses with GT 288A and 67A. Observations on F_1 hybrids indicated that (i) the fertility restorers are available in both germplasm and advanced breeding lines but their frequency is low, (ii) frequency of fertility restorers is less in short duration lines than medium and late, (iii) most lines appear to be heterogeneous for fertility restoration gene(s) and selection within a line is essential to identify pure fertility restorers, (iv) amount of pollen production and duration of pollen dehiscence are not satisfactory though the pollen fertility examined under the microscope ranged from 30 to 80%, and (v) pod set in F_1s under open pollination is as good as normal fertile counterparts whereas it is poor under strict selfing. Efforts are being made to search out new restorers as well as selection and purification of parental lines of already identified restorers through plant-to-plant mating and testing the progenies (F_1) of individual plant for fertility restoration. The lines already identified with CMS lines GT 288 A and 67 A are being used to produce test crosses with newly converted CMS lines to test their fertility restoration ability and heterotic response.

Self Pollinated Crops

Mungbean

Component Breeding

Based upon correlation analysis of various yield components, selections have been based mainly on number of pods per plant, seeds per pod and 100-seed weight and sometimes indicate 'r' values in brackets also on number of pods per bunch and branches per plant or podding per unit area usually called 'Pod Index'. Pods per plant is by far the most important yield component and almost all the workers have found it having positive correlation with seed yield. It is the best selection index for seed yield and could be increased by increasing number of branches per plant or number of bunches per plant or by increasing the number of pods per bunch. Most of the work has shown that branch number per plant is negatively correlated with seed yield, but bunches per plant has mostly been found to be positively correlated with seed yield. Ramanujam (1975) and others have found that pods per bunch and bunches per plant both are positively correlated with yield. Increasing pods per bunch is physiologically constrained in grain legumes lowing to fall of flowers and unripe or partially filled pods as pulses are very sensitive to moisture stress. It seems the most feasible path to increase seed yield is through increasing number of bunches per plant.

Conventional Breeding Methods

Most of the high yielding varieties of mungbean bred and released so far have been developed through single cross pedigree method of selection. The single plant selections made in the early generations restrict carrying forward the bulk of created variability, which gets lost quickly giving way to homozygosity with each succeeding generation. This method has served the mungbean improvement programme well in the past, but lately no productivity advances are materializing due to the inherent genetic limitation of the method. The intermating of selected F_2 plants and selections in the late generations will help to harness most of the fixed desirable genes. Many workers have found single seed decent method to be more efficient in carrying forward more variability and in generating families with higher mean values for grain yield in the later generations.

Mungbean breeding programmes in India and at AVRDC have been following bulk method of breeding with selection for agronomic characters up to F_4 - F_5 generation. Thereafter, pure line selection is followed. Both these methods rapidly fix homozygous genotypes and limit free exchange of favorable genes, thus restricting desirable combinations. Intermating F_2 plants allows accumulation of favorable components of genetic variability *i.e.*, additive and additive x additive gene effects and at the same time allowing breakage of unfavourable repulsion phase gene linkages. India has a wide range of agroclimatic diversity with varying photoperiods and temperatures. Testing of mungbean across spring, summer, *kharif* and *rabi* seasons allows disruptive seasonal selection of lines adapted to diverse agroclimatic conditions and with resistance or tolerance to both biotic and abiotic factors. Backcross and backcross inbred methods have special place in breeding of mungbean in transfer of some specific traits like resistance to powdery mildew and MYMV which are governed by a few major genes. Verma and Singh (1998) reported two recessive genes for MYMV resistance. Inheritance of resistance to powdery mildew has been reported as a monogenic dominance (Chaitieng et al., 2003), two indepenent dominance genes, Pm 1' and Pm 2 (Reddy et al., 1994), three duplicate recessive genes (Kute et al., 2003) and quantitative gene with additive and non additive gene actions (Chaitieng et al., 2003, Gawande and Patil 2003, Sorajjapinum et al 2005. This method may also be employed for improvement in some quantitative and quality traits like methionine and tryptophan contents, seed weathering, *etc.*, in the recurrent parent.

Early Generation Testing

The early generation yield trials allow early identification of better performing crosses and F_2 derived lines within the individual crosses. However, selection emphasis is given in later generations only. The F_2 derived family selection is very appropriate in mungbean, which is prone to high GE interactions and low seed increase ratio, which renders pedigree, bulk and single seed decent methods inefficient. The time required is less and emphasis is on grain yield in replicated progeny tests. It was developed in Canada as a modification of bulk method. The F_2 derived family selection takes benefits of early generation yield testing to eliminate efficiently all the undesired materials both between and within the hybrid populations. Replicated yield trials are conducted across locations/environments for early generation selections among and within populations. These selections are further evaluated and final selections for high yield are made in only the best of families or populations. Due emphasis must be given to make site specific selections for different agro-climatic and production systems. Also depending upon the demands of the location and system, input responsiveness of the selections under high management conditions may be tested. Singh and Satyanarayana (1997) suggested that the segregating generations may be subjected to the selection in different seasons and to select segregants for specific requirements. Different plant types may be evaluated under different cropping systems. The F_3 derived family method is *a* modification of the above method and so is F_4 derived family method. The latter method is used at ICARDA. A common crossing programme with parents identified from different centres and offseason nursery facility can promotes the development of basic materials for selection and varieties desirable for dissemination in a fast track mode.

Mutation Breeding

Both physical and chemical mutagens can be employed in improvement of mungbean crop in India to enhance the variability in the basic population. The main drawbacks of this method are that the frequencies of desirable mutants are very less, necessitating evaluation of very large population and the difficulty in identification and scoring of micromutations. Tickoo and Chandra (1999) using both physical and chemical mutagens could induce significantly higher variability in mungbean for characters like yield per plant, pods per plant, seed number per pod, seed weight, days to flower and harvest index in M_2 generation. Mean values of all the characters had a negative shift in M_2 but after selection changed to positive direction in M_3 but were still associated with significantly higher interfamily and overall variances than the control populations. This indicates further scope for selection and improvement of the characters governed by polygenes. Most of the selected mutants usually have improvement in one or two important agronomical characters. Such characters may be incorporated into the cultivated varieties by backcross method also. Many a time the selected mutants have been released as new varieties as such for cultivation. Fourteen such varieties have been released in India including Pant Mung 2, Co 4, Dhauli, TAP 7, BM 4 and MUM 2 and five in Pakistan including NM 51 and NM 54. The latter two large seeded varieties resistant to MYMV have been developed by hybridization and irradiation of the F_1 seeds.

Work on variability parameters of various characters in mungbean have shown that there is enough genetic diversity in the crop and thus employment of mutagenesis as a tool to create genetic variability in this crop is not essential. However lack of resistant sources to bruchids and abiotic factors like pre-harvest seed damage, drought, water logging, and high temperatures during summer in Indo-Gangetic Plains, *etc.* demand special breeding strategies.

Breeding for Nitrogen Fixing Ability

Mungbean has generated renewed interest as an essential part of the various cropping systems owing to the declining total factor productivity of cereal crops. To improve its atmospheric nitrogen fixing ability (40-60 kg N ha"¹), it is essential to take up that research work on identification and selection of efficient nitrogen fixing genotypes in order to incorporate genes associated with nitrogen fixing ability in cultivars. Very limited work has been done on this aspect in mungbean in India and elsewhere. Hobbs and Mohan (1982) studied the variance in atmospheric nitrogen fixation in pea and found high heritability 0.76 to 0.85 for this character. Holl (1975) found two recessive genes *sym2* and *sym3* responsible for non-nodulation and non-fixing of nitrogen in wild peas in Afghanistan. This is needed to plan breeding programmes in mungbean for efficient fixation of atmospheric nitrogen during the crop growth.

Breeding for Quality Parameters

Indian consumers prefer large seeds with shining green seed coat. In West Bengal and Bangladesh, they prefer yellow small seeded varieties for its dal-flavour, although the variety "Sona mung" itself is very low yielding, late in maturity and susceptible to MYMV. Mungbean seeds generally have about 24% protein which is easily digestible, but is deficient in essential sulphur containing amino acids, methionine and tryptophan and, therefore, limiting the protein quality. Improvement in these amino acids without adversely affecting the easy digestibility of the seeds is the most important quality related objective. The protein and amino acid inheritance is governed by polygenes that show low heritability. Detailed genetic analysis of protein content undertaken in four selected high yielding crosses of mungbean by Chandra and Tickoo (1999) found that the presence of large non-additive gene action makes the task of improving this trait very difficult. *Vigna mungo* and *V. silvestris* (TC 2208, 2209, 2210 and 2211) have been found to contain high levels of methionine and crossing these with mungbean, selections of mungbean with high free methionine (Y-glutamyl-methionine content) than in mungbean have been made at AVRDC (Fernandez and Shanmugasundaram 1988). They have reported the chemical score of 32% of egg protein in mungbean. Backcross method of breeding is generally employed for improvement of quality characters.

Urdbean

Breeding Objectives

Urdbean is largely grown either as a subsistence intercrop during *kharif* season, and as a monocrop on residual moisture in *rabi* (rice fallow) or in spring/ summer season in between two main crops. Therefore, breeding for each growing situations of immense importance. Breeding for early maturity (65-75 days) provides suitability for its cultivation in summer/spring season. Initially, breeding for resistance to diseases of economic importance e.g., MYMV was emphasized. This was done because the disease not only causes severe yield loss but also poses difficulty to be controlled by any other control measure. With the introduction of crop in rice fallow of coastal regions, breeding for resistance to powdery mildew has also assumed importance. Breeding for thrips was also taken up subsequently. Breeding for resistance to bacterial leaf spot, macrophomina blight and crinkle virus and tolerance to hairy caterpillar, galerucid beetle and blister beetle are some other objectives of crop improvement. Breeding for green seeds and large seed has also been undertaken in some states. Nevertheless, increased seed yield has been the primary objective in almost all breeding programmes.

Breeding Strategies

Conventional methods have been used almost exclusively in urdbean breeding programmes. These include mass selection, pedigree selection, backcross method and mutation breeding. Nearly, 50% of the cultivars released so far are selections from landraces. Therefore, selection has been the predominant method of urdbean improvement. With the crop occupying area under non-traditional production system *e.g.*, summer/spring cultivation and in rice fallows, need of combining two or more traits in a cultivar has gradually been felt. As a result, during the last two decades, more than 60% of the varieties released for commercial cultivation have been developed through hybridization. Single, double, three-way and back crosses were used for pyramiding the desirable traits in the recombinants. Top crosses, inter-matings in early segregating generations and recurrent selections are not popular at the moment in urdbean improvement programme. However, their relative importance is under investigation. Early generation yield testing (Dahiya and Singh 1986) has been suggested in some cases where seed yield is the main selection criterion. While pedigree and single pod descent (SPD) methods are found efficient in maintaining mean performance, the SPD and bulk population methods are better in maintaining variability through generations and providing a base for selection (Arya *et al* 1995).

A modest beginning was made for use of mutagens in creating desired variability for early maturity (Sinha 1988) and non dormancy (Apparao and Jana 1974) traits. Three early maturing cultivars *viz.*, Co 4, Sarla and Vamban 2 have been developed through mutation breeding. Sarla and Vamban 2 are MYMV resistant mutants of T 9 while Co 4 is a large seeded mutant of Co 1. For other desired traits *e.g.*, erectness, compact plant type, seed coat colour, *etc.*, the method is yet to be applied extensively.

Wide Hybridization

Wild species are rich reservoir for some of the important traits (Table 3). Inter-specific crosses were made with the objective to broaden the genetic base of cultivated germplasm, incorporation of biotic and abiotic stresses, higher protein content and development of cytoplasmic genetic male sterile lines (van der Maesen 1986). Pundir and Singh (1987) and Moss *et al* (1988) have listed out potentially useful characters available in wild relatives. Verulkar *et al* (1997) have reported that the species *C. scarabaeoides* possesses both physical and antibiosis mechaisms of resistance to pod borers. *Cajanus sericeus* and *C. albicans* are rich in protein,and *C. reticulatus* var. *grandifolius* is hardy and fire tolerant *C. albicans* is tolerant to soil salinity. Several wild relatives were used in hybridization programme with *C. cajan* and male sterile plants were isolated from the segregating populations. Ariyanayagam *et al* (1995) crossed *C. sericeus* with *C. cajan* and isolated male sterile plants from BC_3F_1 population. Tikka *et al* (1997) developed CMS line GT 288 A using *C. scarabaeoides* cytoplasm. Male sterile plants were also isolated from inter-specific cross of *C. cajanifolius* with *C. volubilis*. Saxena and Kumar (2003) developed a cytoplasmic nuclear male sterile line cms 88039A using *C. scarabaeoides* (ICPW 89) and an early maturing line of *C. cajan* (ICPL 88039).

Table 2: Potential wild species of *Cajanus* possessing useful characteristics for wide hybridization in pigeonpea

Species	Useful characteristics	Status of utilization
C. acutifolius	High protein content	Crossable
C. albicans	Resistant to SM and tolerant to salinity	Crossable
C. cajanifolius	High protein content	Crossable
C. lanceolatus	High protein content	Crossable
C. confertiflorus	High protein content	Crossable
C. latisepalus	High protein content	Crossable
C. lineatu	Resistant to sterility mosaic	Crossable
C. reticulates	Wider adaptability	Crossable
C. scarabaeoides	Good for induction of male sterility	Crossable
C. sericeus	Resistant to sterility mosaic and blight	Crossable
C. platycarpus	Extra early and resistant to blight	Not crossable
C. mollis	High protein content	Not crossable

Potential Source of *Alien* Variation

The nature of research reported on wild germplasm related to grain legumes in comparison to cultivars and landraces has contributed to increased opinions of the value of wild germplasm. Exploratory studies suggest that about 150 species of *Vigna* exist in nature and 22 *Vigna* species including cultivated and wild are found in Indian gene centre. Various reports indicate that many desirable traits having great potential for mungbean improvement are available in wild or related species (Table 4). However, most of these species remain unutilized due to complications such as crossing barriers and linkage drag. A wild accession of *Vigna radiata* var. *sublobata,* PLN 15, is found to be the potential donor for more number of pods per plant and seeds per pod (Reddy and Singh 1990).

Crossability Studies

The crossability of cultivars with wild or related species is a pre-requisite for *alien* gene introgression. Therefore, an understanding of the extent of crossability is essential for successful production of hybrids and their derivatives. Crossability studies made in the cultivars of mungbean with their wild and cultivated relatives are given in Table 5.

Table 3. Potential *Vigna* species and their useful traits for introgression

Related species	Useful trait	Status of utilization
Vigna radiata var. *sublobata*	Resistance to MYMV and bruchids tolerance to soil and salinity, alkalinity, cold weather damage	Gene flow is possible. Very limited attempts have been made.
Vigna mungo	Tolerance to shattering, resistance to fungal diseases and MYMV	Gene flow is possible. Amphidiploids have also been produced
Vigna angularis	High productivity, high harvest index, early maturity, determinate growth habit	Limited attempts have been made without success. A mungbean x urdbean derivative was crossed successfully with it.
Vigna umbellate	High productivity, response to irrigation and high inputs, high degree of resistance to MYMV and bruchids	Success has been achieved in raising FI plants which were either completely sterile/produced a few shriveled seed which did not germinate. Amphidiploids have been produced which can be improved for fertility or can be backcrossed with mungbean.
Vigna glabrescens (tetraploid, 2n=44)	Immune to beanfly *(Ophiomya* sp.)	F1 hybrid obtained through embryo rescue technique. The triploid showed low frequency of bivalent formation and was sterile.
Vigna aconitifolia	Tolerance to high temperature and limited moisture	Not utilized

Table 4. Crossability studies involving mungbean with related *Vigna* species *

Wide cross	Reference
V. radiata x *V. mungo*	Sen and Ghosh (I960), Chavan *et al* (1966) Dana (1966), De and Krishnan (1966), Chowdhury and Chowdhury (1974, 1975), Singh *et al* (1975), Verma (1977), Ahn and Hartman (1978a. b), Chen *et al.* (1978), Subramanian (1980), Lukoki and Marechal (1981), Miyazaki (1982), Chen *et al* (1983), Gill *et al* (1983), Miyazaki *et al* (1984), Shanmungam *et al* (1983) Shanmungam *et al* (1984), Verma and Singh (1986), Verma and Yadav (1986), Singh *et al* (1996), Singh and Singh (1998), Subramanian and Muthiah (2001), Singh and Dikshit (2002)

Contd...

V. mungo x *V. radiata*	Gosal and Bajaj (1983a, 1983b), Verma and Singh (1986)
V. radiata x *V. angularis*	Sawa (1973), Ahn and Hartman (1978a, 1978b), Chen *et al* (1983)
V. radiata x *V. glabrescens*	Krishnan and De (1968), Chen *et al* (1989)
V. radiata x *V. grandis*	Chavan *et al* (1966)
V. radiata x *V. mungo var. Silvestris*	Singh and Ahuja (1977), Lukoki and Marechal (1981), Miyazaki (1982)
V. radiata x *V. trilobata*	Chavan *et al* (1966), Dana (1966)
V. radiata x *V. umbellata*	Dana (1966), Sawa (1973), Machado *et al* (1982), Satyan *et al* (1982), Chen *et al* (1983), Chowdhury and Chowdhury (1983)
V. radiata var. *sublobata* x *V. mungo*	Biswas and Dana (1975), Lukoki and Marechal (1981), Miyazaki (1982), Miyazaki *et al* (1984)
V. radiata var. *sublobata* x *V. mungo* var. *silvestris*	Lukoki and Marechal (1981), Miyazaki (1982), Miyazaki *et al* (1984)
V. radiata x *V. radiata* var. *sublobata*	Satija and Kaur (1999)

* Adapted from Masood Ali and Shiv Kumar. (2005). Advances in Pigeonpea Research (Edited). Indian Institute of Pulses Research, Kanpur, India. 356p.

Masood Ali and Shiv Kumar. (2006). Advances in Mungbean and Urdbean (Edited). Indian Institute of Pulses Research, Kanpur, India. 462 p.

A number of studies undertaken on crossability among the different *Vigna* species. Most of the reports have indicated that *V. radiata* produced successful hybrids when used as female parent with *V. mungo, V. umbellata* and *V. angularis.* However, the reciprocal hybrids were not viable.

Crossability Barriers and Methods to Overcome

Cross incompatibility, inviability of F_1 hybrid, and sterility of F_1 hybrid and subsequent progenies are the most common barriers to wide hybridization. Cross incompatibility between parent species occurs when pollen grain does not germinate or pollen tube does not reach ovary or male gamete does not fuse with female gamete (Chowdhury and Chowdhury, 1983). The ploidy level and style length differences may not be a major barrier as the long styled female parent *V. radiata* was successfully crossed with short style male parent *V. trilobata* and crosses between diploid x tetraploid *(V. radiata* x *V. glabrescens)* and tetraploid x diploid *(V. glabrescens* x *V. umbellata)* were also successful. Attempts have been made to overcome incompatibility through the choice of parental lines, manipulation of emasculation and pollination techniques. A large variation for crossability exists among the lines/accessions of cultivated and wild relatives of mungbean. This variation can be exploited to increase the chances of success for obtaining wide crosses. Attempting wide hybridisation using only a few pollinations with a small set of genotypes fail to draw any conclusion about crossability, since the crossability was found to be genotype dependent in a number of studies.

Gene Pool Approach for Improvement

The concept of gene pool given by Harlan and de Wet (1971) has been proved very useful to plant breeders for creating desirable variation for crop improvement. Accessions of species in the primary gene pool of a crop cross easily and produce viable and fertile hybrids. The species in the secondary gene pool can cross with the other with some manipulation and their hybrids may be partially fertile. The species belonging to tertiary gene pool can be crossed by only using special techniques such as embryo rescue, and their hybrids are partially or completely sterile.

Primary, secondary and tertiary gene pool in mungbean and urdbean

Crop	Gene Pool	
Primary	**Secondary**	**Tertiary**
Mungbean *V. radiata* var. *Radiata*	*V. mungo* var. *mungo*	*V. angularis*
V. *radiata* var. *sublobata*	*V. mungo* var. *silvestris*	*V. umbdlata*
V. radiata var. *setulosa*	*V. aconitifolia*	*V. dalzelliana*
	V. trilobata	*V. glabrescens*
		V. grandis
		V. vexillata
Urdbean *V. mungo* var. *Mungo*	*V. radiata* var. *radiata*	*V. angularis*
V. mungo var. *silvestris*	*V. radiata* var. *sublobata*	*V. umbdlata*
	V. radiata var. *setulosa*	*V. dalzelliana*
	V. aconitifolia	*V. glabrescens*
	V. trilobata	*V. grandis*
		V. vexillata

Biotechnological Approaches : A number of innovative biotechnological approaches have shown to enhance the efficiency of pigeonpea breeding programmes. The tools considered are:

Somatic Embryogenesis and Organogenesis

Development of efficient and reproducible *in vitro* regeneration protocol is a pre-requisite for genetic manipulation at cellular levels. Regeneration from embryo and cotyledon has been reported (Sarangi and Gleba 1991). These advances make it possible to employ genetic engineering for manipulation of economically important quantitative as well as qualitative traits.

Anther Culture

Use of haploids in breeding programme helps in reducing the time of breeding cycle by obtaining homozygosity in a single generation. The recalcitrance of pigeonpea pollen has been posing problems for genetic manipulation. The response of pollen grains is restricted to a few divisions resulting in unorganized calli of varying ploidy levels. Kumar *et al* (1983) cultured whole anthers on MS medium + 2,4-D resulting into callus. However, true haploids of pigeonpea are yet to be produced.

Embryo Rescue Technique

Desirable traits from the secondary gene pool have been transferred using embryo rescue technique. Cultures of excised ovaries, oracles or embryos often help to overcome the post fertilization incompatibilities. The first study to include culturing immature pigeonpea embryos was that of Kumar *et al* (1983) who raised callus and plantlets from embryos. Moss *et al* (1988) reported that inter-specific embryos of C. *platycarpus x C. acutifolius,* C. *cajanifolius* and C. *scarabaeoides* var. *scarabaeoides* as small as 0.7-1.0 mm long could be cultured into plantlets by using B5 medium with hormones, together with nurse tissue and dark treatment.

Protoplast and Somatic Hybridization

Techniques have been perfected to isolate protoplasts from pigeonpea leaves radicles (Xu *et al* 1985) and pollen tetrads. These methods utilize enzymatic digestion of the tissue. Production of pigeonpea protoplasts may find application in the hybridization of this crop with presently cross-incompatible species of *Cajanus.*

Somoclonal Variation

Genetic variation in the progeny of tissue culture raised plants is referred to as somaclonal variation. Chintapalli *et al* (1997) exploited somaclonal variation for damage due to pod borer in pigeonpea and indicated the possibility of additional gene for pest tolerance. This technique can be used for screening of a large number of cells in petri dishes particularly for tolerance to drought, temperature, salinity and those diseases where the toxin produced by the pathogen is directly available or a compound related to the toxin is identified.

Gene Transfer Technology

With the advent of genetic transformation techniques, it has become possible to clone and insert alien genes into the crop plants. For example, genes for three types of potentially useful anti-fungal plant-derived proteins can be introduced into pulse crops through transformation technology. These three types of genes are the genes for the phytoalexin synthesizing enzyme stilbene synthetase; the gene for an inhibitor of an important cell wall degrading enzyme of fungal pathogens, polygalacturonase inhibiting protein (PGIP); and a combination of the genes for osmotin and endo-chitinase. However, pulses including pigeonpea are one of the least amenable groups to transformation amongst dicotyledonous crops. Most of the pulse crops can be infected by *Agrobacterium* and produce tumors. In pigeonpea most of the transformations are limited to the transfer of marker genes. The most successful transformation systems in pulses leading to the recovery of transgenic plants are those which have exploited embryonic axes, stem nodal segments or cotyledon hypocotyl sections and apical explants. All such explants have terminal or axillary meristems and hence a high shoot regeneration capacity. Integration of genes governing desirable traits such as resistance to herbicides and insect pests and high methionine content (Muntz *et al* 1998) has been reported. Resistance to insect pests has been demonstrated in transgenic plants expressing genes for a-endotoxin from *Bacillus thuringiensis* (Bt), proteinase inhibitors, alpha-amylase inhibitor, chitinase and plant lectins.

Although most efforts have been centred upon the use of *Agrobacterium,* a natural gene delivery system for introducing genes into pulse crops, there are reports on the use of biolistics. However, the transformation frequency is usually low with biolistics compared to *Agrobacterium* mediated system. The particle bombardment may be preferred for gene introduction into large seeded pulses (including pigeonpea) circumventing the host specificity of many pulse crops to infection by *Agrobacterium.* However, the rooting of transgenic shoots combined with poor shoot growth following rooting have been identified as the major hurdle which remains common to the transformation of some pulse crops including pigeonpea and which needs to be resolved as soon as possible.

Marker Assisted Selection

Molecular markers (RFLP, RAPD, AFLP, microsatellites) are helpful in locating genes of interest and their precise transfer through marker-aided selection. They also help in characterizing genetic variability, identification of cultivars and germplasm and establishing relationships among the species. Identification of genes controlling resistance to diseases and insect pests, salt tolerance, resistance to drought and extreme temperatures and response to day length is possible with marker technology. Ratnaparkhe *et al* (1995) used Random Amplified Polymorphic DNA (RAPD) markers for identification of pigeonpea cultivars and eight of their related wild species. The use of single primers of arbitrary nucleotide sequence resulted in the sensitive amplification of the DNA fragments that were unique to individual accessions. The level of polymorphism among the wild species was extremely high, while little polymorphism was detected within C. *cajan* accessions. All of the cultivars and wild species under study could be easily distinguished with the help of different primers, thereby indicating the immense potential of RAPD in the DNA fingerprinting of pigeonpea.

Total DNA from three putative cytoplasmic male sterile (CMS) progenies derived from crosses between wild, C. *sericeus* and cultivated species (seven accessions of C. *cajan* including two genetic male sterile lines) plus one accession of C. *sericeus* was compared for their Restricted Fragment Length Polymorphism (RFLP) patterns vising maize mitochondrial DNA (mtDNA) specific probes. Three putative CMS progenies from the multiple cross genome transfer of pigeonpea lines (CMS *7-1*, CMS 12-3 and CMS 33-1) showed hybridization patterns identical to that of C. *sericeus* when DNA was designated with Eco RI and Hind III and probed with maize mtDNA clones. The results suggested that these putative CMS progenies were similar in all the restriction enzymes-probe combinations except Hind III at p 6. The genetic male sterile lines, MS Prabhat and QMS-1, differed from each other in their hybridization pattern. The genomic DNA hybridization pattern of Hind III designated DNA from ICPL 87 differed from the other pigeonpea lines when probed with the maize mtDNA clones. The cluster analysis of the hybridization data suggests occurrence of variation in the mitochondrial genome even among the cultivated species.

Tyagi (1997) used RAPD markers for investigating quantitative traits loci (QTLs) in two strains of pigeonpea, and their F, and F_2 progenies. DNA was extracted from the young leaves and subjected to polymerase chain reaction (PCR) for amplification. The use of single primers of arbitrary nucleotide sequence resulted in the selective amplification of DNA fragments that were unique to parents, F_1 and F_2 progenies. However, the level of polymorphism among parents was very low. F_1 was intermediate between the two parents, but F_2 showed little variation indicating that both parents were different morphologically, but with little genetic variation at DNA level.

Linkage Maps

Linkage maps provide the platform for mapping and tagging of useful traits using the information of tightly linked markers. Lambrides *et al.* (2000) reported two genetic linkage maps of mungbean derived from the cross Berken x ACC 41. The map constructed from 67 F_2 individuals consists of 110 markers (52 RFLP and 56 RAPD) that grouped into 12 linkage groups. The linked markers span a total map distance of 758.3 cM. An additional linkage map derived from a recombinant inbred lines (RILs) population, consists of 115 markers in 12

linkage groups. The linked markers span a total map distance of 691.7 cM. Using a framework set of RFLP markers, the F_2 map when compared with another F_2 mungbean map constructed in Minnesota, the order of these markers is found to be consistent between maps.

To facilitate transfer of bruchid resistance, a genetic linkage map has been constructed based on an inter-specific F_2 mapping population between *V. umbellata* and *V. nakashimae* consisting of 74 plants (Somta *et al* 2006). A total of 175 DNA markers (74 RFLPs and 101 SSRs) have been mapped on 11 linkage groups spanning a total length of 652 cM. Comparison of the genome map of azuki bean (*V. angularis*) and this inter-specific genome map shows that 114 (94.2%) markers are located on the same linkage groups in both maps. The marker order is also highly conserved between the two maps.

Recently, Chaitieng *et al* (2006) has published the first genetic linkage map of urdbean using a BC_1F_1 population consisting of 180 individuals. After analyzing the population with 61 SSR primer pairs, 56 RFLP probes, 27 AFLP loci and 1 morphological marker, 148 marker loci have been assigned to 11 linkage groups, which correspond to the haploid chromosome number of urdbean. The linkage groups cover a total of 783 cM distance in the genome with an average of 6 to 23 markers per linkage group. It has also been established that the azuki bean SSR markers can be widely used for Asian *Vigna* species.

Future Directions

Past breeding efforts in the development of pulse varieties have utilized only a fraction of the vast available diversity. Perusal of the pedigree of the released varieties confirms this observation and the narrow genetic base of the released cultivars. Pre-breeding or genetic enhancement needs emphasis for transfer or introgression of genes and gene combinations from unadapted sources into more usable breeding material. There are indications that novel and useful traits can be successfully combined from related species.

There are ample evidences that suggest for source sink manipulation required for yield improvement. Net assimilation rate (NAR) at flowering is an important parameter for which genetic variation exists in the germplasm. Development of relatively photo and thermo - insensitive and early flowering verieties could be another strategy for increasing production. Such varieties can also fit as intercrops during non traditional seasons or as catch crop in existing rotations without affecting the economically dominant crops. Therefore, genetics of flowering response need to be properly understood and photo thermo insensitive lines used in breeding programmes. Efficient screening technique for evaluating resistance for biotic and abiotic stresses is the urgent need for resistance breeding and collaborative program of breeders Entomologists, Pathologists and Physiologists should be charted out. The project should aim for pyramiding genes for durable and multiple resistance.

Proper evaluation and utilization of germplasm from secondary and tertiary gene pools by conventional and biotechnological tools are the immediate priorities. Breeding for efficient fixing of atmospheric nitrogen is a priority area for higher response to the applied nitrogen. Efforts should be made for improving the grain quality, protein percentage and sulphur containing amino acid and also to reduce antinutritional factors.

Stable cytoplasmic genic male sterile lines have been developed in pigeonpea. For these male sterile lines suitable restorers are also identified. Now priority must be given to the development of multiple disease resistant and high yielding hybrids especially in early and medium duration groups.

Use of molecular markers and genomics has emerged as an applied research option during the past two decades and promises to assist breeders in finding solution to many agricultural problems unsolved hitherto. A detailed genome map for pulse crops should be developed using molecular markers. Which will help in identifying and locating genes of interest and their precise transfer through marker aided selection as well as transgenic technologies.

REFERENCES

Apparao, S. and Jana M.K. (1974). Alteration of seed characters in blackgram. *Indian Journal of Agricultural Sciences* 44: 657-660.

Arya, S., Ram Dhari., Dhanda, S.S. and Tomer, Y.S. (1995). Relative efficacy of pedigree and single pod descent method in urdbean *(V.mungo* (L.) Hepper). *Annals of Biology Lidhiana* 11: 59-63.

Ariyanayagam, R.P., Rao, A.N. and Zaveri, P.P. (1995). Cytoplasmic genic male sterility in interspecific mating of *Cajanus. Crop Science 35: 981-985.*

Chowdhury, R.K., and Chowdhury, J.B. (1983). Compatibility between *Vigna radiata* (L.) Wilczek and *Vigna umbellata* (Thumb) Ohwi and Ohashi. *Genetica Agraroa* 37: 257-266.

Chintapalli, P.L., Moss, J.P., Sharma, K.K. and Bhalla, J.K. (1997). *In vitro* culture provides additional variation for pigeonpea *Cajanus cajan* (L.) *Crop Improvement* 33: 30-37.

Chaitieng, B., Laosuwan, P. and Wongkaew, S. (2003). Inheritance of powdery mildew resistance in mungbean (*Vigna radiata*). *Thai journal of Agriculture Science* 36: 73-79.

Chaitieng, B., Kaga, A., Tomooka, N., Isemura, T., Kuroda, Y. and Vaughan, D.A. (2006). Development of a blackgram (*Vigna mungo* (L.) Hepper] linkage map and its comparison with an azuki bean [*Vigna angularis*(Willd) Ohwi and Ohashi] linkage map. *Theoretical & Applied Genetics* 113: 1261-1269.

Chandra, N. and Tickoo, J.L. (1999). Genetic analysis of protein content in mungbean (*Vigna radiata* L. Wilczek). Indian Journal of Genetics & Plant Breeding 58: 475-478.

Dahiya, B.N. and Singh, V.P. (1986). Comparison of single seed descent, selective intermating and mass selection for seed size in greengram (*Vigna radiata* (L.) Wilxzek). *Theoretical & Applied Geetucs* 72: 678-681.

Fernandez, G.C.J. and Shanmugasundaram, S. (1988). Tha AVRDC mungbean improvement programme the past, present and future. Pages 58-70. In: *Mungbean: Processing of the Second International Symposium* (S Shanmugasundaram and BT Mckeabmeds). Shanhua, Taiwan: Asian Vegetable Research and Development Centre.

Gawande, V.L. and Patil, J.V. (2003). Genetics of powdery mildew (*Erysiphe polygoni* DC) resistance in mungbean (*Vigna radiata* (L.). Wilczek). *Crop Protection* 22: 567-571.

Harlan, J.R. and De Wet, M.J. (1971). Towards a rational classification of crop plant. *Taxonomy* 20: 509-517.

Holl, F.B. (1975). Host plant control of inheritance of the nitrogen fixation in the *Pisum-Rhizobium* symbiosis. *Euphytica* 24: 767-770.

Hobbs, S.L.A. and Mohan, J.D. (1982). Heritability of N2 (C2H2) fixation rates and related characters in peas (*Pisum sativum* L.) *Canadian journal of Plant Sciences* 62 : 265-276.

Kumar, P.S., Subrahmanyam, N.C. and Faris, D.G. (1983). Studies on developing pigeonpea haploids. *International Pigeonpea Newsletter* 2: 14-15.

Kute, N.S., Deshmukh, R.B. and Patil, J.V. (2003). Inheritance of resistance to powdery mildew (*Erysiphe polygoni* D.C.) in mungbean (*Vigna radiata* L. Wilczek). *Legume Research* 26: 24-27.

Lambrides, C.J., Lawn, R.J., Godwin, I.D, Manners, J. and Imrie, B.C. (2000). Two genetic linkage maps of mungbean using RFLP and RAPD markers. *Australian journal of Agricultural Sciences 51:* 415-425.

Muntz, K., Christav, V., Sallbach, G., Sallbach, I., Waddel, D., Pickardt, T., Schieder, O. and Wustenhagen, O. (1998). Genetic engineering for high methionine grain legumes. *Nahrung Food* 42: 125-127.

Moss, J.P., Singh, A.K., Sastri, D.C. and Dundas, I.S. (1988). Wide hybridization in legumes at ICRISAT. Pp 87-95.

Masood Ali and Shiv Kumar. (2005). Advnaces in Pigeonpea Research (Edited). Indian Institute of Pulses Research, Kanpur, India. 356p.

Masood Ali and Shiv Kumar. (2006). Advnaces in Mungbean and Urdbean (Edited). Indian Institute of Pulses Research, Kanpur, India. 462 p.

Pundir, R.P.S., and Singh, R.B. (1987). Possibility of genetic improvement of pigeonpea (*Cajanus cajan* (L.) Millsp.) utilizing wild gene sources. *Euphytica 36:* 33-37.

Ramanujam. S. (1975). Genetic diversity, stability and plant type in pulse crops. Pages 167-176. In: *Proceedings of the First International Mungbean symposium*. Shanhua, Taiwan: Asian Vegetable Research and Development Centre.

Reddy, K.R. and Singh, D.P. (1990). The variation and transgreesive segregation in the wide and varietal crosses of mungbean. *Madras Agricultural Journal 77 :* 12-14.

Reddy, C.R., Sekhar, M.R. and Reddy, K.R. (1994). Character association and coefficient analysis in greengram (*Vigna radiata* (L.) Wilczek). *Annals of Agricultural Research* 15: 423-427.

Ratnaparkhe, M.B., Gupta, V.S., Murthy, R.M.V. and Ranjekar, P.K. (1995). Genetic fingerprinting of pigeonpea *(Cajanus cajan* (L.) Millsp). Amd its wild relatives using RAPD markers. *Theoretical and Applied Genetics* 91: 893-898.

Sinha, R.P. (1988). Early maturity, dwarf mutant of urdbean *(V.mungo* (L.) Hepper). *Journal of Nuclear Agriculture & Biology 17:* 61-62.

Sarangi, B.K. and Gleba, Y.Y. (1991). *Cajanus cajan* (Millsp). Pigeonpea embryo, cotyledon based culture and propagation. *Acta Horticultura* 289: 149-150.

Singh, D.P. and Sathyanarayana.A. (1997). Creating higher genetic yield potential in urdbean Pages 157-172. In : Recent Advances in Pulses Research. (AN Asthana and Masood Ali, eds). Kanpur, India: Indian Society of Pulses Research and Development.

Saxena, K.B. and Kumar, R.V. (2003). Development of a cytoplasmic nuclear male sterility system in pigeonpea using *C.scarabaeoides* (L.) Thours. *Indian Journal of Genetics and Plant Breeding* 63(3): 225-229.

Sorajjapinum, W., Rewthongchum, S., Koizumi, M. and Srinives, P. (2005). Quantitative inheritance of resistance to powdery mildew disease in mungbean (*Vigna radiata* (L.) Wilczek). SABRAO Journal 37: 91-96.

Somta, P.A., Kaga, Tomooka, N., Kashiwaba, K. Isemura, T, Chaitieng, B. Srinives, P. and Vaughan, D.A. (2006). Development of an interspecific *Vigna* linkage map between

Vigna umbellata (Thunb.) Ohwi & Ohashi and *V. nakashimae* (Ohwi) Ohwi & Ohashi and its use in analysis of bruchid resistance and comparative genomics. *Plant Breeding* 125: 77-84.

Tyagi, A.P. H.S. (1997). Investigation of quantitative trait loci (QTLs) in two diverse strains of pigeonpea (*Cajanus cajan*) using RAPD markers, *Journal of South Pacific Agriculture* 4 (1-2): 92-98.

Tikka, S.B.S., Parmer, L.D. and Chauhan, R.M. (1997). First record of cytoplasmic genic male sterility system in pigeonpea (*Cajanus cajan* (L.) Millsp.). *GAU Research Journal* 22: 160-162.

Tickoo, J.L. and Chandra, N. (1999). Mutagen induced polygenic variability in mungbean (*V. radiata* (L.) Wilczek). *Indian Journal of Genetics & Plant Breeding* 53: 395-398.

Van der Maesen, L.J.G. (1986). *Cajanus* DC and *Atylosia* W. and A. (Leguminosae). Agricultural University Wagenengin Papers 85-4, Agricultural University, Wageningen, the Netherlands. 225p.

Verulkar, S.B., Singh, D.P. and Bhattacharya, A.K. (1997). Inheritance of resistance to podfly and pod borer in the interspecific cross of pigeonpea. *Theoretical Applied Genetics* 95: 506-508.

Xu, Z., Davey, M.R. and Cocking, E.C. (1985). Root protoplast isolation and culture in higher plants. *Scientia Sinica* 28 (4): 386-392.

□□□

Chapter 3

Transgenic Approach for Pulses Improvement

NADARAJAN. N AND P. JAYAMANI
Department of Pulses, Centre for Plant Breeding and Genetics
Tamil Nadu Agricultural University, Coimbatore

Introduction

Globally grain legumes play a crucial and important role in the sustainability of agricultural system, maintaining soil fertility and protein-enriched foodstuff particularly in developing countries. Several constraints that limit genetic improvement for crop yield, nutritional quality and productivity have been overcome by conventional breeding, agronomic practices, enhanced management, but there are situations where existing germplasm or cultivars lack the required desired traits for productivity, grain yield, nitrogen use efficiency, resistance to pathogens including insect-pests and maturation duration. Consequence of these limitations and barriers for inter - and intra specific hybridization, the genetic improvement programme in grain legume is restricted for progressive gains reflecting overall static grain yield in major legume for last two decades.

Grain legume crops like pigeonpea, chickpea, mungbean, groundnut and soybean are extensively grown in the rainfed and dryland areas of India. They have a narrow genetic base since they are essentially self pollinated. Thus, there is need to widen the genetic base and incorporate desirable characters. There is an urgent need to use transgenic technologies for improvement of leguminous crops. Routine transformation protocols are limited in most grain legumes. The low success has been attributed to poor regeneration ability (especially via callus) and lack of compatible gene delivery methods, although some success has been achieved in soybean.

Although many different techniques (electroporation of intact tissues, silicon carbide whiskers, etc.) have been tested for gene delivery to plant cells, two major methods, namely *Agrobacterium-mediated* and particle bombardment, have been extensively employed for genetic transformation of crop plants. Regeneration via the callus lends itself easily (compared to explants regenerating directly) to *Agrobacterium-mediated* transformations, while direct regeneration is more amenable for particle bombardment. Organogenesis from an unorganized callus (pathway I) has only been reported in soybean and pea, where shoots were recovered from callus tissues at a low frequency. Thus, this pathway, although amenable to *Agro*-bacterium transformation, could not be used extensively for the recovery of transgenic plants in these crops.

The efficient production of transgenic plants requires stringent selection procedures supported by a selectable marker gene that confers resistance to agents such as antibiotics or herbicides. Several such selection systems have recently been described for grain legumes, based on the marker genes neomycin phosphotransferase II *(nptII)*, hygromycin phosphotransferase *(hph, aph* IV or *hyg)*, phosphinotricin acetyltransferase *(bar* or *pat)*, conferring resistance to kanamycin, hygromycin and the herbicide phosphinotricin (BASTA™), respectively.

Why Transform Legumes?

The answer to this question is, of course, that some legume species are much more difficult to transform than others. Legume transformation systems, like transformation in all organisms, require development of: (a) a source of totipotent cells or gametes that serve as recipients of delivered DNA, (b) a means of delivering DNA into the target cells, and (c) a system for selecting or identifying transformed cells. For legumes that have been regarded as recalcitrant to transformation, regeneration *in vitro* is highly genotype specific and only rarely are cultivated varieties amenable to regeneration. In these cases, plant regeneration remains an "art" that requires considerable training of the practitioner to develop the skills needed to generate sufficient transgenic plants. In addition, regeneration is often slow and the frequency of transformation (no. of transformed plants generated from each explant) is often low. In species that are amenable to *in vitro* somatic embryogenesis such as alfalfa, (lucerne; *Medicago sativa),* relatively rapid and efficient transformation methods have been developed based on cocultivation of tissue pieces (explants) with *Agrobacterium tumefaciens.* Because inducing somatic embryogenesis or organogenesis in many legume species is difficult, a variety of transformation methods have been reported that use cultures of meristematic cells as sources of totipotent cells. Most commonly, transformation has been based on infection by *A. tumefaciens* although *Agrobacterium rhizogenes* is used for transformation of some species. Regeneration of shoots from the cotyledonary node or from other meristematic explants after *Agrobacterium* infection is emerging as a rapid and relatively efficient method of transformation in a number of legume species including soybean (Olhoft and Somers, 2001), *Lotus japonicus* (Oger et al., 1996), barrel medic *(Medicago truncatula;* Trieu and Harrison, 1996), and *Trifolium repens* (Larkin et al., 1996). A number of legume species also have been transformed by direct DNA transfer methods including microinjection, electroporation, and microprojectile bombardment (Christou, 1997; Atkins and Smith, 1997; Babaoglu et al., 2000) (Table 1).

Table 1. Promising methods for gene transfer to selected grain legumes

Legume species, cultivar	Transform ation method	Expressed gene(s)	Evidence	Reproduc ibility and robustness
Common bean (*P.vulgaris* Olathe	At	uidA neo, 25 albumin nptII, uidA, arc 5-1	Transmission to T_1 (Southern blot)	Low
Chickpea (*C.arietinum*) Turkey and Chafa	At	Pat, nptII	Transmission to T_1 (PCR)	Low
Cowpea (*V.unguiculata*) C 152	At	Hpt	Transgenic T_0 (Southern blot), no germination of T_1 seeds	Low
Pigeonpea (*C.cajan*) ICPL 87	At	nptII, uidA	Transmission T_1 (PCR)	TETE
Lentil (*L.culinaris*) Laird	MPB	Als	Transmission T_1 (PCR, Southern blot)	TETE
Field pea (*P.sativum*) Greenfeast, Rondo	At	bar, nptI, AI	Transmission and expression in advanced generations	Good
Faba bean (*V.faba*) Mythos	At	uidA, nptII, sfa8, lysC	Transmission and expression in advanced generations	TETE
Mungbean (*V.radiata*) K 851	At	nptII, uidA	Stable integration and expression in To plants	TETE
Blackgram (*V.mungo*)	At	nptII, uidA	Stable integration and expression in To plants, transmission to T_1 progeny	TETE

At : *Agrobacterium tumefaciens* MPB : Micro particle bombardment TETE : To early to evaluate

Common Bean (*Phaseolus vulgaris* Linn.)

Common bean can be regenerated *in vitro* from different explants, including shoot tips, leaf petioles, seedlings, embryonic axes, cotyledons, seedling nodes or meristimatic calli. Susceptibility to Agrobacterium has been shown to be genotype dependent. The first attempts to produce stable transgenic *Phaseolus* plants resulted in non-regenerable transgenic callus or chimeric tissues. There is no confirmed report of stably transgenic *P. vulgaris* plants using an *A. tumefaciens* based system.

On the other hand, *P.acutifolius* can be routinely transformed using Agrobacterium. The wild *P.acutifolius* line NI 576 and the large seeded cultivar, TB1, have been used successfully . Transformation is based on callus explants, and between 5 and 10 independent, fertile transformants can be obtained in a routine experiment starting from 50 bean seeds. Since *P.acutifolius* can be crossed with *P. vulgaris* and fertile plants can be obtained by embryo rescue , this may be the most efficient way to improve the common bean with gene technology. In contrast to the situation with Agrobacterium, a system to generate transgenic common beans has been developed using particle bombardment of apical meristems. Although this often results in chimeric plants, the progeny can be screened for stable integration of the transgene.

Chickpea (*Cicer arietinum* L.)

Plant regeneration via somatic embryogenesis after callus induction with 2,4-D or other synthetic auxins was an early success. For instance, "cotyledon like structures" can be induced by combining zeatin and low levels of indole-3-acetic acid (IAA). Transformation experiments relying on callus failed due to limited shoot regeneration but demonstrated the potential of *A. tumefaciens* as a transformation vector for chickpea.

The first report of chickpea transformation also provided molecular evidence for the transgenic nature of sexual progeny of chickpea after co-cultivation of embryonic axes with *A tumefaciens.* Genes encoding (GUS) and neomycin phosphotransferase (NPTII) were expressed and the transgenes were transmitted up to the generation. Using similar experimental protocols the formation of multiple shoots from different genotypes and the production of primary transgenic plants has been reported. Multiple shoot formation was achieved on MS medium supplemented with 6-benzylaminopurine (BAP), BAP and otnaphtalene acetic acid (NAA) or BAP, NAA and kinetin. Transgenic plants were selected via multiple culture cycles on media containing kanamycin or phosphinotricin. Transformation frequencies were low and reproducibility was a limiting factor.

There are two reports of trangenic plants expressing potentially useful transgenes in chickpeas. Transgenic chickpea plants produced via biolistic transformation expressed the bacterial crylAc gene from *B.thuringiensis* together with *nptll* as the selectable marker. Insect feeding trials with one primary transgenic plant demonstrated an inhibitory effect on growth of larvae of the chickpea pod borer *H.armigera*. Transmission to progeny was demonstrated although further analysis has not been reported.

Agrobacterium was used to transform desi type chickpeas with a seed specific alpha-amylase inhibitor *(aAIt)* gene from P. *vulgaris* and the *nptII* gene as selectable marker. Stable transmission and expression of the transgenes in subsequent generations was demonstrated. The high level of expression of the protected chickpea seeds from insect damage by severely inhibiting the development of cowpea weevils (C. *maculatus)* and adzuki bean weevils (C. *chinensis)*.

Cowpea (*Vigna unguiculata* (L.))

The introduction of genes for insect resistance into cowpea by genetic transformation has the potential to address the problem and could have an important impact on food security. Regeneration of cowpeas via somatic embryogenesis has been attempted using different tissues including cell suspensions derived from callus and callus derived from young leaves, mature embryonic axes and immature cotyledons. Cowpea callus failed to regenerate plants at an acceptable frequency. Therefore, organogenic regeneration is the current focus for cowpea transformation.

Direct organogenesis from tissue derived from axenic hypocotyls, epicotyl, cotyledon, hypocotyls and mature cotyledons has been achieved with moderate levels of cytokinins. The easy to culture cotyledonary node explants seem the most promising with reproducible responses reported by several laboratories. Recently, fertile cowpea plants were regenerated from cotyledonary node thin cell layer explants by application of the cytokinin analogue, thidiazuron. If a high proportion of the shoot buds give rise to independent, phenotypically normal plants, this system could be an attractive route to transformation.

Several *Agrobacterium* strains were shown to infect cowpeas although no regenerated cowpea plants were obtained. By avoiding the callus regeneration step and using mature de-embryonated cotyledons on hygromycin selection, four plants were obtained that produced seeds. Southern blot analysis of one transgenic plant was positive but the T_1 seeds failed to germinate. The particle gun has been used for cowpea transformation and although transformation may have occurred, Mendelian transmission of the transgene could not be confirmed. In summary, to date there is no evidence for a robust transformation system in cowpeas despite the great need for such a development.

Pigeonpea (*Cajanus cajan (L.) Millsp)*

Pigeonpea regeneration requires considerable training to develop the skills in *in vitro* manipulation. In addition, regeneration is often slow and the frequency of transformation is low. A variety of transformation methods have been reported that vise cotyledonary node, leaf disc and apical shoot meristem as explants. *A.tumefaciens* mediated method is the most commonly used method for transformation Genetic improvement of pigeonpea through genetic engineering has been envisaged for the biotic and abiotic stresses and the improvement of protein quality. The control of lepidopteran insects by insect control proteins from a prokaryotic source specially, *Bacillus thuringiensis* (Bt) has been considered to be a potential alternative for insect resistance. Transfer of *Bt* gene to pigeonpea will provide to the latter the desired resistance. In addition, the major seed proteins of leguminous crop species including pigeonpea are deficient in the essential sulfur containing amino acid, methionine. The introduction of DNA sequences encoding methionine will make the seed rich in respect of essential amino acids. For successful genetic modification and production of transgenic plants, effective transformation and regeneration system is imperative.

With the establishment of efficient and reproducible regeneration systems, transformation in pigeonpea has become wide open (Geetha *et al* 1998, Rathore *et al* 1999, Lawrence and Koundal 2001, Manoj Kumar *et al* 2002). Geetha *et al* (1999) reported the development of a reproducible system for production of transgenic plants via *Agrobacterium* mediated transformation of pigeonpea. They achieved transformation by co-cultivating shoot apices and cotyledonary explants with *Agrobacterium tumefaciens* strain LBA 4404 harbouring a binary vector carrying the reporter gene, p-glucuronidase *(uidA)* and the marker gene neomycin phosphotransferase *(nptlT)*. Transgenic plants were established in pots and presence of *uidA* and *nptll* genes in the transgenic plants was verified by PCR analysis. Lawrence and Koundal (2001) achieved transformation of pigeonpea using *Agrobacterium tumefaciens* strain GV 2260 containing the construct of isolated cowpea protease inhibitor gene (pCPI; Accession no.: AJ 271752). The gene was driven by CaMV 35S promoter containing kanamycin resistance as plant selection marker. Molecular analysis of the transformed plants was done by northern hybridization techniques. Manoj Kumar *et al* (2002) have standardized the optimal parameters for the genetic transformation of the pigeonpea explants by particle bombardment as well as *Agrobacterium* mediated method with GUS gene construct. A protocol for efficient plant regeneration from leaf explants was developed by Dayal *et al* (2003). The procedure is very simple, efficient and reproducible and is applicable across diverse genotypes of pigeonpea. These workers in biolistics-mediated gene transfer have demonstrated the genetic transformation of pigeonpea by using *nptll* and *uidA* as marker genes, where 50% of the selected plants showed gene integration and expression. With the optimization of protocols for the development of transgenic plants with appropriate vectors and promoters, researchers will soon be able to produce transgenic plants routinely in pigeonpea incorporating the genes of interest. Pigeonpea plants with *cry lAb* and *soybean trypsin inhibitor* (SBTI) genes have been developed and are being tested against *H. armigera* (Gopalswamy *et al* 2003).

Greengram (*Vigna radiata (L.) Wilczek)* and Blackgram *(Vigna mungo (L.) Hepper)*

Although efficient plant regeneration and genetic transformation protocols have been developed in mungbean and urdbean, improvement of these crops via genetic transformation has been limited. Transformation frequencies as high as 23% with LBA 4404 and 10% with EHA105 strains of *Agrobacterium tumefaciens* for leaf discs of urdbean have been reported (Karthikeyan *et al* 1996). However, transformed calli did not give rise to any shoot. Microprojectile bombardment can be performed with any tissue of most species; however, the process is relatively inefficient because few cells are stably transformed. Transient expression of *gus* gene in meristematic zones and recovery of stably transformed plantlets and complete seedlings of bombarded embryo axes of mature seeds of mungbean, urdbean and *V. aconitifolia* have also been achieved. Transgenic mungbean plants has been developed through an efficient *Agrobacterium* mediated genetic transformation method using primary leaf explants that could be induced to directly regenerate shoots via a rapid, reliable and genotype independent protocol.

In mungbean and urdbean, generation of transgenics through recombinant DNA technology has emerged as a priority for containing disease like MYMV. The complete genome of MYMV has been sequenced and the viral regulatory elements and the protein coding regions are well characterized. Three different approaches have been followed to obtain resistance. In induced resistance, the gene for enterotoxin dianthin has been cloned under the inducible rightward primer of VMYMV. This promoter is induced by AC 2, a product of leftward promoter. Upon infection, the virus induces the enterotoxin transgene causing death of invaded cell and thus stopping further spreading of the virus (Poogin *et al* 2003). In case of silencing, simultaneous presence of sense and anti-sense RNA sequences represses the expression from transgenes, endogens and viral genes. Silencing can occur in the transcriptional level by promoter inactivation *e.g.,* by DNA methylation and in the cytoplasm at post transcriptional level by RNA degradation. Transgenic plants of urdbean are being developed at Madurai Kamraj University by transcribing VMYMV promoter and coding sequence in sense and anti-sense orientation. Attempts are also being made to analyze the virus protein-protein interaction and to determine the interactive domains. Based on this information, development of transgenic plants should be possible using genes for competitive inhibition (Hanley-Bowdoin *et al.,* 1999).

Lentil (*Lens Culinaris Medik)*

Tissue culture of lentils has been achieved with shoot apical meristem tips, nodal segments and intact seedlings. Cotyledonary nodes showed the highest frequency and genotype independence for shoot regeneration. T-DNA was integrated into the lentil genome via axenic tumors on in vivo and *in vitro* inoculated stems. Transient expression of the reporter gene GUS was achieved by electroporation, particle bombardment and by Agrobacterium transfer methods.

Fertile transgenic lentil plants were produced after bombarding cotyledonary nodes with a mutant of the tobacco acetolactate synthase gene, which confers resistance to sulfonylurea herbicides. Shoots were micrografted on non-transgenic plantlets. Seven transgenic plants survived leaf-painting tests and were phenotypically normal. Southern hybridisation of the T_1 progeny provided evidence for the transmission of the transgenes.

Soybean *(Glycine max* (L.) Merrill)

Soybean is one of the most important leguminous seed crops as it is an economic source of both oil and protein. Developing an efficient transformation system should be helpful to investigating the functional soybean genes, as well as to improving soybean cultivars. Transgenic soybean plants have been produced using various DNA deliveries and plant tissue cultivation, including microprojectile bombardment of shoot meristems, embryogenic suspension cultures, *Agrobacterium tumefaciens* mediated T-DNA delivery into immature cotyledons and axillary meristematic issue located in seedling cotyledonary nodes.

Roundup Ready Soybean

Monsanto's Roundup is the world's biggest selling herbicide, approved for use in over 100 countries. The first herbicide-tolerant food crop to be commercially released was Monsanto's Roundup Ready soybean. The first field trials of this soybean were conducted in 1989-90. The transgenic plant showed tolerance, but it was only after 1991 that it demonstrated commercially useful levels coupled with no loss of yield. The first major commercial planting took place in 1996, and accounted for 2% of the total USA crop. This increased to 15% in 1997 and around 30% in 1998. The transgenic soybean is now also being grown in Canada and Argentina.

Roundup ready (RR) soybeans had a copy of a gene from *Agrobacterium* sp. strain CP 4, inserted into its genome by means of a gene gun. The survival transgenic plants were sprayed by the on-selective herbicide, Roundup, which kills conventional soybeans.

CONCLUSION

Somatic embryogenic and organogenic tissue cultures are the primary sources of totipotent target cells used in legume transformation systems. For some species a range of genotypes, including cultivars, are amenable to a specific tissue culture type, whereas in others, only a specific genotype or wild relative can be used for tissue culture initiation and, therefore, transformation. Expanding the range of genotypes within a species that undergo the requisite tissue culture process would provide a major contribution to improving the transformation system. This may be accomplished by the conventional empirical approach of manipulating culture media composition, phytohormones, explant source, and tissue culture environment. Finally, extending the current trend in legume transformation of using meristems as sources of totipotent cells is also likely to be productive. Development of soybean transformation systems from meristems provides an interesting case study illustrating this point.

There is a current trend toward increasing the use of *A.tumifaciens* for DNA delivery in crop improvement programs compared with micro projectile bombardment. This is driven by recent development of highly virulent strains and binary vectors that are useful for legume transformation and its ease of use and researcher familiarity. There is also the consensus that because *A.tumifaciens* generally only delivers the T-DNA, transgene loci resulting from *A.tumifaciens* infection are less complex than those produced via direct DNA delivery methods.

The optimization of selection and identification systems is crucial for improving transformation efficiency. For example, in soybean, development of a selection system based on hygromycin B greatly increased transgenic plant production and reduced both the number of non-transformed escapes and time in culture (Olhoft et al., 2003). Extensive evaluation of selection systems for legumes is reflected in the array of selectable marker genes and selective agents. There is no overall trend evident from reviewing the data, suggesting strong interactions between the selection system, culture type, and genotype within a species that require substantial experimentation to optimize.

Certainly, eliminating the requisite tissue culture step for legume transformation would be a great boost in progress toward development of high-throughput systems. However, is the development of non-tissue culture systems for legumes feasible? There are several reports describing legume transformation systems that require reduced or no tissue culture. At least two approaches for development of non-tissue culture transformation systems can be pursued. Either the floral dip method for Arabidopsis is adapted to legumes or a novel legume-specific system is developed.

Novel legume-specific non-tissue culture systems are already being developed in a number of species. The recalcitrance of many legumes to tissue culture initiation and plant regeneration has driven researchers to develop transformation systems that target apical and axillary meristems in the embryonic axis as sources of totipotent target cells. Further development of meristem culture systems and experience in production of transgenic plants from them will likely provide researchers with insights to bypass the tissue culture phase. Minimal tissue culture is required in the meristem multiplication method described by McCabe et al. (1988) for soybean as is the *A.tumefaciens based* method described for peanut by Rohini and Rao (2000). Further research in developing such new sources of totipotent cells as targets for transformation, especially those that are less dedifferentiated, will require concomitant improvements in DNA delivery and methods for selection or identification of transgenic plants. Substantial progress in those areas has been achieved in legumes, suggesting that non-tissue culture methods for most legumes based on meristem cultures may be feasible. Recent progress in legume transformation suggests that some systems will achieve the transformation efficiencies required for functional genomics applications in the near future.

REFERENCES

Atkins, C.A. and Smith, P.M.C. (1997). Genetic transformation and regeneration of legumes. In A Legocki, H Bothe, A Puhler, eds, Biological Fixation of Nitrogen for Ecology and Sustainable Agriculture. Springer-Verlag, Berlin, pp 283-304.

Babaoglu, M., Devey, M.R. and Power, J.B. (2000). Genetic engineering of grain legumes: Key transformation events. *Ag Biotech Net* 2000, Vol 2, June ABN 050.

Christou, P. (1997). Biotechnology applied to grain legumes, *Field Crops Research* 53 : 83-97.

Dayal, S., Lavanya, M., Devi, P. and Sharma, K.K. (2003). An efficient protocol for shoot regeneration and genetic transformation of pigeonpea (*Cajanus cajan*(L.) Millsp.) using leaf explants. *Plant Cell Reports* 21: 1072-1079.

Geetha, N., Venkatachalam, P. and Lakshmi Sita, G. (1998). High frequency induction of multiple shoots and plant regeneration from seedling explants of pigeonpea (*Cajanus cajan* L.) *Current Science* 75: 1036-1041.

Geetha, N., Venkatachalam, P. and Lakshmi Sita, G. (1999). Agrobacterium - mediated genetic transformation of pigeonpea (*Cajanus cajan* L.) and development of transgenic plants via direct organogenesis. *Plant Biotechnology* 16: 213-218.

Gopalswamy, S.V.S., Kumar, S., Subaratnam, G.V., Sharma, H.S. and Sharma, K.K. (2003). Transgenic pigeonpea. A new tool to manage *Helicoverpa armigera.* Poster presented at the National Symposium an Bioresources, Biotechnology and Bioenterprise, 19-20 November 2003, Osmania University, Hyderabad, India

Hanley, B., Settlage, L.S.B., Orozco. B. M., Nagar, S. and Robertson, D. (1999). Gemini viruses: models for plant DNA replication, transcription and cell cycle regulation. *Critical Reviews in Plant Sciences* 18: 71-106.

Karthikeyan, A.S., Sharma, K.S. and Veluthambi, K. (1996). Agrobacterium tumefaciens - mediated transformation of *Vigna mungo* (L.) Heppere. Plant Cell Reports 15 : 328-331.

Larkin, P.J., Gibson, J.M., Mathesius, U., Weinman, J.J., Gartner, E., Hall, E., Tanner, G.J., Rolfe, B.G., Djordjevic. M.A. (1996). Trangenic white clover. Studies with the auxin responsive promoter, GH 3, in root gravitropism and lateral root development. *Transgenic Res.* 5 : 325-335.

Lawrence, P.K. and Koundal, K.R. (2001). Agrobacterium tumefaciens - mediated transformation of pigeonpea (*Cajanus cajan* L.Millsp.) and molecular analysis of regenerated plants. *Current Science* 80 (11) : 1428 – 1432.

Manoj Kumar, S. Syamanta, D., Sharma, K.K. and Devi. P. (2002). Protocol of efficient plant regeneration and Agrobacterium tumefaciens - mediated genetic transformation of pigeonpea (*Cajanus cajan* L.Millsp.). *Indian Journal of Genetics and Plant Breeding* 63 (4): 289-294.

McCabe, D.E., Swain, W.F., Martinell, B.J., Christou, P. (1988). Stable transformationof soybean (*Glycine max*) by particle acceleration. *Bio/Technology* 6: 923-926.

Oger, P., Petit, A. and Dessaux, Y. (1996). A simple technique for direct transformation and regeneration of the diploid legume species *Lotus japonicus Plant Sci.* 116: 159-168.

Olhoft, P.M. and Somers, D.A. (2001). L: Cysteine increase *Agrobacterium* medicated T-DNA delivery into soybean cotyledonary node cells. *Plant Cell Rep* 20: 706-711.

Olhoft, P.M., Flagel, L.E., Donovan, C.M. and Somers, D.A. (2003). Efficient soybean transformation using hygromycin B selection in the cotyledonary node method, *Planta* 216: 723-735.

Poogin, M.P.V., Shivaprasad, K.V. and Hohn. T. (2003). RNAi targeting of DNA virus in plants. *Nature Biotechnology* 21 : 131-132.

Rathore, R.S., Chand, L. and Garg, G.K. (1999). Regeneration of pigeonpea (*Cajanus cajan* (L.) Millsp.) from cotyledonary node. *Physiolosgy & Molecular Biology of Plants* 5:. 167-171.

Rohini, V.K. and Rao, K.S. (2000). Transformation of peanut (*Arachis hypogaea* L): a non-tissue culture based approach for generating transgenic plants. *Plants Sci.* 150: 41-49.

Trieu, A.T. and Harrison, M.J. (1996). Rapid transformation of *Medicago truncatula*: regeneration via shoot organogenesis, *Plant Cell Rep* 16 : 6-11.

□□□

Chapter 4

Genetic Analysis for Physiological Traits in Mungbean (*Vigna radiata* L. Wilczek)

SUNIL KUMAR, B., MURUGAN, S. and M. PRAKASH
Department of Agricultural Botany, Faculty of Agriculture
Annamalai University, Annamalai Nagar

Introduction

Mungbean (*Vigna radiata* (L.) Wilczek) is an important grain legume crop grown under diverse situations throughout the length and breadth of the country. The farmers rarely raise this pulse crop as a sole crop due to its low yield potential. Secondly, it is known for its biological value, digestibility and for its photo-thermo insensitive nature. Seed yield is a complex character and is found to be controlled by large number of genes.

Mungbean, being a self-pollinated crop and the scope of exploitation of heterosis will always depend on the direction and magnitude of biological feasibility and the type of gene action involved. The genetic potential of heterosis has been exploited by many scientists but very little information is available on the physiological basis of heterosis in mungbean. Recent development of new cultivars has been taken into consideration with biochemical characters only. Whereas information regarding the heterosis of physiological characters like photosynthetic rate, transpiration rate, stomatal conductance and dry matter production is lacking. Hence, the present investigation was carried out to assess the heterosis for physiological characters and their relationship with seed yield.

Materials and Methods

The material consisted of a set of five hybrids of mungbean raised in Randomized Block Design (RBD) with three replications at the Experimental Farm, Department of Agricultural Botany, Faculty of Agriculture, Annamalai University during summer 2003. The distance between rows was 30 cm and between plants was 10cm. Standard agronomic and protection measures were followed as per the recommendations needed for mungbean cultivation.

Observations were recorded on five randomly selected competitive plants at 45 Days after sowing (DAS). Photosynthetic rate, transpiration rate and stomatal conductance were measured using portable photosynthesis system (LCA 2). The instrument was setup as per the manufacturer's recommended instructions. After photosynthesis measurement, leaf was excised and the leaf area was measured. Then the leaf was dried at 70°C for 48 hours and dry weight was determined for Specific Leaf Area (SLA). At harvest, number of branches per plant, number of clusters per plant and dry matter production were also recorded. Heterosis was estimated for all the characters over Mid Parent (MP) and Better Parent (BP) as per the standard procedure and significance for heterosis was tested by using 't' test (Wynne *et al.*, 1970). Correlation and other statistical analysis were carried out using standard statistical software.

Result and Discussion

Mean values of five hybrids and their related characters are presented in Table 1. Based on the mean performance, the best hybrids found for each character were VRMGG 1/ VBN 1 (Number of clusters per plant, 100 seed weight and seed yield per plant), EC 30072/ VBN 1 (Number of branches per plant) and VRMGG Local/ PMB 27 (Number of pods per plant). Aher *et al.* (1998) and Khattak *et al.* (2002) observed similar high mean values for yield related characters. The results regarding heterosis are presented in Table 2. Among the three types of heterosis, as far as breeder is concerned-standard heterosis is more important because the hybrid to be released is expected to outperform the superior local variety or the local hybrid (standard check). For photosynthetic rate, all the hybrids recorded positive and significant percent heterobeltiosis and standard heterosis. For transpiration rate, the maximum heterobeltiosis was exhibited by VRMGG 1/ VBN 1, EC 30072/ VBN 1 and VRMGG Local/ PMB 27 recorded negatively significant standard heterosis. The hybrid LCG 460/ PMB 27 recorded highest percent of heterobeltiosis (35.32%) and standard heterosis (33.94%) for stomatal conductance. Similar high percent of heterobeltiosis and standard heterosis for biophysical and morpho-physiological traits were observed by Sunil Kumar *et al.* (2006), Jirali *et al.* (2003) and Ashwathama and Patil (2003). All the hybrids except LGG 450/ LGG 410 showed positive and significant heterobeltiosis and standard heterosis for dry matter production and specific leaf area. Wallace *et al.* (1972) stressed the importance of specific leaf area, whereas Chimmad and Kamatar (2003) reported negatively significant heterobeltiosis and standard heterosis for transpiration rate.

Table 1. Yield and other important characters among five hybrids in mungbean

Crosses	Number of branches per plant	Number of clusters per plant	Number of pods per plant	100 seed weight (g)	Seed yield per plant
LGG 450/ LGG 410	4.38	12.00	2.81	3.03	11.68
VRMGG Local/ PMB 27	4.27	13.28	46.44	3.32	12.90
LGG 460 / PMB 27	4.61	13.02	45.29	3.14	12.44
EC 30072 / VBN 1	4.12	12.64	40.38	3.45	12.65
VRMGG 1 / VBN 1	4.17	13.92	45.50	4.12	13.04

Table 2. Heterotic behavior (per cent) over standard parent (di) and better parent (dii) for physiological characters among five hybrids in mungbean

Characters \ Crosses		LGG 450 /LGG 410	VRMGG Local /PMB 27	LGG 460 /PMB 27	EC 30072/ VBN 1	VRMGG 1 / VBN 1
Photosynthetic rate	di	9.09**	38.46**	32.94**	21.95**	44.44**
	dii	4.38	35.00**	36.84**	13.64**	23.81**
Transpiration rate	di	14.31**	34.48**	25.20**	-21.86**	28.28**
	dii	0.21	-28.05**	-25.38**	8.00	-30.28**
Stomatal conductance	di	12.60**	24.84**	35.32**	31.30**	26.12**
	dii	-0.69	20.46**	33.94**	25.83**	30.77**
Dry matter production	di	7.32*	23.33**	41.71**	23.75**	31.22**
	dii	-6.13	26.05**	37.22**	15.26**	14.81**
Specific leaf area	di	17.17**	28.65**	40.01**	23.50**	40.00**
	dii	6.60	16.67**	14.94**	13.00**	17.82**

Correlation analysis revealed (Table 3) that seed yield per plant had positive and significant association with photosynthetic rate, transpiration rate, stomatal conductance and dry matter production, whereas recorded negatively significant association with specific leaf area at both 30 and 45 Days After Sowing (DAS). It was found to be instrumental in the increase or decrease in seed yield per plant. This was in support with Ashwathama and Patil (2003).

Table 3. Correlation between physiological characters and seed yield per plant among five hybrids in mungbean at 30 and 45 days after sowing (DAS)

Characters \ Crosses	Days	LGG 450 /LGG 410	VRMGG Local /PMB 27	LGG 460 /PMB 27	EC 30072/ VBN 1	VRMGG 1 / VBN 1
Photosynthetic rate	30	0.58**	0.89**	0.57**	0.51**	0.84**
	45	0.51**	0.81**	0.66**	0.48**	0.83**
Transpiration rate	30	0.41**	0.61**	0.54**	0.19	0.37**
	45	0.31**	0.58**	0.33**	0.28*	0.36**
Stomatal conductance	30	0.21*	0.72**	0.61**	0.23*	0.41**
	45	0.29*	0.75**	0.74**	0.33**	0.35**
Dry matter production	30	0.42**	0.61**	0.81**	0.45**	0.74**
	45	0.38**	0.62**	0.95**	0.26*	0.56**
Specific leaf area	30	0.42**	0.34**	0.87**	0.44**	0.75**
	45	0.40**	0.39**	0.89**	0.43**	0.82**

In general out of the five hybrids, the hybrids VRMGG 1/ VBN 1 and VRMGG Local/ PMB 27 recorded higher mean performance, high percent of heterobeltiosis and standard heterosis and showed positively significant association of the characters studied with seed yield per plant. Hence, these hybrids could be exploited for higher yields in mungbean.

REFERENCES

Aher, R.P., Kute, N.S., Barhate, B.G. and Sarode. N.D. (1998). Heterosis and combining ability for protein content in green gram. *Madras Agric. J.*, 85(3-4): 197-198.

Ashwathama, V.H. and Patil, B.C. (2003). Studies on heterosis for yield attributes and biophysical traits in interspecific and intra hirsutum cotton hybrids. *2nd Intl. Cong. Plant Physiol.*, Jan 8-12, New Delhi. pp.59.

Chimmad, V.P. and Kamatar, M.Y. (2003). Physiological characterization of sorghum parents of kharif and rabi seasons. 2nd *Intl. Cong. Plant Physiol.*, Jan 8-12, New Delhi. Pp. 27

Jirali,D.I., Hiremath, S.M., Chetti, M.B. and Patil, B.C. (2003). Physiological basis of yield variation in turmetic (*Curcuma longo* L.) genotypes. *2nd Intl. Cong. Plant Physiol.*, Jan 8-12, New Delhi, 69.

Khattak, G.S.S., Haq, M.A., Marwat, E.U.K., Ashraf, M. and Srinives, P. (2002). Heterosis for seed yield and yield components in mungbean (*Vigna radiata* (L.) Wilczek). *Science Asia*, 28: 345-350.

Sunil Kumar, B., Prakash,M. Saravanan,K. and Ganesan, J. (2006). Physiological and morphophysiological basis of heterosis in mungbean (*Vigna radiata* (L.) Wilczek). *Res. On Crops*, 7(1): 318-320.

Wallace, D.M., Osbun, J.L. and Muger, H.M. (1972). Physiological genetics of crop yield. *Adv. Agron.*, 24: 97- 146.

Wynne, J.C., Emery, D.A. and Rice. P.W. (1970). Combining ability estimates for Arachis hypogaea L. 2nd Field performance of F1 hybrids. *Crop Sci.*, 10: 713- 715.

□□□

Chapter 5

Enhancement of Gamma Rays Mutagenicity by Ethidium Bromide and Ethyl Methane – Sulphonate in Cowpea

THIRUGNANA KUMAR, S., S. THIRUMAL RAJ, G. KANAGARAJ, R. BALAMURUGAN AND P.A. VELUSAMY
Department of Agricultural Botany, Faculty of Agriculture
Annamalai University, Annamalai Nagar - 608 002, Tamil Nadu, India

Introduction

Experimental mutagenesis is an important unconventional source to increase the variability in cultivated plants. The application of induced mutagenesis to crop improvement is being made in Tamil Nadu Agricultural University since 1965. It is significant to mention that one fourth of the released mutant varieties are pulse crops (Sree Rangasamy, 1986).

Ionizing radiations, especially, gamma rays have been a very important source of inducing genetic changes either by having direct effect on chromosomes or indirectly by producing free radicals, which on collision with chromosomes bring out mutations. EB - an acridine dye, is an intercalating agent and its potentiality has been well recognized by Burton and Hanna (1976) in pearl millet and by Levy and Ashri (1975) in peanut. It is known that, EMS is an alkylating agent. It causes point mutations. The fact that EB is an inhibitor of DNA polymerase (Meyer and Simpson, 1969) suggests the possibility that treatment in combination with gamma rays may have synergistic effect in the induction of mutation. Similarly, as both gamma rays and EMS are shown to affect only nuclear mutations, it is expected to cause additive and/or synergistic effect on the induction of mutations. In this paper, the comparative effects of combination treatments of gamma rays with EB and EMS on the yield of chlorophyll, viable and non-viable mutations is reported.

Materials and Methods

Well filled selfed seeds of cowpea (*Vigna unguiculata* (L.) Walp.) culture Co. Vu. 623 were used. EB and EMS were obtained from Sigma Chemical Company, USA.

Dry seeds were irradiated with, 40, 45, 50, 55, 60, 65 and 70 kilo rads of gamma rays from a 60 Co source at an exposure of 54.05 rads/sec., at room temperature. Presoaked seeds for 6 h. were treated with 0.08, 0.10, 0.13, 0.15, 0.18 and 0.20 mM of EB and with 20, 30 mM EMS for 6 h. at 21 ± 1.°C. The mutagen solution was prepared in double distilled water without buffer. The seeds were washed in running tap-water for half an hour after the treatment with mutagen solution. The combination treatments were performed by treating the seeds with gamma rays first then treating with chemical mutagens. The combination treatments were fixed at LD_{50} ± 10 of LD_{50} (germination). Separate untreated dry seeds, pre-soaked for 6 h. and 12 h. were also included in the experiments, as control.

After the treatments, the seeds were sown in the field, in Randomized Block Design with three replications. In M_1 generation, reduction in the survival on 30th day was taken as lethality and reduction in plant height on 30th day was taken as injury and the reduction in seed fertility was considered as sterility. In M_2 generation, chlorophyll mutation was scored during the first two weeks after sowing and viable mutations were scored at the time of flowering and at the time of maturity. The mutagenic effectiveness and efficiency of mutagens were worked out as per the formulae given by Konzak *et al.* (1965).

Results and Discussion

The treatment with each mutagen alone, as well as the combination treatments reduced the survival, plant height and seed fertility. Gamma rays with EMS post-treatment showed maximum lethality (88.99 per cent at 65 kR + 20 mM), whereas the injury was highest with gamma rays plus EB post treatment (45.40 per cent, at 55 kR + 0.08 mM). EB alone at 0.13 mM showed maximum seed sterility (39.87 per cent). A dose dependent increase in lethality was observed with gamma rays post treated with EB, EMS and in sterility with gamma rays post treated with EB. Injury showed inconsistent trend with treatments.

As seen from Table 1, chlorophyll mutant was not observed among the progenies of the seeds treated with EB alone. But as post treatment with gamma rays, the induction of the chlorophyll mutant showed a dose dependent increase. Other treatments showed an inconsistent relationship. The progenies

Table 1. Mutagenic effectiveness and efficiency

Mutagens Dose/conc.	Lethality (L)	Injury (I)	Sterility (S)	Mutation per M_2 plants		Effectiveness $\frac{M \times 100}{kR \text{ or conc. } xt}$		Efficiency $\frac{M \times 100}{L}$		$\frac{M \times 100}{I}$		$\frac{M \times 100}{S}$	
				a	b	a	b	a	b	a	b	a	b
Gamma rays (kR)													
40	30.29	00.99	25.58	1.38	1.11	3.45	02.78	04.56	3.66	139.39	112.12	05.39	4.34
45	28.85	00.00	28.94	1.33	0.68	2.96	01.51	04.61	2.36	-	-	04.60	2.35
50	38.94	00.00	31.58	3.19	1.16	6.38	02.32	08.19	2.98	-	-	10.10	3.67
55	43.27	02.23	33.52	1.96	0.78	3.56	01.42	04.53	1.80	087.89	034.98	05.85	2.33
60	36.74	11.08	29.28	3.14	0.58	5.23	00.97	08.56	1.58	028.34	005.23	10.72	1.98
65	40.87	00.00	29.68	2.68	0.78	4.12	01.20	06.56	1.91	-	-	09.03	2.63
70	52.41	18.86	39.17	4.33	0.73	6.19	01.04	11.87	1.40	032.82	003.87	15.80	1.86
EB (mM)													
0.08	18.22	20.30	33.30	0.0	0.26	-	54.17	-	1.43	-	001.28	-	0.78
0.10	38.41	27.17	38.57	0.0	0.43	-	71.67	-	1.12	-	001.58	-	1.11
0.13	39.71	24.43	39.87	0.0	0.45	-	55.13	-	1.13	-	001.84	-	1.13
0.15	36.34	16.47	30.57	0.0	0.20	-	22.22	-	0.55	-	001.21	-	0.65
0.18	36.34	21.94	32.99	0.0	0.38	-	35.19	-	1.05	-	001.73	-	1.15
0.20	36.86	28.81	28.79	0.0	0.27	-	22.50	-	0.73	-	000.94	-	0.94
EMS (mM)													
20	29.19	35.44	23.46	1.14	0.64	0.95	00.53	03.91	2.19	003.22	001.81	04.86	2.73
30	40.68	40.67	32.20	1.91	0.94	1.06	00.52	04.70	2.31	004.70	002.31	05.91	2.91
Gamma rays + EB													
45 kR + 0.08 mM	43.97	29.57	29.40	1.84	0.71	NE	NE	04.18	1.61	006.22	002.40	06.26	2.41
55 kR + 0.08 mM	56.52	45.40	30.22	2.83	1.00	NE	NE	05.01	1.77	006.23	002.20	09.36	3.31
65 kR + 0.08 mM	65.22	40.22	37.19	2.83	0.67	NE	NE	04.34	1.03	007.04	001.67	07.61	1.80
Gamma rays + EMS													
45 kR + 20 mM	77.30	44.24	35.16	2.02	1.09	NE	NE	02.61	1.41	004.57	002.46	05.75	3.10
55 kR + 20 mM	82.61	37.63	32.41	1.85	0.68	NE	NE	02.24	0.82	004.92	001.81	05.71	2.10
65 kR + 20 mM	88.89	35.61	36.23	3.27	1.20	NE	NE	03.68	1.35	009.18	003.37	09.03	3.31

Note : a : On the basis of chlorophyll mutations

b : On the basis of viable mutations

of the seeds treated with gamma rays exhibited maximum chlorophyll mutation frequency (4.33 per cent at 70 KR). The frequency of viable mutations showed an inconsistent relationship with the gamma rays dosage and EB and EMS both individually and in combination. The maximum frequency of viable mutation was exhibited by the gamma rays with EMS post treatment (1.20).

For gamma rays, the effectiveness for chlorophyll mutations ranged from 2.96% to 6.38% and for viable mutations, it ranged from 0.97% to 2.78%. In respect of EB, the effectiveness for viable mutations ranged form 22.22% to 71.67%. In the case of EMS, the effectiveness for chlorophyll mutations was 0.95% to 1.06%. For viable mutations the effectiveness was almost equal at the two dose levels of 20 and 30 mM. Gamma rays at 50 kR showed the highest effectiveness for chlorophyll mutations, whereas 40 kR showed maximum effectiveness for viable mutations. Combination of gamma rays + EB showed synergistic effect for chlorophyll mutations and less than synergistic effect for viable mutations. Whereas gamma rays post treated with EMS showed invariably a less than synergistic effect. In terms of viable mutations 40 kR showed the highest efficiency on the basis of lethality, injury and sterility basis. EB at 0.08, 0.13 and 0.18 mM showed highest efficiency respectively on lethality, injury and sterility basis.

Gamma rays with EB post treatment at 55 KR + 0.08 mM showed maximum efficiency in terms of chlorophyll and viable mutations on the basis lethality and seed sterility. Gamma rays plus EMS treatment at 65 kR + 20 mM showed the maximum efficiency in terms of lethality, injury and sterility basis and for viable mutations on the basis of injury and sterility. Combination treatments involving gamma rays and EB showed a synergistic effect for chlorophyll mutations and less than synergistic effect for viable mutations. Whereas gamma rays post treated with EMS invariably showed a less than synergistic effect for both chlorophyll and viable mutations.

The treatment with each mutagen solely as well as in combination reduced the survival, plant height and seed fertility. A dose dependent increase in lethality was observed with gamma rays post treatment with EB and EMS and for sterility with gamma rays post treated with EB only. Injury showed inconsistent relationship with almost all the mutagens and their combinations.

The maximum lethality observed with gamma rays post treated with EMS treatment may be due to the joint action of both the mutagens on nuclear DNA. As a cell can tolerate a loss of a whole chromosome but not even a single aberration, the concurrent action of the mutagens may have brought out a large deletion at the gene and chromosome level. Whereas the highest plant injury observed with gamma rays post treated with EB may be due to the independent and joint action of the two mutagens on two different genetic systems. As it is a well established fact that gamma rays can cause aberration in any of the hereditary systems, the knowledge of the system on which EB acts would be worth. It had been shown that EB being an intercalating agent (Sideris *et al.*, 1973 and Waring, 1966) has more affinity towards circular DNA (Mahler and Bastos, 1974) and can block DNA replication by interfering with nucleic acid synthesis *in vivo* (Newton, 1957) through the inhibition of DNA polymerases (Meyer and Simpson, 1969) and transcription by inhibiting DNA dependent RNA polymerases (Elliot, 1963), the independent but a coordinated and integrated action and interaction of these two mutagens might have caused the blockage of the nuclear coded cytosol born mitochondrial proteins or the targeting of the synthesized proteins to the mitochondrial matrix, inner or outer membrane or inter membranous space leading to a modified intracellular protein topogenesis or even the self-synthesizing capacity of the mitochondrial system itself may be affected leading to the loss of their semi-autonomous nature. It had been reported about the loss of the kinetoplast DNA in *Trypanasoma* sp., (Steinert, 1969) and induction of cytoplasmic petite mutations in Bakers Yeast (*Saccharomyces cerevisae*) after treatment with EB (Nagley and Linnane, 1972; Slonimski *et al.*, 1968). Totally the affected cell may had a debilitated mitochondrial system or a reduced population below the threshold level in the dividing cells. The coincidence in occurrence of the dwarfs with delayed flowering confirmed the above view (Thirugnanakumar, 1986).

Vaughn *et al.* (1980) had reported that in rice during the normal course of micro sporagenesis the organelle genomes such as mitochondria and chloroplasts were coagulated and aggregated due to membrane disruption and had debilitated function. In this study, the debilitation of the organelle genomes may be hastened in the early stages of archesporial differentiation in to PMC's, as a cumulative action of the normal and induced by the mutagens. The pollen developed from these affected cells might have received "Zero mitochondria" (mitochondria without respiratory function), which is turn reflected in the high seed sterility observed with EB treatments exclusively.

Chlorophyll mutation frequency was highest with gamma rays treatments, whereas the viable mutation frequency was highest with gamma rays post treated with EMS treatments. The mutation rate induced by gamma rays was enhanced by EB and EMS post treatments but the rate of enhancement of the mutation was found to differ greatly within and among the combination doses.

Konzak *et al.* (1965) proposed the term "effectiveness" as a measure of gene mutation in relation to thé dose and 'efficiency" as an estimate of the mutation rate in relation to other biological effects induced, such as lethality, injury or sterility.

Gamma rays as single treatment showed maximum effectiveness and efficiency in terms of chlorophyll mutations. The effectiveness for viable mutations was high in EB treatment but the efficiency was high with gamma rays, which was in conformity with Konzak *et al.* (1965) who stated that the most effective mutagen need not be the most efficient one. It was also noticed that with the increase in the concentration of EB, there was a decrease in its mutagenic efficiency and it may be due to its high toxic level at its high concentration. The effectiveness of gamma rays in inducing chlorophyll and viable mutations was in agreement with Rathnaswamy (1975) in lablab. But, Sharma (1969) and Palanisamy (1975) reported that EMS was more effective in inducing chlorophyll and viable mutations in cowpea.

The very significant finding of our study was that EB as a sole mutagen failed to induce any chlorophyll mutation. But in combination with gamma rays it showed a synergistic effect. The synergistic effect of EB in combination with gamma rays may be due to the independent genetic changes induced in different genetic systems. Perhaps the mutagenic effects of these chemicals may not be masked or inhibited by the one on the other. The failure of EB to induce chlorophyll mutations in M_2 generation may be due to its genomic specificity. On the other hand the absence of chlorophyll mutations in M_2 may be due to the elimination of the gametes carrying the mutations or due to zygotic inviability. As a mutation event has both time and space reference the failure in chlorophyll mutation hunting in M_2 could not be mistaken for its mutagenicity. The weak mutagenic effect of EB in inducing chlorophyll mutation observed in this study was in agreement with the report of Sarma and Patnaik (1979) in rice, Khalatkar *et al.* (1975) in barley and Prabakhar (1985) in sesame.

On the contrary, the potentiality of EB in inducing viable mutations both as single as well as post treatment with gamma rays was well evidenced in our study and was in corroboration with the findings of Prabakhar (1985) in sesame. Gamma rays post treated with EB showed a less than synergistic effect. So it was conceived that a differential effect of gamma rays post treated with EB may be due to the interaction of the cytoplasmic components and EB, as well as the interaction between the mutated nuclear wild organellar genomes or wild nuclear mutated organellar genomes. While our study proposed the enhancement of gamma rays mutagenicity by EB,

Hasegawa and Inoue (1983) observed no enhancement of chlorophyll mutation in rice.

Gamma rays post treated with EMS showed a less than synergistic effect for both chlorophyll and viable mutations. The less than synergistic effect observed with gamma rays plus EMS may be due to the following reasons. A mutation event at DNA level is a product of pre-mutational changes and its subsequent fixation (Doudney, 1961 and Witkin, 1966). The mutagen applied first and second may acted on different loci, as one on the loci governing morphogenetic potentiality and the other on the loci involved in the repair process. If the above said mutagenic action is correct it would have resulted in a synergistic effect due to the fixation of the mutational lesion by the second mutagen, induced by the first mutagen. The effect of the second mutagen, *i.e.* EMS, on the repair genes may not be absolute or complete. That may led to the deviation from the expected synergism.

The increased chlorophyll and viable mutation frequencies in the combination treatments showed a more or less similar as well as a close relationship. The decreased viable mutation frequencies as compared to chlorophyll mutation frequencies in the present study, perhaps, may be due to the increased size of the mutated sector as stated by Gaul (1961) involving multiple mutations as well as viable mutations. The coincidence in the occurrence of both chlorophyll and viable mutations would lead to an under estimation of viable mutations, as they would be eliminated at seedling stage owing to the lethal chlorophyll mutations.

REFERENCES

Burton, G.W. and Hanna, W.W. (1976). Ethidium bromide induced cytoplasmic male sterility in pearl millet. *Crop Sci.*, 16: 731-732.

Doudney, C.O. (1961). Nucleic acid formation and U.V. induced mutations in bacteria: Some considerations in light of recent advances. *J. Cell Comp. Physiol.*, **58:** 145-150.

Elliot, W.H. (1963). The effect of antimicrobial agents on Deoxyribonucleic acid polymerase. *Biochem. J.*, 86: 562-567.

Gaul, H. (1961). Use of induced mutations in seed propagated species. Mutations in plant breeding. NAS. NRC. 891: 206-252.

Hasegawa, H. and Inoue, M. (1983). Enhancement of sodium azide mutagenicity by ethidium bromide in rice (*Oryza sativa* L.). *Environ. Exp. Bot.*, 23: 53-57.

Khalatkar, A.S., Gopal - Ayengar and Bhatia C.R. (1975). Modification of mutagenic efficiency of ethyl methane sulphonate with ethidium bromide, iodoacetamide and sodium fluoride in barley. *Mutat. Res.*, **27:** 331-340.

Konzak, C.F., Nilan, R.N. and Foster, R.J. (1965). Efficient chemical mutagenesis. The use of induced mutations in plants breeding (Rep. FAO/IAEA Tech. meeting, Rome, 1964). Pergamon Press, 49-70.

Levy, A. and Ashri, A. (1975). Ethidium bromide – an efficient mutagen in higher plants. *Mutat. Res.*, 28: 397-404.

Mahler, H.R. and Bastos, R.N. (1974). Coupling between mitochondrial mutation and energy transduction *Proc. Natn. Acad.* USA. 71: 2241-2245.

Meyer, R.R. and Simpson, M.V. (1969). DNA biosynthesis in mitochondria: Differential inhibition of mitochondrial and nuclear DNA polymerases by the mutagenic dyes ethidium promide and acriflavin. *Biochem. Biophys. Res. Commun.*, 34: 238-244.

Nagley, P. and Linnane, A. (1972). Biogenesis of mitochondria. XXI. Studies on the nature of the mitochondrial genome in yeast : The degenerative effects of ethidium bromide on mitochondrial genetic information in a respiratory competent strain. *J. Mol. Biol.*, 66: 181-193.

Newton, B.A. (1957). The mode of action of phenanthridines : The effect of Ethidium Bromide on cell division and nucleic acid synthesis. *J. Gen. Microbial.*, 17: 718.

Palanisamy, G.A. (1975). Investigations on the induction of mutations in cowpea (*Vigna sinensis* (L.) Savi.) M.Sc. (Ag.) Thesis, Tamil Nadu Agricultural University, Coimbatore, India.

Prabhakar, L.V. (1985). Studies on induced mutagenesis in *Sesamum indicum* L. *M.Sc. (Ag.) Thesis,* Tamil Nadu Agricultural University, Coimbatore, India.

Rathnaswamy, R. (1975). Investigations on induced mutagenesis in lablab (*Lablab niger* Medikus). M.Sc. (Ag.) Thesis, Tamil Nadu, Agricultural University, Coimbatore, India.

Sarma, N.P. and Patnaik, A. (1979). Mutagenicity of sodium azide and ethidium bromide in rice. Symposium on the role of induced mutations in crop improvement Hyderabad, Sep. 1979, 59 (Abstract).

Sharma, B. (1969). Chemically induced mutations in cowpea (*Vigna sinensis* L. Savi.) *Curr. Sci.*, 38: 520-521.

Sideris, E.G., Nilan, R.A. and Bogyo, T.P. (1973). Differential effect of sodium azide on the frequency of radiation induced chromosome aberrations Vs. the frequency of radiation induced chlorophyll mutations in *Hordeum vulgare. Radiat. Bot.*, 13: 315-322.

Slonimski, P.P., Perrodin, G. and Croft, J.H. (1968). Ethidium bromide induced mutation of yeast mitochondria : complete transformation of cells in to respiratory deficient non-chromosomal "petite". *Biochem. Biophys. Res. Commun.*, 30: 232-239.

Sree Rangasamy, S.R. (1986). Applied mutation research in field crops at Tamil Nadu Agricultural University. Mutation Breed. *Newsl.*, 28: 13-14.

Steinert, M. (1969). specific loss of kinetoplastic DNA in Trypanosomatidae treated with ethidium bromide. *Exp. Cell Res.*, 55: 248-252.

Thirugnana Kumar, S. (1986). Studies on induced mutagenesis in cowpea (*Vigna unguiculata* L. Walp.). Unpublished M.Sc. (Ag.) Thesis, submitted to Tamil Nadu Agricultural University, Coimbatore, India.

Vaughan, K.C., Debonte, L.R., Wilson, K.G. and Schaefer, G.W. (1980). Organelle alteration as a mechanism for maternal inheritance. *Science*, 280: 196-198.

Waring, M.J. (1966). Structural requirements for the bindings of ethidium to nucleic acids. *Biochem. Biophys. Acta*, 114: 234-244.

Witkin, E.M. (1966). Modification of mutagenesis induced by U.V. light through post treatment of bacteria with basic dyes. *J. Cell. Physiol.* 58: 135-144.

□□□

Chapter 6

VRI (Gn) 6 – New High Yielding Spanish Bunch Groundnut

VISWANATHAN, PL., P. VINDHIYAVARMAN, G. NALLATHAMBI, D. KUMARESAN, A. MOTHILAL, S. CHIDAMABARAM, K. N. RAGUMOORTHI, K. SACHITHANANTHAM, K. SUBRAMANIAN, K. MANIAN, M. BALUSAMY AND V. MURALIDHARAN
Regional Research Station, Tamil Nadu Agricultural University, Vriddhachalam-606 001

Introduction

Groundnut is an important oilseeds crop in Tamil Nadu and occupies 0.59 m.ha. with the production of 0.91 m.t. The cultivated groundnut has four distinct botanical groups viz., Virginia bunch, Virginia runner, Spanish bunch and Valencia. Among these Spanish bunch groundnut occupies 90% of the total cultivated area followed by semi spreading (Virginia bunch). Spanish bunch type groundnut varieties are early maturing, easy to harvest and have desirable pod features. Hence, it is highly suitable under multiple cropping systems. The high yielding bunch type groundnut varieties grown in the state have low yields due to biotic and abiotic stress. Hence, a research objective was made to develop biotic and abiotic resistant bunch type groundnut better than existing varieties at Regional Research Station, Vriddhachalam. This resulted in the improvement of promising culture VG 9816 and it was released as improved variety as VRI (Gn) 6 during 2007 for cultivation in Tamil Nadu.

Materials and Methods

The Spanish bunch culture VG 9816 was evolved at Regional Research Station, Vriddhachalam. The cross was made between ALR 2 x VG 9513. Elite plants were selected from F_2 onwards and evaluated for their performance and homozygosity and the culture VG 9816 was identified as the best. The culture VG 9816 was evaluated with check varieties at Regional Research Station, Vriddhachalam between 1999-2005. Further evaluations were made in multilocation trials during 2001-2002 and ART during 2004-05 at Trichy, Dindigul, Theni, Thiruvallur, Cuddalore, Coimbatore, Kancheepuram, Tirunelveli, Pudukottai, Perambalur, Madurai, Tanjore, Sivagangai, Karur, Namakkal, Virudhunagar and Villupuram district of Tamil Nadu. This culture was also tested in All India Coordinated trials during Rabi/summer 2003-04.

Results and Discussion

The results of station trials, multilocation trials, adaptive research trials and All India Coordinated trials are presented in Table 1. The culture VG 9816 which was tested in station trials from 1999 to 2005 has recorded an average pod yield of 2379 kg/ha and 2426 kg/ha under rainfed and irrigated conditions respectively. It represents an increased pod yield of 10.5 and 13.5% over VRI 2 in rainfed and irrigated situation. Similarly under multilocation trials, it recorded an average pod yield of 1775 and 2158 kg/ha during kharif and Rabi/summer seasons which is 3.1 and 5.4% higher over the best check VRI (Gn) 5.

Table 1. Overall performance of groundnut culture VG 9816 in various trials for pod yield (kg/ha)

Season	Name of the Trial	No. of locations	VG 9816	VRI 2	VRI (Gn) 5	TAG 24	ICGS 44
Kharif	Station trial	7	2379	1941			
	MLT	15	1775	--	1721		
	ART 2005	30	1756	1631			
	Mean	**52**	**1916**	**1734**	**1721**		
	% over VRI 2		**10.5**				
	% Over VRI (Gn) 5		**3.1**				
Rabi/ **Summer**	Station trial	7	2426	1933			
	MLT	12	2158	--	2048		
	ART 04-06	48	2477	2178	--		
	AICRP Zone III b	6	2296	--	--	2251	--
	AICRP Zone IV	4	1932	--	--		1731
	Mean	**67**	**2403**	**2117**	**2048**	**2251**	**1731**
	% over VRI 2		**13.6**				
	% Over VRI (Gn) 5		**5.4**				
	% over TAG 24		**2.0**				
	% over ICGS 44		**11.6**				

Under advanced varietal trial of All India Coordianted trials, this culture has recorded a pod yield of 2296 kg/ha in Zone III b which is 2.0% increased yield over TAG 24. In Zone IV, this culture recorded a mean pod yield of 1932 kg/ha which is 11.6% increase over the Zonal check ICGS 44. Finally it was tested in adaptive research trials, it recorded a mean pod yield of 1756 and 2477 kg/ha during Kharif and Rabi/summer which is 7.7 and 13.7 per cent superior over the best check VRI 2 (Table 1).

The culture VG 9816 has been identified as early drought resistance type through empirical (yield and harvest index) and trait (proline and relative water content) based screening. The pooled data for three years under early drought situation (Table 2) indicated that VG 9816 had produced a mean pod yield of 1242 kg/ha as compared to VRI 2, which produced only 902 kg/ha. Similarly the harvest index of pooled data for three years indicated that the culture VG 9816 had maximum harvest index of 34.62 per cent as against VRI 2 which had only 27.81 per cent. The three year data on relative water content revealed (Table 3) that VG 9816 had maintained more than 80 per cent (80.69%) relative water content as compared to VRI 2 (79.25%). The proline content was very high to the magnitude of 144 µmol per gram of fresh weight of leaf as compared to VRI 2 (27.71 µmol).

Table 2. Performance of groundnut culture VG 9816 for yield and harvest index to early drought condition

	Pod yield (kg/ha)				Harvest index (%)			
Entries	*Kharif* 04	*Rabi* 04-05	*Kharif* 05	Pooled	*Kharif* 04	*Rabi* 04-05	*Kharif* 05	Pooled
VG 9816	1656	960	111	1242	39.39	25.49	38.97	34.62
VRI 2	850	864	992	902	32.56	18.49	32.38	27.81

Table 3. Performance of groundnut culture VG 9816 for drought traits

Entries	Relative water content (RWC %)				Proline content (Micro molar per gram of fresh leaves)			
	Kharif 04	*Rabi* 04-05	*Kharif* 05	Pooled	*Kharif* 04	*Rabi* 04-05	*Kharif* 05	Pooled
VG 9816	81.38	76.87	83.81	**80.69**	118	182	132	**144**
VRI 2	80.35	75.63	81.78	**79.25**	20	39	25	**28**

Thus the data revealed that under early drought situations VG 9816 had more drought resistance as compared to VRI 2.

The groundnut culture VG 9816 recorded moderate resistance to late leaf spot with a score of 3.1 to 3.3, while the check variety VRI 2 recorded 3.9 to 4.5 score under field condition. Similarly it recorded moderate resistance to rust (Table 4a). The groundnut culture VG 9816 recorded moderate resistance to leaf miner and *Spodoptera* pests (Table 4b).

Table 4a. Disease reaction to foliar disease of groundnut at Regional Research Station, Vriddhachalam

S No	Seasons	Disease incidence in 1-9 scale			
		VG 9816		VRI 2	
		LLS	Rust	LLS	Rust
1	*Kharif* 2004	3.3	3.2	4.1	4.3
2	*Rabi*/summer 2004-05	3.1	3.0	3.9	4.0
3	*Kharif* 2005	3.2	3.1	4.2	4.1
4	*Rabi*/summer 2005-06	3.1	3.2	4.5	4.1

Table 4b. Reaction of VG 9816 to leaf miner and *spodoptera* at Regional Research Station, Vriddhachalam

S. No	Seasons	VG 9816*		VRI 2	
		Leaf miner (%)	*Spodoptera litura* (%)	Leaf miner (%)	*Spodoptera litura* (%)
1	*Kharif* 2004	9.8	7.5	32.1	13.5
2	*Rabi*/summer 2004-05	6.6	16.9	12.8	43.5
3	*Kharif* 2005	13.6	No incidence	48.9	No incidence
4	*Rabi*/summer 2005-06	3.9	19.7	23.9	41.3

* Moderately Resistant

Pedigree of VRI (Gn) 6 Groundnut (Bunch)

1996	- ALR 2 x VG 9513 (Progeny testing, selection of variants and evaluation)
↓	
1999	- Station Trials (VG 9816)
↓	
2001-02	- MLT
↓	
2004-05, 2005-06	- ART & AICRP (IVT) & OFT
↓	
2007	- Variety release

□□□

Chapter 7

Heterosis for Yield and Yield Components in Greengram (*Vigna radiata* (L.) Wilczek)

SARAVANAN, K., T. SABESAN AND M. PRAKASH
Department of Agricultural Botany, Faculty of Agriculture
Annamalai University, Annamalainagar-608 002

Introduction

Greengram (*Vigna radiata* (L.) Wilczek.) is one of the most commonly cultivated pulse crop in India. It is an excellent source of digestible and high value protein in Indian diet. The magnitude of heterosis provides a basis for determining genetic diversity and also serves as a guide to the choice of desirable parents. The information on heterosis involving eight lines and three testers in greengram was studied in order to study the gene action of yield and yield components.

Materials and Methods

The materials for the present study consisted of eight lines *viz.*, LGG 410 (L_1), Chidambaram Local (L_2), VRMGg Local (L_3), LGG 460 (L_4), PMB 27 (L_5), Chidambaram local bold (L_6), EC 30072 (L_7) and VRMGg 1 (L_8) and three testers *viz.*, KM 2 (T_1), Chidambaram local polish (T_2) and VBN 1 (T_3). They were crossed during 2004-2005. The 24 F1s and eleven parents were grown in randomized block design with three replications. Each entry was grown in a single row of 4 m length. The inter and intra row spacing was 40 cm and 15 cm respectively. Observations were recorded on five random plants from each plot for days to first flower, plant height, number of branches per plant, number of clusters per plant, number of pods per plant, number of seeds per pod, hundred seed weight and seed yield per plant. Heterosis was estimated over mid parent (MP) and better parent (BP) and tested for significance as suggested by Wynne *et al* (1970).

Results and Discussion

The estimates of mean squares were highly significant for most of the characters indicating considerable diversity of parents. Most heterotic crosses identified on the basis of range, mean performance, heterosis parameters are presented in Table 1. The highest range was observed for plant height and number of pods per plant, number of clusters per plant, number of branches per plant, plant height and seed yield per plant. A perusal of heterotic crosses revealed that none of the crosses was which cross heterotic for all the traits simultaneously. The former showed higher potential for number of seeds per pod and plant height and the latter for number of pods per cluster and test weight.

From a total of 24 crosses studied, seven most promising combinations were identified and their heterotic expression for different characters analyzed (Table 2). Out of seven promising crosses, five showed positive significant heterosis. This may be attributed to presence of non additive gene effects for grain yield as reported by Chowdhuri (1991). The highest MP (103.48) and BP (53.56) heterosis were recorded for seed yield per plant in the cross $L_8 \times T_3$. The crosses $L_1 \times T_3$, $L_2 \times T_3$, $L_8 \times T_3$, $L_3 \times T_3$ and $L_6 \times T_3$ also displayed high significant positive heterosis for seed yield per plant. The crosses $L_3 \times T_3$ and $L_6 \times T_3$ were noted for days to first flower, plant height, number of branches, number of pods per cluster, number of clusters per plant and number of pods per plant.

In general, high x high and high x low gca crosses, which registered high magnitude of heterosis, indicated that the parents with diverse origin should be used for heterosis breeding. There seems to be a general dominance basis for number of clusters per plant, since five out of 24 crosses showed positive MP heterosis of which 19 significantly exceeded the BP values. MP and BP heterosis was positively significant in 15 an 11 crosses for seed yield per plant respectively. The magnitude and high incidence of heterosis in these crosses were indicative of high degree of dominance or epistasis or both. Similar results were reported by earlier workers (Bhatnagar and Singh, 1964 and Gupta 1980).

Table 1. Range, heterosis and most heterotic crosses for nine characters in blackgram

Characters	Range		Heterosis (%)		Best parent	Best hybrid	
	Parents	Crosses	MP	BP		Based on mean performance	Based on BP heterosis
Days to first flower	30.33 to 38.00	30.07 to 38.20	-7.49 to 5.95	-15.26 to 0.53	L8	L8 X T3	L6 X T3
Plant height	23.13 to 69.19	31.57 to 70.77	-11.84 to 108.98	-28.49 to 60.57	L6	L6 X T3	L3 X T3
Number of branches per plant	1.27 to 3.20	1.27 to 3.93	-28.89 to 83.61	-47.92 to 54.17	L3	L6 X T3	L1 X T3
Number of clusters per plant	6.33 to 15.80	7.73 to 17.00	-20.00 to 95.35	-40.51 to 75.00	L6	L6 X T3	L2 X T3
Number of pods per cluster	2.60 to 4.40	1.93 to 4.73	-50.85 to 43.43	-56.06 to 33.33	L3	L8 X T3	L3 X T3
Number of pods per plant	15.93 to 26.47	10.93 to 41.27	-38.69 to 131.40	-44.59 to 109.12	T3	L6 X T3	L8 X T3
Number of seeds per pod	5.33 to 7.73	5.33 to 7.73	-17.95 to 30.18	-28.57 to 27.91	L6	L3 X T3	L3 X T3
Hundred seed weight	4.39 to 5.36	3.30 to 5.85	-27.19 to 20.04	-29.52 to 13.50	L3	L6 X T3	L3 X T3
Seed yield per plant	3.88 to 10.41	4.38 to 13.39	-22.59 to 103.48	-35.78 to 61.87	L6	L8 X T3	L8 X T3

The positive heterosis observed in some of the crosses may be attributed to non allelic interaction which can either increase or decrease the expression of heterosis (Table 2), in most of the cases. Significant positive heterosis for seed yield per plant was associated with heterosis for number of pods per plant. This indicate that heterosis for yield was through heterosis of their yield components. Similar observation on high heterosis for grain yield was reported by Singh and Jain (1970) and Loganathan (2001).

Table 2. Heterosis for nine characters in promising crosses of greengram

Characters	$L_1 X T_3$		$L_2 X T_3$		$L_3 X T_3$		$L_5 X T_3$	
	MP	BP	MP	BP	MP	BP	MP	BP
Days to first flower	0.59**	-0.96**	-0.48	-0.58	-5.32**	-9.47**	-5.75	-15.56**
Plant height	-5.65	-23.76**	43.53**	42.23**	23.16**	9.55**	0.85**	-28.49*
Number of branches per plant	16.92**	-7.32	10.00**	-19.51	-5.62**	-12.50	-26.47	-47.92**
Number of clusters per plant	44.83**	11.11**	37.86**	12.70**	9.37**	7.69**	-20.00**	-40.51**
Number of pods per cluster	-4.26	-13.46	11.11**	7.14**	-14.85**	-27.12	-50.40**	-53.03**
Number of pods per plant	20.75**	4.79	12.69**	-6.05	14.29**	-0.25*	-38.69**	-44.59**
Number of seeds per pod	17.10**	5.61	15.15**	-1.79	11.58**	1.92	17.65	5.77**
Hundred seed weight	-5.26	-8.37	-12.05**	13.50**	3.31	-0.01	20.04*	9.14
Seed yield per plant	36.23**	26.02**	28.88**	22.43	33.16**	30.91**	-16.09**	-35.93

In general, the genotype L_3 and T_3, were found promising parents which resulted in high heterosis for seed yield per plant. The present study suggested that exploitation of L_6 x T_3, L_8 x T_3 and L_3 x T_3 could be more rewarding for breeding programme of green gram.

REFERENCES

Bhatnagar, C.P. and Singh, B. (1964). Genetic analysis of quantitative characters in blackgram. *J. Genet.*, 24: 89-91.

Choudhary, R.K. (1991). Combining ability and heterosis studies in blackgram. *Agric. Sci. Digest.*, 11: 175-177.

Gupta, S.N. (1980). Heterosis for yield and other quantitative characters in blackgram (*Vigna mungo* (L.) Hepper). *Indian J. Hered.*, 12: 49-55.

Loganathan, P (2001). Heterosis for yield and yield components in black gram (*Vigna mungo* (L.) Hepper). *Legume Res.*, 24: 77-81.

Singh, K.B. and Jain, R.P. (1970). Heterosis and combining ability in greengram. *Indian J. Genet.*, 30:244-250.

Wynne, J.C., Emery, D.A. and Rice, P.W. (1970). Combining ability estimates of *Arachis hypogaea* L. II Field performance of F_1 hybrids, *Crop Sci.*, 10: 713-715.

□□□

Chapter 8

Combining Ability and Heterosis for Yield and Yield Component Characters in Greengram (*Vigna Radiata* (L.) Wilczek.)

ANITHA VASLINE, Y. AND V. SUGUNA*
Department of Agricultural Botany, Faculty of Agriculture
Annamalai University, Annamalainagar - 608 002
*Department of Millets, TNAU, Coimbatore - 641 003

Introduction

Mungbean is one of the important pulse crops as it is an excellent source of easily digestable protein. Improvement of yield in greengram has been so far achieved through pure line selection by exploiting the existing variability. An yield break through has been achieved by hybridization. Breeding for high yielding varieties largely depends on the choice of parents. Knowledge of combining ability and heterosis helps in the selection of appropriate parents for hybridization programme. Hence the present study of line x tester design was undertaken to investigate the combining ability of 10 parents and to quantify the extent of heterosis for yield and yield component characters in greengram.

Materials and Methods

Seven lines *viz.*, VRMGG 1, LGG 410, VGG 8, LGG 450, VRMGG Local, PMB 27 and CDM Local and three testers *viz.*, ADT 3, VBN 1 and KM 2 were crossed in a L x T fashion in the Plant Breeding Farm, Faculty of Agriculture, Annamalai University during two seasons January to March, 2005 and June to August 2005. They were grown in a randomized block design with two replications by adopting a spacing of 30 cm x 10 cm. Recommended cultural and plant protection measures were adopted. Observations were recorded on number of clusters per plant, number of pods per cluster, number of pods per plant, number of seeds per pod, pod length, 100 seed weight and seed yield per plant. Combining ability analysis was worked out following the method suggested by Kempthorne (1957) and the significance of heterosis was tested using the formula suggested by Wynee *et al.* (1970).

Results and Discussion

Analysis of variance revealed significant difference for all the characters studied. The estimates of GCA and SCA variances revealed that the SCA variance were higher for all the characters indicating the preponderance of non-additive gene action. This results are similar with the earlier findings (Gopi, 2005).

The estimates of *gca* effects are presented in Table 1. Combining ability is an important parameter to evaluate the genetic potential of the material

Table 1. Estimates of general combining ability effects

Parents	Number of clusters/ plant	Number of pods / cluster	Number of pods / plant	Number of seeds / pod	Pod length	100 seed weight	Seed yield / plant
Lines							
VRMGG 1	0.33	-0.05	0.98*	-0.57*	-0.33**	0.20**	-1.87**
LGG 410	-1.17**	0.62**	-1.52**	-0.24*	-0.15**	-0.17**	-1.89**
VGG 8	-0.33	-0.05	-1.69**	0.60**	0.50**	-0.17**	-0.83**
LGG 450	1.17**	0.29**	0.98*	0.43**	0.33**	0.52**	4.67**
VRMGG Local	1.00**	-0.05	0.48	0.60**	0.17**	-0.02	1.57**
PMB 27	0.17	-0.38**	-0.36	-0.57**	-0.35**	-0.10*	-1.53**
CDM Local	0.17	-0.05	1.14*	-0.24*	-0.17**	-0.27*	-0.12
Testers							
ADT 3	0.36*	0.29**	1.68**	1.07**	0.16**	0.22**	0.68**
VBN 1	-0.07	-0.14**	1.48**	0.00	-0.04**	-0.30**	-0.76**
KM 2	-0.29*	-0.14**	-3.17**	-0.07	-0.12**	0.80*	0.63**

* Significant at 5 per cent level
** Significant at 1 per cent level

handled. Dhillon (1975) pointed out that the combining ability gives useful information on the choice of parents. In the present study, the *gca* effects of lines indicated that the line LGG 450 and VRMGG Local were good general combiners for all the yield attributing characters. This is in accordance with Saxena and Sharma (1993) and Gopi (2005). Among the testers, ADT 3 possessed desirable *gca* effects for all the yield attributing characters.

The estimates of *sca* effects are presented in Table 2. The *sca* effect is an important criterion for the evaluation of hybrids. Among the hybrids, LGG 450/ ADT 3 showed the highest *sca* effect for seed yield and its component traits. This is in accordance with Gopi (2005). The other promising hybrids like VRMGG Local / ADT 3 performed creditably well for number of pods / cluster, number of seeds per pod, pod length, 100 seed weight and seed yield per plant.

Diwakar and Singh (1993) suggested that the crosses with high *gca* effects generally gave high *sca* effects. In the present study, LGG 450/ ADT 3 was outstanding for yield and its component traits. This hybrid showed superior *sca* effect since their parents possessed high x high combination of *gca* effect. High *sca* effects were not only produced by high x high but also with high x low or low x high combinations of *gca* effects as observed in the hybrid like VGG 8 / KM 2 for hundred seed weight and seed yield per plant. Such potential cross from high x low combination have attributed to interaction between dominant alleles from poor combiner (Dubey, 1975). This is in accordance with Singh (1988).

A good hybrid manifest high amount of heterosis for commercial exploitation. Devaraj and Nadarajan (1995) stressed the need for computing standard heterosis for commercial exploitation of hybrid vigour. Hence the hybrids are evaluated based on standard heterosis. Among the hybrids, LGG 450 / ADT 3 and VRMGG Local / ADT 3 exhibited significant and positive standard heterosis for all the yield attributing traits (Table 3). This is in conformity with the findings by Aher *et al.* (2001) and Gopi (2005).

The selection of hybrids is based on high *sca* and heterotic effects. It may be concluded that the hybrids LGG 450 / ADT 3 and VRMGG Local / ADT 3 satisfies all these criteria and so they could be exploited for further crop improvement.

Table 2. Estimates of specific combining ability effects

Crosses	Number of clusters /plant	Number of pods / cluster	Number of pods / plant	Number of seeds / pod	Pod length	100 seed weight	Seed yield / plant
VRMGG 1 / ADT 3	-1.19**	0.05	-3.69**	-0.07	-0.16**	0.18*	-0.34
VRMGG 1/ VBN 1	1.24*	-0.52**	0.02	-1.00**	-0.64**	0.15	-2.77**
VRMGG 1 / KM 2	-0.05	0.48**	3.68**	1.07**	0.62**	-0.33**	3.11**
LGG 410 / ADT 3	-0.69	0.38**	-2.19**	1.10**	0.66**	0.25**	1.43**
LGG 410 / VBN 1	0.24	-0.19*	-0.48	0.17	-0.09	-0.43**	-0.17
LGG 410 / KM 2	0.45	-0.19*	2.67**	-1.26**	-0.56**	0.19*	-1.26**
VGG 8 / ADT 3	2.98**	-0.95**	7.98**	-0.24	0.01	-0.35**	-3.43**
VGG 8 / VBN 1	-1.60**	0.48**	-0.81	-0.17	0.21**	-0.23**	2.36**
VGG 8 / KM 2	-1.38**	0.48**	-7.17**	0.40*	-0.21**	0.59**	1.07**
LGG 450 / ADT 3	1.31**	1.95**	10.19**	1.07**	0.33**	0.66**	3.78**
LGG 450 / VBN 1	-0.76*	0.48**	9.52**	0.00	-0.13*	-0.17*	1.58**
LGG 450 / KM 2	-0.55	0.48**	-9.33**	1.07**	0.45**	-0.50**	-3.36**
VRMGG Local / ADT 3	-0.36	1.05**	5.81**	-0.24	-0.16**	-0.25**	0.06
VRMGG Local / VBN 1	2.07**	-0.52**	0.52	0.33	0.04	0.22**	-0.49
VRMGG Local / KM 2	-1.71**	-0.52**	-6.33**	-0.10	0.12*	0.04	0.44
PMB / ADT 3	-1.02**	0.38**	-1.86*	-0.07	-0.19**	-0.62**	1.95**
PMB / VBN 1	-0.60	-0.19*	-3.64**	1.00**	0.56**	0.30**	0.10
PMB / KM 2	1.62**	-0.19*	5.50**	-0.93**	-0.36**	0.32**	1.85**
CDM Local / ADT 3	-1.02**	0.05	-5.86**	0.60**	0.17**	0.15	2.45**
CDM Local / VBN 1	-0.60	0.48**	-5.14**	-0.33	-0.13*	0.17*	-0.60
CDM Local / KM 2	1.62**	-0.52**	11.00**	-0.05	-0.05	-0.31**	-1.85**

* Significant at 5 per cent level ** Significant at 1 per cent level

Table 3. Percentage of heterosis

Cross combination	No. of clusters per plant		No. of pods / cluster		No. of pods / plant		No. of seeds / pod		Pod length		100 seed weight		Seed yield / plant	
	d_{ii}	d_{iii}	d_{ii}	d_{iii}	d_{ii}	d_{iii}	d_{ii}	d_{iii}	d_{ii}	d_{iii}	d_{ii}	d_{iii}	d_{ii}	d_{iii}
VRMGG 1 × ADT 3	18.18	30.00**	25.00**	25.00**	51.22**	51.22**	-8.33**	10.00**	-13.33**	-13.33**	5.77*	11.11*	-38.54**	-21.30**
LGG 410 × ADT 3	0.00	10.00	20.00**	50.00**	42.86**	46.34**	4.17	25.00**	20.00**	0.00	0.10	5.05*	-27.82**	-7.57
VGG 8 × ADT 3	81.82**	100.00**	0.00	0.00	105.13**	68.29**	0.00	20.00**	15.38**	0.00	-11.54**	-7.07**	-51.10**	-37.39**
LGG 450 × ADT 3	30.77**	70.00**	-20.00**	50.00**	16.95**	95.12**	-8.33**	25.00**	-6.67**	-.667**	21.15**	27.27**	14.52**	46.65**
VRMGG Local × ADT 3	45.45**	60.00**	50.00**	25.00**	105.13**	95.12**	0.00	25.00**	0.00	-6.67**	-6.73**	-2.02	-15.01**	28.83*
PMB 27 ×ADT 3	18.18	30.00**	0.00	25.00**	57.50**	5366**	-8.33**	10.00**	-7.86**	-14.00**	-15.38	-11.11**	-46.32**	-31.27**
CDM Local × ADT 3	8.33	30.00**	0.00	25.00**	18.37**	41.46**	0.00	20.00**	0.00	-6.67**	-3.85	1.01	-10.72**	14.32**
VRMGG 1 × VBN 1	70.00**	70.00**	0.00	25.00**	50.00**	68.29**	0.00	0.00	-20.00**	-20.00**	0.00	0.00	-47.35**	-47.35**
LGG 410 ×VBN 1	20.00	20.00	0.00	25.00**	36.96**	53.66**	4.55	15.00**	-6.43**	-12.67**	-14.89**	-19.19**	-27.00**	-27.00**
VGG 8 × VBN 1	0.00	0.00	25.00**	25.00**	34.78**	51.22**	20.00**	20.00**	7.14**	0.00	-10.64**	-15.15**	-13.73**	1.10
LGG 450 × VBN 1	-7.69	20.00	25.00**	11.11**	49.15**	114.63**	9.09**	20.00**	-6.67**	-6.67**	-1.00	0.00	18.87**	38.13**
VRMGG Local × ADT 3	81.82**	100.00**	0.00	0.00	50.00**	68.29**	25.00**	25.00**	0.00	-6.67**	2.13	-3.03	-2.43	-2.47
PMB 27	30.00**	30.00**	0.00	-11.11**	28.26**	43.90**	9.09**	20.00**	0.00	-6.67**	2.13	-3.03	-22.14**	-22.17**
CDM Local × ADT 3	8.33	30.00**	25.00**	-11.11**	20.41**	43.90**	-8.33**	10.00**	-7.14**	13.33**	-10.00**	-9.09**	-16.89**	-16.59**
VRMGG 1 × KM 2	7.69	40.00**	25.00**	25.00**	63.41**	63.41**	0.00	20.00**	-6.67**	-6.67**	-10.19**	-2.02	9.79*	9.77*
LGG 410 × KM 2	-7.69	20.00	0.00	25.00**	42.86**	46.34**	-16.67**	0.00	-7.69**	-20.00**	-7.41**	1.01	-16.45**	-24.68**
VGG 8 × KM 2	-23.08**	0.00	25.00**	25.00**	2.56**	-2.44	4.17	25.00**	7.69**	-6.67**	0.00	9.09**	13.02**	1.92
LGG 450 × KM 2	-7.69	20.00	0.00	25.00**	-30.51**	0.00	8.33**	30.00**	0.00	0.00	-7.41**	1.01	-5.06	10.32*
VRMGGG ×KM 2	-7.69	20.00	0.00	0.00	17.95**	12.20*	0.00	20.00**	0.00	6.67**	-7.41**	1.01	40.26**	15.77**
PMN 27× KM 2	30.77**	70.00**	-20.00**	0.00	70.00**	65.85**	-16.67**	0.00	-14.29**	-20.00**	-3.70	5.05*	4.98	2.55
CDM Local × KM 2	30.77**	70.00**	-20.00**	0.00	67.35**	100.00**	-8.33**	10.00**	-7.14**	-13.33**	-18.52**	-11.11**	-15.72**	15.42**

* Significant at 5 per cent level ** Significant at 1 per cent level

REFERENCES

Aher, R.P., Dahat, D.V. and Surve, P.P. (2001). Diallel analysis for yield contributing characters in mungbean (*Vigna radiata* (L.) Wilczek). *Legume Res.*, 24(2):124-126.

Devaraj, M. and Nadarajan. N. (1995). Gene action for seven bio metrical traits in rice. *Ann. Agric. Res.*, 16(3): 360-361.

Dhillon, B.S. (1975). The application of partial diallel crosses in plant breeding review. *Crop Improv.*, 2: 1-7.

Diwakar, N.C. and Singh, A.K. (1993). Combining ability for oil content and yield attributes in yellow seeded Indian mustard (*Brassica juncea* L. Zerncross). *Ann. Agric. Res.*, 14: 194-198.

Dubey, R.S. (1975). Combining ability of cigar filter tobacco. *Indian J. Genet.*, 35: 76-82.

Gopi, P. (2005). Studies on genetics of certain economic characters through line x tester analysis in greengram (*Vigna radiata* (L.) Wilczek). *M.Sc. (Ag.) Thesis,* Annamalai Univ., Annamalainagar.

Kempthorne, O. (1957). An introduction to genetic statistics. John Wiley and Sons Inc., New York.

Saxena, S.D. and Sharma, R.K. (1993). Combining ability for yield in greengram (*Vigna radiata* (L.) Wilczek). *Legume Res.*, 16: 13-16.

Singh, K.M. (1988). Combining ability in wheat in normal and sodic soils. *Indian J. Genet.*, 48: 99-102.

Wynne, J.C., Emery, D.A. and Rice, D.W. (1970). Combining ability estimates in *Arachis hypogaea* L. II yield performance of F_1 hybrids. *Crop. Sci.*, 19(6): 713-715.

□□□

Chapter **9**

Interspecific Hybridization in Vigna Species

RAMAMOORTHI, N., M.PANDIYAN, SK.GANESH AND S. JEBARAJ
Tamil Nadu Agricultural University, National Pulses Research Centre
Vamban Colony, Pudukkottai, District

Importance of Pulses

Pulses play a vital role in furthering sustainable Agriculture and they are the major sources of dietary protein in the vegetarian diet. Importance of pulses is relatively more in Indian diet than that in Asia and world as a whole. Each plant of a pulses crop is virtually a nature's mini Nitrogen fertilizer factory which enables it to meet its own requirement as well as that of succeeding cereal crop. To alleviate protein-energy malnutrition, a minimum of 50 g pulses/capita/ day is required in addition to other sources of proteins such as cereals, milk, meat, eggs, etc. Today, more than 25 per cent of the babies born in the country are of low birth weight. Nearly, 50 per cent of the children under five years of age are stunted and approximately 70-90 per cent of the pregnant women are anaemic. This picture has not changed for the last 20 years.

Globally 60 million tonnes of pulses are produced annually from 70 million hectares. India is the largest producer with 33% of global area and produced

25 % of world's production. Normally the area under total pulses in the country is around 22.76 million hectares with a production of 13.12 million tonnes. The average productivity of the country is 577 kg/ha which is below the average global productivity of 857 kg/ha.

In Tamil Nadu, the total area under pulses is around 5.9 lakh ha with a production of 2.46 lakh tons. The average productivity of pulses in the state is around 410 Kg/ha which is far below the average productivity of the country as well as that of the global productivity. The area under blackgram in the state is around 2.34 lakh ha in the year 2004- 05 with a production of 1.02 lakh tons which accounts for an average productivity of 430 kg/ha (Directorate of Economics and Statistics, M/A, GOI). The increase in area and production is attributed to the development of high yielding and YMV resistant varieties suitable for cultivation in rabi season in rice fallows. The average productivity of blackgram in the state is just above the national average productivity of 419 Kg/ha. In the case of greengram the area was around 1.23 lakh ha in the year 2004- 05 with a production of 0.533 lakh tons which works out to an average productivity of 317 kg/ha. It is also just above the national average productivity of 435 kg/ha.

About 19.6 million tonnes of pulses are required by 2008, which is expected to touch 28 million tonnes by 2020. This can only be achieved by improving productivity and using additional areas for pulse production through diversification coupled with favourable Government policies.

Production Constraints in Pulses

- Lack of high yielding varieties adapted to diverse growing condition
- Large area under rainfed cultivation (88%)
- Biotic and abiotic stresses (up to 30% losses)
- Poor plant stand
- Poor response to high input conditions and better management
- Moisture stress at terminal growth stage
- Inadequate seed replacement rate
- Emerging deficiencies of secondary and micronutrients
- Low risk bearing capacity
- Resource poor farmers
- Poor crop management

There is an urgent need for a technology mission for boosting the production of pulses and legumes. This is a neglected area and should now receive priority.

Strategies for Increasing Pulses Production

There are three options to rectify this problem. They are,

- Genetic options
- Management options
- Developmental efforts

Genetic Options

Improved plant type
Exploitation of hybrid vigour
Exploitation of untapped yield genes
Transgenics in pulses
Pyramiding of resistance genes
Development of nutrient responsive and nutrient use efficient genotypes
Improvement in quality of protein

Management Options

Integrated nutrient management
Enhancing biological nitrogen fixation
IPM module against pod borer complex
Drought management
Mechanization

Developmental Options

Bringing Additional Area under Pulses (Horizontal Expansion)
Policy Support
Policy Issues
Post-harvest Technology and Value Addition

Apart from above options to achieve the yield increase, the yellow mosaic virus problem is major thrust area now in greengram and blackgram cultivation because these two crops occupied major area under cultivation in Tamil Nadu and also in India level. The wild species of these crops, in general are resistant to biotic and/or abiotic stresses and hence have been used to transfer the resistance/tolerance to cultivated types.

Even though yellow mosaic resistant varieties in grengram and black gram are available for cultivation now, these varieties are not showing complete resistance in all seasons. The varieties show varying degrees of resistance from season to season. The Priority should be given for research on development of durable resistant genotypes against yellow mosaic virus for which interspecific hybridization with available wild *Vigna* species is the better option.

Vigna Subgenus Ceraotropis

The genus *Vigna* is a large pan - tropical genus with 82 described species distributed among 7 subgenera (Marechal *et al.*, 1978).

Among the subgenera in the genus *Vigna* only subgenus Ceratrotropis has its center of species diversity in Asia.

Family	Leguminosae
Subfamily	Papilionoideae
Tribe	Phaseoleae
Subtribe	Phaseolinae
Genus	Vigna

Subgenera Ceratotropis, Haydonia, Lasiospron, Macrorhyncha, Plectotropis, Sigmoidtropis, and Vigna

The Asian *Vigna* consists of 21 species and remarkably, eight of these are used for human or animal food. The major cultivated species of genus *Vigna* are mung bean or green gram (*Vigna radiata* (L) Wilczek, urd bean or black gram (*V. mungo* (L) Hepper, cowpea (*V. unugiculata* (L) Walp.) and minor cultivated species moth bean (*V. aconitifolia* (Jacq.) Marechal), rice bean (*V. umbellata* (Thunb.) Ohwi and Ohashi), adzuki bean (*V. angularis* (Wild) Ohawi and Ohashi), *V. settulosa*, and *V. hainiana.*

The wild species are *V. radiata var. sublobata, V.mungo var. silvestris, V. trilobata, V. acconotifolia, V.glabrescense, V. hainiana, V. pilosa, V. bournea, V. marina, V. minima, V. khandalensis, V. dalzalliana and V. vexillata.*

Apart from cerototropis sub genus we have plectotropis species also (*V.vexillata* (L.) A. Rich).

Importance of Asian *Vigna*

Asian Vigna have importance in various traits that may make this group attractive.

- Very high nutritional value and high harvest index (*V. umbellata*) Herklots, 1972)

- Grain legume and cover crop adapted to very dry regions (*V.aconitifolia*) (Duke, 1981)
- Forage legume adapted to dry regions (*V. trilobata*) (Duke, 1981)
- Grain legume for sweet paste production (*V. angularis*) (Lumpkin and McClary, 1994)
- Grain legume for nutritious bean sprouts (*V. radiata*) (Smartt, 1990)

Interspecific Hybridization of Vigna Species Research in India

In national level, Vigna species are exploitated limitedly. Singh and Singh (2007) have attempted crosses in *V. radiata* x *V. mungo*, *V. mungo* x *V. radiata*, *V. radiata* var sublobata and *V. mungo* var. silvestris and *V. radiata* x *V. Umbellata* crosses.

V. radiata var sublobata and V. mungo var. silvestris have been used successfully (as male) in transfer of YMV resistance gene(s) to cultivated species. The crosses involving some other wild species with mungbean are reported to be successful but have not been used as an effective tool for improvement of mungbean.

The wide hybridization involving cultivated species, mungbean x black gram, blackgram x mungbean, mungbean x ricebean have been reported to be successful.

Blackgram x mungbean crosses were reported successful following embryo resue technique and success in mungbean x rice bean cross was reported with the use of an immunosuppressant, e-aminocaprioc acid (eaca)

A bridging species V. radiata var. Sublobata in mungbean x blackgram crosses have most frequently been successful and resulted into the development and release of commercial varieties.

This has provided scope for combining the desirable characteristics of mungbean and blackgram. Some accessions of blackgram possess durable resistance to mungbean yellow mosaic virus (MYMV) and are resistant to Cercospora leaf spot (cls), in addition blackgram cultivars are synchronous in maturity. Mungbean varieties are early in maturity, erect in growth habit and have been released. These are; pant mung-4 (type 44 x upu 2), meha (pant mung 2 x amp-36), hum1 and pant mung 6 (Pant M-2x AMP-36).

Sharma and Singh (2007) studied the wide cross involving mungbean (cv. BOYR-1) as female and blackgram (cv. OPU 8831) as male was successful.

Interspecific Hybridization of Vigna Species Research in Tamil Nadu

In Tamil Nadu we are exploiting more number of *Vigna* species to develop resistant genotypes in greengram and blackgram against Yellow Mosaic Virus. They are *V. radiata* x *V. mungo, V. mungo* x *V. radiata, V.mungo* x *V. unguiculata, Vigna radiata* x *Vigna radiata var. sublobata, V. radiata* x *Vigna mungo var. Silvestris, Vigna radiata* x *V.hainiana, V. radiata* x *V. umbellata, V. radiata* x *V. trilobata, V. radiata* x *V. acconitifolia, V. radiata* x *V. vexillata, V. radiata* x *V. glabresense, V. radiata var sublobata* x *V. mungo var. silvestris, Vigna radiata var. sublobata* x *V. radiata, Vigna mungo var. Silvestris* x *Vigna radiata, V.hainiana* x *Vigna radiata, V. umbellata* x *Vigna radiata, V. trilobata* x *Vigna radiata, V. acconitifolia* x *Vigna radiata, V. vexillata* x *Vigna radiata, V. glabresense* x *Vigna radiata.*

Research Programme Involving Wide Hybridization at TNAU

In TNAU, the cross involving V. radiata x V. mungo was found to be successful (Muralidharan, 1986, Ganeshram, 1993, Subramanian and Muthiah, 2000 and Pandiyan 2005).

Umamaheshwari (2002), Angel Priya (2004) and Pandiyan (2005) have studied the Vigna species crosses in *V. radiata* x *V. unguiculata, V. radiata* v *V.sublobata, V. radiata* x *V.umbellata* and found worthy in exploiting hybrid vigour and to get array of segregants.

National Pulses Research Centre is a lead centre for pulses crop research in Tamil Nadu and one of the main centres in national level. The main mandate of the centre is to evolve high yielding, biotic and abiotic resistant pulses varieties suitable for different ecosystems.

Maintenance and utilization of gene pool including wild species is of paramount importance in creating genetic variability. Keeping it in view we are maintaining more than 13 *Vigna* species namely *V. radiata, V. mungo, V. unguiculata, V. radiata var sublobata, V. mungo var. silvestris, V. hainiana, V. umbellata, V. trilobata, V. acconitifolia, V. vexillata, V. glabresense, V. dalzelliana,* and *V. khandalensis.* Every Vigna species have it own desirable characters as follows

S.No	Species	Desirable characters
1.	*Vigna radiata*	Short duration cultivated species and nutritional value, cosmetic field
2.	*Vigna mungo*	Short duration cultivated species
3	*Vigna unguiculata*	Short duration cultivated species

Contd...

4	*Vigna radiata* var *sublobata*	High number of pods per plant, MYMV resistance
5	*Vigna mungo var. silvestris*	YMV resistance and Bruchid resistance
6	*Vigna hainiana*	Short duration species, higher number of pods
7	*Vigna vexillata*	Lengthy pod, high number of seeds per pod & bruchids resistance
8	*Vigna umbellata*	Immune to MYMVand bruchid resistance
9	*V. trilobata*	Early ness, higher number of pods, fodder value
10	*V. acconitifolia*	Drought tolerant and dew tolerant
11	*V. glabresense*	Bruchid resistance
12	*V. dalzelliana*	YMV resistance
13	*V. khandalensis*	YMV resistance

Development of high yield coupled with MYMV resistance in mungbean is the need of the hour. To forward into the desired direction we are attempting wide crosses with wild Vigna species as detailed below

V. radiata x *V. mungo, V. mungo* x *V. radiata, V.mungo* x *V. unguiculata, Vigna radiata* x *Vigna radiata var . sublobata, V. radiata* x *Vigna mungo var. silvestris, Vigna radiata* x *V.hainiana, V. radiata* x *V. umbellata, V. radiata* x *V. trilobata, V. radiata* x *V. acconitifolia, V. radiata* x *V. vexillata, V. radiata* x *V. glabresense, V. radiata var sublobata* x *V. mungo var. silvestris.* In addition V. mungo with V. mungo var. silvestris crosses were also attempted.

Vigna radiata var. sublobata x *V. radiata, Vigna mungo var. Silvestris* x *Vigna radiata, V.hainiana* x *Vigna radiata, V. umbellata* x *Vigna radiata, V. trilobata* x *Vigna radiata, V. acconitifolia* x *Vigna radiata, V. vexillata* x *Vigna radiata* and *V. glabresense* x *Vigna radiata.* Both direct and reciprocal crosses were effected and hybrids recovered.

The highest pod set of 21.92 per cent was recorded in the cross *V. radiata* x *V.radiata* var. *sublobata* while *V. vexillata* x *V. radiata* possessed low crossability of 2.85 per cent. Among crosses F_1 hybrid of V. *radiata* x *V. hainiana* recorded high germination of 80.00 per cent. Even though crossability barriers were predominant, it was possible to recover interspecific hybrids from all the twelve direct and reciprocal crosses. High leathality and inviability was also noticed in F_1 hybrids. The highest pollen germination of 65.30 per cent was recorded in *V. radiata* x *V. radiata* var. *sublobata.* Pollen fertility estimate was sufficient enough to recover F_2 segregants in all the crosses except *V. radiata* x *V. umbellata.*

For quantitative traits the interspecific cross *V. umbellata* x *V. radiata* exhibited high *per se* performance for many traits viz., plant height, number of branches per plant, length of branch, number of clusters per branch, number of clusters per plant, hundred seed weight, and dry matter production. For seed yield , the cross *V. radiata* x *V. trilobata* registered the highest mean value among the direct crosses.

Promising F_1S were advanced further for evaluation. In F_2 generation, for all the traits in majority of the crosses the skewness was positive indicating that predominance of dominant alleles. Among the direct crosses *V. radiata* x *V. mungo* var. *silvestris* exhibited highest mean per se value for four traits viz., number of clusters per plant, number of clusters per branch, single plant yield and dry matter production. The cross *V. radiata* x *V. vexillata* showed better performance for the traits viz., length of pod and number of seeds per pod in both direction and also for other two traits viz., single plant yield and hundred seed weight only in reciprocal direction. The cross *V. trilobata* x *V. radiata* recorded best performance for three traits viz., number of clusters per branch, number of clusters per plant and number of pods per plant.

We have developed many YMV resistance entries both in greengram and blackgram through interspecific hybridization. The PYT, AYT (NPRC, VAMBAN), MLT and ART (Tamil Nadu) entries are in advanced stages which are derived from interspecific hybridization involving wild species.

Greengram

PYT and AYT

VGG04-006 (VRM(Gg)1 x *V. silvestris*), VGG04-015 (VRM(Gg)1 x V. silvestris), VGG04-018 (VRM(Gg)1 x *V. sublobata*), VGG04-20 (VRM(Gg)1 x *V. silvestris* 1), VGG04-021(VRM(Gg)1 x V. sublobata shiny seed), VGG04-022 (VRM(Gg)1 x V. silvestris 2), VGG04-024 (VRM(Gg)1 x V. silvestris 1),VGG 04-025, VGG 04-026, VGG 04-027, VGG05-002, VGG05-003, VGG05-004, VGG05-005, VGG05-006, VGG05-007, VGG05-008, VGG05-009, VGG05-010, VGG06-001, (VRM(Gg)1 x *V.umbellata*) VGG06-002 (VRM(Gg)1 x V. Glabresense), VGG06-003, VGG05-004 (VRM(Gg)1 x *V. glabresense*).

MLT

VGG 04-021 (VRM(Gg)1 x *V. sublobata* shiny seed)

ART

VGG04-024 (KM4-44 x VRM(Gg)1 x V. silvestris 1)

Blackgram

PYT and AYT

VBG 04-002 (*Vigna mungo* var silvestris x VM(Gg)1), VBG 04-003 (*Vigna mungo* var silvestris x *Vigna radiata* CO6), VBG 04-004 (*Vigna mungo* var silvestris x Vigna radiata VBN1), VBG 04-005 (*Vigna mungo* var silvestris x *Vigna radiata* var. Sublobata early), VBG 04-006 (*Vigna mungo* var silvestris x *Vigna glabresense*), VBG 04-007 Yellow seeds (*Vigna mungo* var silvestris x Vigna radiata var. Sublobata), VBG 04-009 Yellow seeds (*Vigna mungo var. silvestris* 2 x *V. radiata (VRM(Gg)1)*, VBG 04-014 (*V. mungo x Vigna mungo var. silvestris* 2),

VBG 05-001, VBG 05-002, VBG 05-003, VBG 05-004,VBG 05-005, VBG 05-006,VBG 05-007

VBG 05-008, VBG 05-009, VBG 05-010, VBG 05-011, VBG 05-012, VBG 05-013,VBG 05-014, VBG 05-015,VBG 06-001, VBG 06-002, VBG 06-003, VBG 06-004,VBG 06-005, VBG 06-006,VBG 06-007, VBG 06-008, VBG 06-009, VBG 06-010, VBG 06-011, VBG 06-012, VBG 06-013,VBG 06-014, VBG 06-015, ,VBG 06-016,VBG 06-017,VBG 06-018,VBG 06-019,VBG 06-020,VBG 06-021

MLT

VBG 04-014 (V.mungo *x Vigna mungo var. silvestris* 2)

ART

VBG 04-001 (Vigna mungo var silvestris x Vigna radiata var. sublobata)

VBG 04-008 (*Vigna mungo var. silvestris* 2 x *V. radiata (VRM(Gg)1)*

Among the entries derived from *V. radiata, V. mungo* with wild *Vigna* species, the *Vigna radiata* entry VGG04-024 (KM4-44 x VRM(Gg)1 x V. silvestris 1) is in Adaptive Research Trial. It matures in 65 to 70 days duration, terminal podding and medium bold type with MYMV resistance and also with desirable nutritional quality parameters viz., protein and phosphorus content.

Development of ideotype is one of the primary objectives of pulses breeding programme. For blackgram, the entries VBG 04-008, VBG 04-009 (Yellow blackgram) and VBG04-014 are evolved with such objective. These entries are highly resistant to Yellow Mosaic Virus when grown throughout the year.

It is highlighted that VBG 04-009 is yellow grain type in blackgram which is interspecific derivative of V. mungo x V. mungo var silvestris. This entry has high amount of protein, minerals and carbohydrates. It is highly suitable for jelabi and jangry making in bakery. The current mission mode technology is not freedom from hunger but prevention from under nutrition. The development of quality protein may envisage the desired goal.

Future Challenges

In Addition to YMV Development

- To improve the nutritional status of pulse consumers, it is essential to develop micro nutrient - dense varieties. The most important micronutrients, which can be improved through breeding, are iron and zinc.
- Marker Assisted Selection has had dramatic impacts, particularly in the private sector, on evolution of varieties with inbuilt resistance to either biotic or abiotic stress. A combination of conventional and molecular breeding approaches can lead to new advances in legume productivity for agricultural development and improved livelihoods of the rural poor.
- Integrated Crop Management (ICM) technologies should focus on the entire cropping system with emphasis on year - round and multi- year management.
- There is a need to evaluate the challenging scenario of pests in a crop sequence as a whole, and detailed information on the ecology, epidemiology and life cycle pattern of key pests is essential to develop, validate and promote cost effective IPM modules for pod borer complex and effective storage systems for Bruchid management.
- Development of disease prediction models needs to be initiated. The precision and accuracy of these models need to be tested and validated at Multi Location for future suitability especially for the devastating disease, YMV.
- The potentials of local antagonists and bio- agents should be tapped and the most suitable ones should be involved as components of IPM programmes.
- The markets for pulses are thin and fragmented, new niches need to be identified and appropriately supported by improved technologies and policies.
- Although government regularly announces minimum support prices for pulses but their procurement is not effective as for major cereals. The government should take measures to ensure that growers get remunerative price for their produce.

- Three approaches are to be adopted to enhance nutritional quality of pulses. First, biotechnology would help to alter methionine contents of legumes through the use of methionine - rich or methionine - enriched protein genes driven by seed specific promoters. But, they should be devoid of allergens. Secondly, improving the digestibility of proteins and starch, reducing flatulence, altering fatty acid composition, minimizing formation of undesirable flavors and aroma should also be the future strategy. Finally, processing technologies should help to transform raw grains into useful products with minimum nutritional loss. Innovative fortification of cereals with legumes to improve protein content quality should ensure nutritional security.

❑❑❑

crossing. In the next season, selfed seeds of F_1 were harvested for generation building.

All the five generations of six crosses were raised during summer 2001 in randomized block design with three replications. The crosses were [illegible] generations were randomized within [illegible] were raised in a plot with a spacing of 30 cm between rows and 10 cm between plants. Ten plants each [illegible] in F_1 and 40 plants in F_2 generation per replication were randomly selected for observations. The A, B, C and D scaling test for testing adequacy of additive-dominance model was computed using the scales [illegible] Mather (1949). Genetic components of generation mean were estimated by following five-parameter model suggested by Hayman (1958).

Results and Discussion

Pusa Garima, being the most used [illegible] variety, with desirable yield [illegible] et al. [illegible] and Joshi [illegible] crosses involving one of the parents as good general combiner can be expected to throw desirable segregants. In each of the [illegible] crosses, [illegible] P_1, P_2, F_1, F_2 and F_3 were considered [illegible] important because [illegible] additive and dominance components [illegible] gene interaction. The scaling test (Mather, 1949) showed that in all six crosses with five generations, either C or D was significant for all the characters studied, indicating the presence of epistatic interactions of either additive x additive or dominance x dominance [illegible] both C and D were significant indicating the presence of both types of epistatic interactions. The mean performance of five generations of cross viz. P_1, P_2, F_1, F_2 and F_3 involving Pusa Garima as one of the parents are depicted in (Table 1). The additive (*d*), dominance (*h*) and epistatic interaction effects (*i*, *l*) and the scales C and D estimated from the means of five generations (Table 2). The significant components (*d*), (*h*), (*i*) and (*l*) in crosses 2, 4 and 5 indicated involvement of all three types of gene action like additive, dominance and epistasis in days to first flowering. Complementary type of gene action was observed in crosses 1, 3 and 6 and duplicate gene action was observed in crosses 2, 4 and 5. In cross 1 and 6, dominance x dominance (*l*) interaction alone was significant. Though dominant x dominant interaction was significant in crosses 1 and 6, no significant heterosis was observed. It may be due to the fact that, dominance effect did not deviate from zero as denoted by non-significant '*h*' values in

Chapter **10**

Genetic Analysis for Yield Attributes in Vegetable Cowpea (*Vigna Unguiculata* (L.) Walp.)

MANIVANNAN, R.[1], DR. K. SEKAR[2] AND DR. MOHAN ANDRE SAVERY[3]
1. Dept. of Horticulture, Pondicherry University, Puducherry - 605 014
2. Dept. of Horticulture, Faculty of Agriculture, Annamalai University Annamalainagar - 608 002. 3. Krishi Vigyan Kendra, Puducherry - 605 009

Introduction

Studies on the nature of gene action governing the yield and its component traits in cowpea are available while information on vegetable cowpea are limited. Hence, an attempt was made to unravel the genetic architecture involving vegetable cowpea varieties.

Materials and Methods

Five high yielding vegetable cowpea lines viz., EC 240741, EC 244018, EC 367690, IC 201087, IC 201099 and IC 202853 were crossed with Arks Garima during June, 2000 at the Horticulture Dept., Faculty of Agriculture, Annamalai University, Annamalai Nagar to effect fresh crosses to obtain enough F_1 seeds. Selfing of parents were also done to get enough seeds for generation building. The selfed seeds of selected hybrids obtained from line x tester analysis were also raised to study F_2 after reserving enough seeds for raising in the next season. Selfed seeds of F_2 were harvested for generation building.

All the five generations of six crosses were raised during January, 2001 in randomized block design with three replications. The crosses were first randomized in each block and the different generations were randomized within a cross. Plants were raised by adopting a spacing of 60 cm between rows and 30 cm between plants. Ten plants in each of P_1, P_2, and F_1; 60 plants in F_2 and 40 plants in F_3 generation per replication were randomly selected to record observations. The adequacy of the data for a simple additive - dominance model was tested utilizing the scales C and D as suggested by Mather (1949). Genetic components of generation mean were estimated by following five-parameter model suggested by Hayman (1958).

Results and Discussion

Arka Garima, being the broad based ruling variety, with desirable yield and quality characteristics of vegetable cowpea, six crosses involving it as one of the parents were selected for generation mean analysis (Table 1). Paroda and Joshi (1970) and Gill *et al.* (1972) also opined that crosses involving one of the parents as good general combiner can be expected to throw desirable segregants. In each of the six cross combinations, five generations viz., P_1, P_2, F_1, F_2 and F_3 were constituted and studied. The test of additivity of scales is important because in most of the cases the estimation of additive and dominance components of variance is made assuming the absence of gene interaction. The scaling test (Mather, 1949) showed that in all six crosses with five generations either scale C or D was significant for all the characters studied, indicating the presence of genetic interactions of either additive x additive or dominant x dominant interactions. In some cases both C and D were significant indicating the presence of both types of genetic interactions. The mean performance of five generations of crosses viz., P_1, P_2, F_1, F_2 and F_3 involving Arka Garima as one of the parents are depicted in (Table 1). The additive (*d*), dominance (*h*) and epistatic interaction effects (*i* , *l*) and the scales C and D estimated from the means of five generations (Table 2). The significant components (*d*), (*h*), (*i*) and (*l*) in crosses 3, 4 and 5 indicated involvement of all three types of gene action like additive, dominance and epistasis in days to first to first flowering. Complementary type of gene action was observed in crosses 1, 5 and 6 and duplicate gene action was observed in crosses 2, 3 and 4. In cross 1 and 6, dominant x dominant (*l*) interaction alone was significant. Though dominant x dominant interaction was significant in crosses 1 and 6, no significant heterosis was observed. It may be due to the fact that, dominance effect did not deviate from zero as denoted by non-significant '*h*' values in crosses 1 and 6 for the trait, days to first flowering. This is in agreement with the earlier reports by Thiagarajan *et al.* (1990), Mallikarjun *et al.* (1995), Ranvir Singh and Dabas *et al.* (1992), Bhor and Dumbre (1998) and Anbuselvam (1999). Cross 4 showed significant (*d*), (*h*), (*i*) and (*l*) components indicating the presence of all three types of gene action in controlling the trait, plant height. Presence of additive, dominance and epistasis gene action was observed in cross 4 for number of branches per plant which had significant '*d*', '*h*', '*i*' and '*l*' effects, whereas crosses 1 and 5 showed significant '*d*', '*h*', and '*i*'. All three types of gene action were indicated by the significance of '*d*', '*h*', '*i*' and '*l*' in crosses 1, 4 and 5 and '*d*', '*h*' and '*l*' in cross 2 for number of branches per plant. The result goes in support of the findings of Rahman and Saad (2000). Complementary type of epistatic gene action was observed in crosses 1, 2, 4 and 6 while it was duplicatory in crosses 3 and 5 for this trait. In pod length, the crosses 1 and 3 expressed significant *d, h* and *l* component. The *d, h* and *l* were significant in cross 2 whereas additive and epistatic gene action was observed in cross 4. Additive component alone was significant in cross 5 and the effects '*d*' and '*i*' were significant in cross 6. The results fall in line with the findings of Nagaraj *et al.* (2002) who also indicated additive x additive (*i*) epistatic interaction for this trait. The trait pod weight showed additive, dominance and epistatic gene action in cross 3 whereas (*d*), (*i*) and (*l*) were significant in crosses 1, 2 and 5. In cross 4, additive, dominance and additive x additive gene action were observed. Additive gene effects (*d*) and (*i*) alone was significant in cross 6. Complementary type of gene action was found in all crosses except cross 3 for this trait. Involvement of all types of gene action viz., additive, dominance and epistasis was observed in all crosses for the trait, green pod yield per plant. Rahman and Saad (2000) observed both additive x additive (*i*) and dominance (*l*) contributed more for pod yield per plant. Complementary type of epistatic gene action was observed in all crosses. Nagaraj *et al.* (2002) also found (*i*) type of epistasis for this trait. The trait, protein content expressed all types of gene action indicated by significant effects of *d, h, i* and *l* in all crosses except cross 5. Additive (*d*) and epistatic gene action (*i*) and (*l*) were significant in cross 5. Complementary type of gene action was found in all crosses for this trait. Crude fibre content was observed to be controlled by additive, dominance and epistasis gene action in crosses 1 and 4. Significant '*d*', '*i*' and '*l*' in crosses 5 and 6 indicated additive and epistatic gene action. The dominance and epistatic components '*h*', '*i*' and '*l*' were significant in cross 2. Epistatic gene effects (*i*) and (*l*) were significant in cross 2. Complementary gene action was observed in crosses 1 and 2 and it was duplicatory in other crosses.

Table 1. Generation means value for various traits

a. Days to first flowering

Parameter	Cross 1	Cross 2	Cross 3	Cross 4	Cross 5	Cross 6
P_1	3.980 ± 0.74	41.71 ± 1.13	40.70 ± 0.84	36.97 ± 0.46	38.33 ± 0.43	37.20 ± 0.72
P_2	38.33 ± 0.77	38.33 ± 0.77	38.33 ± 0.77	38.33 ± 0.77	38.33 ± 0.77	38.33 ± 0.77
F_1	36.37 ± 0.73	40.90 ± 1.03	37.65 ± 0.78	35.73 ± 0.60	38.37 ± 0.58	37.14 ± 0.68
F_2	39.00 ± 0.31	44.82 ± 0.46	38.60 ± 0.32	36.62 ± 0.25	41.27 ± 0.25	39.92 ± 0.30
F_3	38.61 ± 0.38	43.12 ± 0.53	42.51 ± 0.43	39.49 ± 0.33	41.59 ± 0.30	39.27 ± 0.36

b. Plant height (cm)

Parameter	Cross 1	Cross 2	Cross 3	Cross 4	Cross 5	Cross 6
P_1	54.78 ± 1.10	53.78 ± 1.35	60.18 ± 1.24	56.33 ± 0.95	62.94 ± 0.94	60.86 ± 1.13
P_2	60.71 ± 1.22	60.71 ± 1.22	60.71 ± 1.22	60.71 ± 1.22	60.71 ± 1.22	60.71 ± 1.22
F_1	54.76 ± 1.10	53.47 ± 1.35	62.06 ± 1.28	52.59 ± 0.89	71.12 ± 1.07	59.61 ± 1.10
F_2	56.89 ± 0.46	57.54 ± 0.85	70.24 ± 0.58	53.03 ± 0.36	73.10 ± 0.44	62.92 ± 0.47
F_3	55.30 ± 0.55	57.43 ± 0.71	70.33 ± 0.72	54.96 ± 0.46	80.41 0.60	65.33 0.60

c. Number of branches per plant

Parameter	Cross 1	Cross 2	Cross 3	Cross 4	Cross 5	Cross 6
P_1	3.19 ± 0.06	4.15 ± 0.10	3.86 ± 0.08	4.96 ± 0.08	4.65 ± 0.07	4.39 ± 0.08
P_2	3.90 ± 0.08	3.90 ± 0.08	3.90 ± 0.08	3.90 ± 0.08	3.90 ± 0.08	3.90 ± 0.08
F_1	3.39 ± 0.06	3.65 ± 0.09	3.10 ± 0.06	4.87 ± 0.08	4.88 ± 0.07	3.96 ± 0.07
F_2	2.86 ± 0.02	3.82 ± 0.16	2.97 ± 0.02	4.44 ± 0.03	4.70 ± 0.03	3.67 ± 0.03
F_3	2.62 ± 0.03	3.46 ± 0.04	3.04 ± 0.03	4.41 ± 0.04	4.62 ± 0.03	3.98 ± 0.04

d. Number of pods per plant

Parameter	Cross 1	Cross 2	Cross 3	Cross 4	Cross 5	Cross 6
P_1	26.70 ± 0.54	20.16 ± 0.51	24.02 ± 0.50	25.49 ± 0.43	28.94 ± 0.43	24.32 ± 0.45
P_2	23.02 ± 0.46	23.02 ± 0.46	23.02 ± 0.46	23.02 ± 0.46	23.02 ± 0.46	23.02 ± 0.46
F_1	23.12 ± 0.46	21.88 ± 0.55	26.49 ± 0.55	26.91 ± 0.45	26.09 ± 0.39	25.97 ± 0.48
F_2	15.43 ± 0.12	17.92 ± 0.18	24.28 ± 0.20	24.69 ± 0.17	25.53 ± 0.15	22.90 ± 0.17
F_3	14.36 ± 0.14	18.14 ± 0.22	22.49 ± 0.23	24.52 ± 0.20	26.62 ± 0.20	23.15 ± 0.21

e. Pod length (cm)

Parameter	Cross 1	Cross 2	Cross 3	Cross 4	Cross 5	Cross 6
P_1	18.51 ± 0.36	17.38 ± 0.44	19.11 ± 0.39	15.01 ± 0.25	27.32 ± 0.41	15.19 ± 0.28
P_2	22.72 ± 0.46	22.72 ± 0.46	22.72 ± 0.46	22.72 ± 0.46	22.72 ± 0.46	22.72 ± 0.46
F_1	17.79 ± 0.36	16.55 ± 0.42	19.26 ± 0.40	19.21 ± 0.33	28.50 ± 0.43	18.09 ± 0.33
F_2	14.87 ± 0.12	14.71 ± 0.17	16.50 ± 0.14	14.63 ± 0.10	28.50 ± 0.17	18.09 ± 0.13
F_3	14.62 ± 0.14	13.94 ± 0.17	16.70 ± 0.17	15.70 ± 0.13	28.73 ± 0.21	17.72 ± 0.16

f. pod weight (g)

Parameter	Cross 1	Cross 2	Cross 3	Cross 4	Cross 5	Cross 6
P_1	5.73 ± 0.11	5.55 ± 0.14	4.43 ± 0.09	4.11 ± 0.07	7.94 ± 0.12	5.92 ± 0.11
P_2	10.28 ± 0.21	10.28 ± 0.21	10.28 ± 0.21	10.28 ± 0.21	10.28 ± 0.21	10.28 ± 0.21
F_1	7.10 ± 0.14	7.00 ± 0.18	5.62 ± 0.12	5.06 ± 0.09	10.56 ± 0.16	6.38 ± 0.12
F_2	6.59 ± 0.05	7.89 ± 1.03	5.35 ± 0.04	4.85 ± 0.03	10.08 ± 0.06	6.30 ± 0.05
F_3	6.63 ± 0.07	6.16 ± 0.08	6.11 ± 0.06	4.80 ± 0.04	10.12 ± 0.08	6.29 ± 0.06

g. Green pod yield per plant (g)

Parameter	Cross 1	Cross 2	Cross 3	Cross 4	Cross 5	Cross 6
P_1	125.80 ± 2.52	125.60 ± 3.15	110.84 ± 2.28	103.80 ± 1.75	227.74 ± 3.41	149.74 ± 2.77
P_2	260.17 ± 5.22	260.17 ± 5.22	260.17 ± 5.22	260.17 ± 5.22	260.17 ± 5.22	260.17 ± 5.22
F_1	161.65 ± 3.25	147.24 ± 3.62	135.85 ± 2.81	149.95 ± 2.53	347.32 ± 5.22	143.85 ± 2.63
F_2	109.72 ± 0.89	117.45 ± 1.63	103.20 ± 0.86	114.89 ± 0.78	294.46 ± 1.77	107.98 ± 0.83
F_3	107.62 ± 1.06	114.37 ± 1.42	102.28 ± 1.04	108.91± 0.91	295.63 ± 2.24	107.69 ± 0.98

h. Protein content

Parameter	Cross 1	Cross 2	Cross 3	Cross 4	Cross 5	Cross 6
P_1	6.02 ± 0.12	4.63 ± 0.12	4.82 ± 0.10	5.39 ± 0.09	5.85 ± 0.04	3.86 ± 0.07
P_2	4.92 ± 0.05	4.92 ± 0.05	4.92 ± 0.05	4.92 ± 0.05	4.92 ± 0.05	4.92 ± 0.05
F_1	4.65 ± 0.09	4.33 ± 0.11	5.00 ± 0.10	4.82 ± 0.08	5.23 ± 0.03	5.15 ± 0.09
F_2	3.96 ± 0.03	3.81 ± 0.15	3.66 ± 0.03	3.52 ± 0.02	3.82 ± 0.01	3.30 ± 0.02
F_3	3.79 ± 0.04	3.43 ± 0.04	3.74 ± 0.04	3.44 ± 0.03	3.73 ± 0.02	3.09 ± 0.03

i. Crude fibre

Parameter	Cross 1	Cross 2	Cross 3	Cross 4	Cross 5	Cross 6
P_1	21.15 ± 0.42	20.48 ± 0.51	18.97 ± 0.39	24.08 ± 0.41	20.43 ± 0.31	21.13 ± 0.39
P_2	18.52 ± 0.37	18.52 ± 0.37	18.52 ± 0.37	18.52 ± 0.37	18.52 ± 0.37	18.52 ± 0.37
F_1	20.75 ± 0.42	21.11 ± 0.53	20.02 ± 0.41	23.24 ± 0.39	16.71 ± 0.25	19.23 ± 0.35
F_2	12.74 ± 0.10	20.09 ± 0.31	19.31 ± 0.16	23.24 ± 0.16	14.62 ± 0.09	17.82 ± 0.13
F_3	13.99 ± 0.14	19.61 ± 0.24	19.99 ± 0.20	22.14 ± 0.18	15.25 ± 0.11	17.85 ± 0.16

Table 2. Scaling test and genetic effects for various traits

a. Days to first flowering

Parameters	Cross 1	Cross 2	Cross 3	Cross 4	Cross 5	Cross 6
Scale C	7.95*	-81.46*	0.07	9.28*	21.58*	9.87*
Scale D	1.13	-41.80*	13.81*	18.98*	17.06*	1.71
(*m*)	39.00* ± 0.31	44.54*± 0.45	38.60*± 0.32	36.62*± 0.25	41.27* ± 0.25	39.92* ± 0.30
(*d*)	-0.68 ± 0.53	1.69* ± 0.69	1.18* ± 0.57	-5.46*± 0.45	-4.95* ± 0.44	-0.57 ± 0.53
(*h*)	-0.71 ± 1.65	1.36 ± 1.83	-11.06* ± 1.42	-8.25*± 1.09	-2.79* ± 1.03	-0.12 ± 1.22
(*i*)	-0.78 ± 1.49	3.86 ± 2.16	-6.83* ± 1.63	-22.03* ±1.17	-17.68*± 1.12	-0.63 ± 1.43
(l)	-9.09* ± 3.77	-17.28* ± 5.39	18.32* ± 4.03	12.93*± 3.10	-6.03* ± 2.99	-10.88* ± 3.56
Epistasis	Complementary	Duplicate	Duplicate	Duplicate	Complementary	Complementary

b. Plant height

Parameters	Cross 1	Cross 2	Cross 3	Cross 4	Cross 5	Cross 6
Scale C	2.55	-42.13*	35.95*	-10.10*	26.51*	10.89*
Scale D	-8.07*	-31.66*	19.95*	-3.26	51.79*	13.91*
(*m*)	56.89* ± 0.46	58.65* ± 0.60	70.24* ± 0.58	53.03*± 0.36	73.10* ± 0.44	62.92* ± 0.13
(*d*)	-2.97* ± 0.82	-3.47* ± 0.93	-0.27 ± 0.87	-2.19 ± 0.77	1.12 ± 0.77	-0.08 ± 0.83
(*h*)	2.82 ± 1.87	-2.00 ± 2.43	-5.69* ± 2.40	-5.44*± 1.54	-20.81*± 1.96	-8.63* ± 1.99
(*I*)	-0.13 ± 2.21	-3.36 ± 2.78	-7.84* ± 2.66	-3.89*± 1.86	-27.87*± 2.13	-7.31* ± 2.28
(l)	-14.16* ± 5.52	-20.32* ± 7.12	-21.33* ± 6.93	9.12* ± 4.46	33.71* ± 5.54	4.03 ± 5.72
Epistasis	Duplicate	Complementary	Complementary	Duplicate	Duplicate	Duplicate

Number of branches per plant

Parameters	Cross 1	Cross 2	Cross 3	Cross 4	Cross 5	Cross 6
Scale C	-2.43*	214.80*	-2.08*	-0.84*	0.49*	-1.53*
Scale D	-2.33*	106.59*	-1.54*	-0.10	0.53*	0.29
(*m*)	2.86* ± 0.02	3.51* ± 0.04	2.97* ± 0.02	4.44* ± 0.03	4.70* ± 0.03	3.67* ± 0.03
(*d*)	-0.36* ± 0.05	0.13* ± 0.06	-0.02 ± 0.06	0.53* ± 0.06	0.36* ± 0.05	0.25* ± 0.06
(*h*)	0.99* ± 0.09	-0.23 ± 0.15	-0.10 ± 0.11	0.37* ± 0.13	0.33* ± 0.12	-0.63* ± 0.12
(*I*)	0.44* ± 0.12	0.85* ± 0.19	0.64* ± 0.14	0.99* ± 0.16	0.46* ± 0.14	0.04 ± 0.15
(l)	0.13 ± 0.29	0.11 ± 0.44	0.72* ± 0.31	0.99* ± 0.38	0.05 ± 0.35	2.43* ± 0.35
Epistasis	Complementary	Duplicate	Duplicate	Complementary	Complementary	Duplicate

c. Pod length

Parameters	Cross 1	Cross 2	Cross 3	Cross 4	Cross 5	Cross 6
Scale C	-17.33*	-1.51	-14.35*	-17.63*	-6.96*	-1.73
Scale D	-12.49*	-3.38*	-8.03*	-4.19*	-7.88*	-3.21*
(*m*)	14.87* ± 0.12	14.85* ± 0.15	16.50* ± 0.14	14.36*± 0.10	28.50* ± 0.17	18.09* ± 0.13
(*d*)	-2.11* ± 0.29	-2.67* ± 0.32	-1.81* ± 0.30	-3.86* ± 0.26	2.30* ± 0.31	-3.77* ± 0.27
(*h*)	2.61* ± 0.51	3.56* ± 0.62	1.31* ± 0.59	0.20 ± 0.46	-0.61 ± 0.72	0.99 ± 0.56
(I)	1.23 ± 0.67	1.72* ± 0.80	-0.65 ± 0.74	-7.86* ± 0.56	0.51 ± 0.84	-5.68* ± 0.65
(l)	6.45* ± 1.56	0.32 ± 1.89	8.43* ± 1.78	17.92*± 1.38	1.23 ± 2.12	-1.97 ± 1.65
Epistasis	Complementary	Complementary	Complementary	Complementary	Duplicate	Duplicate

d. Pod weight

Parameters	Cross 1	Cross 2	Cross 3	Cross 4	Cross 5	Cross 6
Scale C	-3.85*	29.01*	-4.55*	-5.11*	0.98	-3.76*
Scale D	-2.67*	10.51*	-0.97*	-4.89*	2.10*	-3.64*
(*m*)	6.59* ± 0.05	6.09* ± 6.06	5.35* ± 0.04	4.85* ± 0.03	10.08* ± 0.06	6.30* ± 0.05
(*d*)	-2.28* ± 0.12	-2.37* ± 0.13	-2.93* ± 0.11	-3.09* ± 0.11	-1.17* ± 0.12	-2.18* ± 0.12
(*h*)	0.23 ± 0.23	0.42 ± 0.27	-1.85* ± 0.20	0.27* ± 0.14	0.21 ± 0.26	0.08 ± 0.20
(I)	-3.41* ± 0.27	-3.39* ± 0.31	-5.96* ± 0.23	-3.76* ± 0.18	-3.58*± 0.29	-2.56* ± 0.24
(l)	1.56* ± 0.67	2.79* ± 0.80	4.77* ± 0.58	0.29 ± 0.41	1.49* ± 0.76	0.60 ± 0.50
Epistasis	Complementary	Complementary	Duplicate	Complementary	Complementary	Complementary

e. Green pod yield per plant

Parameters	Cross 1	Cross 2	Cross 3	Cross 4	Cross 5	Cross 6
Scale C	270.39*	648.71*	229.91*	-204.31*	-4.71	-265.46*
Scale D	-174.93*	-376.90*	-168.29*	-158.11*	105.69*	-194.88*
(*m*)	109.72*± 0.89	119.29* ± 1.21	103.20*± 0.86	114.89* ± 0.78	294.46*± 1.77	107.98*± 0.83
(*d*)	-67.19*± 2.90	-67.28* ± 3.14	-74.67* ± 2.85	-78.19*± 2.76	-16.22*± 3.12	-55.33* ± 2.96
(*h*)	40.22* ± 3.98	31.75* ± 5.13	24.22* ± 3.77	39.32*± 3.34	32.12*± 7.77	24.68* ± 3.57
(*I*)	-62.82*± 5.42	-57.17* ± 6.61	-75.46* ± 5.10	-85.02* ± 4.56	-103.67* ± 8.56	-24.98* ± 5.28
(l)	127.28*± 12.54	48.29* ± 15.74	82.16*± 11.57	61.60± 10.39	147.20*± 23.19	94.11*± 11.01
Epistasis	Complementary	Complementary	Complementary	Complementary	Complementary	Complementary

f. Protein content

Parameters	Cross 1	Cross 2	Cross 3	Cross 4	Cross 5	Cross 6
Scale C	-1.99*	454.02*	-2.69*	-3.46*	-1.36*	-3.47*
Scale D	-1.29*	215.43*	0.31	-1.18*	-1.08*	-0.61*
(*m*)	3.96* ± 0.03	3.56* ± 0.04	3.66* ± 0.03	3.52* ± 0.02	2.15* ± 0.01	3.30* ± 0.02
(*d*)	1.76* ± 0.07	1.06* ± 0.06	1.16* ± 0.06	1.44* ± 0.05	0.18* ± 0.03	0.68* ± 0.04
(*h*)	0.91* ± 0.13	0.86* ± 0.15	0.68* ± 0.14	1.08* ± 0.10	0.09 ± 0.06	1.79* ± 0.11
(*I*)	4.04* ± 0.19	2.22* ± 0.20	1.66* ± 0.17	3.09* ± 0.14	0.90* ± 0.07	1.18* ± 0.14
(l)	0.93* ± 0.41	1.36* ± 0.47	4.00* ± 0.42	3.04* ± 0.33	0.37* ± 0.16	3.81* ± 0.35
Epistasis	Complementary	Complementary	Complementary	Complementary	Complementary	Complementary

g. Crude fibre

Parameters	Cross 1	Cross 2	Cross 3	Cross 4	Cross 5	Cross 6
Scale C	-30.21*	-65.96*	-0.29	3.88*	-13.89*	-6.83*
Scale D	-9.19*	-32.91*	3.85*	-0.52	-7.19*	-3.89*
(*m*)	12.74* ± 0.10	19.61* ± 0.19	19.31* ± 0.16	23.24*± 0.16	14.62* ± 0.09	17.82* ± 0.13
(*d*)	1.32* ± 0.28	0.98 ± 0.32	-0.23 ± 0.27	2.78* ± 0.28	0.96* ± 0.41	1.31* ± 0.27
(*h*)	2.01* ± 0.51	12.56* ± 0.48	-1.34* ± 0.69	2.93* ± 0.64	-0.29 ± 0.39	0.86 ± 0.56
(*I*)	3.72* ± 0.70	1.35* ± 0.99	-2.17* ± 0.79	6.55* ± 0.77	4.39* ± 0.53	4.07* ± 0.70
(l)	28.083*± 1.57	3.99* ± 2.51	5.52* ± 2.01	-5.87*± 1.91	8.93* ± 1.15	-3.92* ± 1.67
Epistasis	Complementary	Complementary	Duplicate	Duplicate	Duplicate	Duplicate

Generation mean analysis showed that the cross 5 gave the highest pod yield than others. The superiority of the cross 5 might have resulted from its high mean for number of branches per plant, number of pods per plant, pod length and pod weight which are all major yield components. Apart from this, cross 5 also had quality characters like protein content and crude fibre. Thought the cross 5 is adjudged the best on F_1 mean, other crosses could also throw transgressive segregants in later generations. In addition, Complementary type of epistasis gene action could be deduced from the same sign of '*h*' and '*l*' components. Hence, one or two cycles of intermating of selected segregants may be adopted to accumulate favourable genes through reciprocal recurrent selection.

REFERENCES

Anbuselvam, Y. (1999). Studies on genetics of certain quantitative traits in cowpea (*Vigna unguiculata* (L.) Walp.). Ph. D. Thesis. Annamalai University, Annamalainagar.

Bhor, TJ., and Dumbre, AD. (1998). Gene action of some characters in cowpea. *Legume Res.* 21 ($^3/_4$): 177-182.

Gill, KS., Dhillon, SS., and Bains, KS. (1972). Combining ability and inheritance of yield components in crosses involving indian and exotic germplasm. *Indian J. Genet.* 32: 421-430.

Hayman, B. (1958). The separation of epistatic from additive and dominance variation in generation means. *Heredity,* 12: 371-390.

Mallikarjun, SN., Arathi, HS., Gangappa, E. and Ramesh, S. (1995). Gene action for yield attributes in cowpea (*Vigna unguiculata* (L.) Walp.). *Mysore J. Agric. Sci.* 29: 289-292.

Mather, K. (1949). Biometrical Genetics. Dover publications, Inc. New york.

Nagaraj, KM., Savithramma, DL., and Ramesh, S. (2002). Triple Test cross analysis in two crosses of vegetable cowpea (*Vigna unguiculata* (L.) Walp.). *South Indian Hort.,* 50 (1/3): 98-104.

Paroda, RS. and Joshi, AB. (1970). Combining ability in wheat. *Indian J. Genet.* 30: 630 - 637.

Rahman, MA. and Saad, MS. (2000). Estimation of additive, dominance and digenic epistatic interaction effects for certain yield characters in *Vigna sesquipedalis* Fruw. *Euphytica.* 114 (1): 61-66.

Ranvir Singh and Dabas, BS. (1992). Inheritance of yield and its components in cowpea (*Vigna unguiculata* (L.) Walp.). *Int. J. Trop. Agri.* 10 (2): 161-164.

Thiagarajan, K., Natarajan, C. and Rathinaswamy, R. (1990). Combining ability and inheritance studies in cowpea (*Vigna unguiculata* (L.) Walp.). *Madras Agric. J.* 77 (7&8): 305-309.

□□□

Chapter 11

Future of Pulse Breeding in the Post PPV & FRA, 2001 Era

SABESAN, T.
Department of Agricultural Botany, Faculty of Agriculture
Annamalai University, Annamalainagar - 608 002

Introduction

The concept of ownership of biodiversity has changed radically during the last 25 years. Until 1980s genetic diversity was considered the common heritage of humankind. Now, it is increasingly being subjected to laws and treaties that allow institutions and companies employing plant breeders to claim ownership on cultivars. The full effect of this change on plant breeding innovation, germplasm accessibility, and exchange remains to be determined, but it is limiting the exchange of plant materials among breeding programs because of the increase in ownership claims such as patents, plant variety protection (PVP) certificates, and material transfer agreements (MTAs). Overall, public research and development (R&D) investment has decreased but private R&D investment has increased (Fowler and Lower, 2005), which mirrors the employment trends mentioned earlier. Major issues that might influence the breeding of crops in general and pulses in particular are discussed.

Production and Availability of Pulses in India

India's pulses production was 13.1 million tonnes in 2005-06. In fact, production 15 years ago was 14.3 million tonnes. Efforts to attain a production of 15 million tonnes have proved unsuccessful. Though production has stagnated at around 13 million tonnes, consumption has been hovering around 17 million tonnes a year. The per capita availability of pulses has dropped from 22.5 kg per annum in 1965-66 to 10.6 kg in 2003-04 (Punnathara, 2007).

The productivity of pulses has remained virtually stagnant for the last 45 years between 539 kg per hectare in 1961 to 585 kg in 2005-06. The per capita availability at present is 38 g/person/day as against 75 g/person/day in 1960 (Anon, 2001). The productivity remains stagnant for nearly 45 years while the consumption dwindled to one half. Owing to their floral morphology and pollination behavior pulses like blackgram, greengram are often neglected by private seed companies. The role of public and private sector institutions in breeding of pulses, seed production, marketing, and issues on pricing with reference to the newly enacted PPV&FR Act, 2001 will not only affect the fate of the farmers but also the food security of our country.

Public and Private Plant Breeding

Plant Variety Protection (PVP) will provide not only a legislative framework but also a suitable environment for plant breeding and biotechnology research. The products developed by research and development efforts in public and private research establishments can be protected and unethical practices during contract seed production can be prevented. The immediate effect of PPV&FR Act will be in terms of increased access to seeds developed by Multi National Companies (MNCs) in other countries. MNCs may establish joint research ventures with Indian Seed Companies and sell seed on their own or tie up with national seed companies for multiplication and marketing of their material. Increased plant breeding activity by private sector will increase availability of improved hybrids in some pulses e.g. pigeonpea.

Since, the private sector is not expected to invest in the research and seed production of self pollinated crops in a significant way, public sector research institutes and seed organizations will continue to develop varieties and produce seeds for the major food crops like pulses, soybean, groundnut, rice and wheat which are the life line of food security in our country. The rainfed, hilly, lowland and other sub-optimal ecologies also should be catered by the public sector institutions (Frey, 1997). Private sector seed companies feel that clauses pertaining to sale of seeds of protected varieties by the farmers and benefit sharing in the PPV&FR Act may not induce the private sector to seek protection and to increase research investment in breeding varieties in self-pollinated crops.

Availability of Germplasm and Use of Protected Varieties for Plant Breeding

Free and unrestricted availability of germplasm for crop breeding led to the development of outstanding varieties and hybrids in green and post green revolution era. During this period, most of the developed countries opted for protection of crop varieties and there was transition from government funded plant breeding to private plant breeding and involvement of large MNCs in biotechnology, variety development, seed production and marketing. After the WTO, more and more countries including those from developing world enacted the Plant Variety Protection legislation. Along with these developments due to Convention on Biodiversity (CBD), which has been signed by most countries of the world, germplasm is no longer heritage of mankind and thus, not freely available for crop breeding. Its ownership now rests with the sovereign nations and access depends on multilateral or bilateral agreements.

The clauses related to breeders' exemption in the PVP legislations of different countries ensure free availability and use of protected varieties for research and further breeding. For example the release and use of new protected varieties for cultivation has increased from 35 per cent in the year 1995 to 94 per cent in Argentina (UPOV, 2005).

Our PPV&FR legislation also provides the plant breeders, use of a protected variety for creating new varieties without prior approval of Plant Breeders Rights (PBR) holder. However, the clauses related to benefit sharing of the Act may act as impediment for the use of both protected varieties in plant breeding. Since in the post PPV era, there will be restriction in the free availability of germplasm, steps should be taken to ensure that crop breeding programmes do not suffer. This will require initiatives facilitating exchange of germplasm under multilateral agreement as well as devising simple and transparent material transfer agreements within the country for access to germplasm.

DUS Tests and Nucleus Seed Production

As per the provisions of PPV&FR Act, the protection will be granted to a new variety based on distinctiveness, uniformity and stability (DUS) tests. The applicant has to submit pure nuclear seed of variety along with the application in sufficient quantities for two years of DUS trials and for keeping representative sample of the variety for another seven years in reference collection. Nucleus seed production will have to be started when the particular variety is entered in IET (Sharma, 2008). DUS test guidelines for six crops *viz.*, lentil, kidney bean, chickpea, greengram, blackgram and pigeonpea have been released so far.

Rationale for Strengthening Breeding of Pulses in Post PPV&FRA Era

India accounts for nearly 25 per cent of the total global production of pulses, 27 per cent of consumption, and 34 per cent of global use of pulses in food. The right trigger can revive pulses production and productivity in India, reducing rural poverty and sustain the momentum of growth in agriculture. For example, Canada has brought large area under pulses and its exports to India increased from 73,000 tonnes in 1996 to 2,20,000 tonnes in 2006. To strengthen its position, Canada recently set up a Pulses Crop Research Laboratory at the University of Saskatchewan. As the Foreign Agricultural Service of the United States Department of Agriculture (USDA) states: "There is every indication that India's huge market for pulses will continue to grow. Not only is India's population increasing by 18 million every year but production of pulses has languished as government research and price supports have favoured cereal grains over other foodstuffs" (Punnathara, 2007).

Food Security

Plant breeding decisions determine the future of the world's food supply. Placing the responsibility for the world's crop germplasm and plant improvement in the hands of a few companies is bad public policy. The primary goal of private corporations is to make profit, and even in the case of the most civic-minded corporations, this goal will be at odds with certain public needs. Even if we assume that the one or two companies controlling a crop were completely altruistic, it is extremely dangerous to have so few people making decisions that will determine the future of a crop. Even well intentioned people make mistakes. The future of our food supply requires genetic diversity but also demands a diversity of decision makers i.e. plant breeders (Tracy, 2004). The article "A dying breed" in *Nature* 2003 also highlights the importance of public sector plant breeding efforts for food security. Furthermore, the GM crops so far developed involve only the addition of a single gene. Whether complex traits involving multiple genes will be amenable to manipulation through genetic engineering is unclear which may deter or steer the MNCs away owing to long gestation period. If unprepared, the public sector effort in plant breeding may be ill equipped to take up the slack (Jonathan Knight, 2003).

CONCLUSION

The objective of PPV&FR Act is to provide incentive for increased investment in plant breeding. Since public sector institutions have the social responsibility of ensuring access of seeds of improved varieties, support to public sector plant breeding programmes should continue in case of pulses to meet the requirements of farmers and to avoid monopoly of few seed companies. Access of germplasm to breeders should be ensured. Suitable incentives to public plant breeders for developing new varieties in pulses should be given. Special incentives for variety development of pulses for difficult ecologies should also be granted. Present breeder seed production system should continue with suitable modifications. Public sector seed organizations should be strengthened for seed production and supply of public bred varieties for stability of seed price. Appropriate government policy to regulate availability of sufficient quantity of seeds at proper price and effective implementation of compulsory licensing stated in PPV&FR Act should be ensured. Investment by private sector on contract research in public sector should be initiated in case of pulses as the expected requirement is around 30 million tonnes by 2025 assuming the per capita availability of 55g/day. As the world population is predicted to increase by as much as two billion people by 2030 (IFPRI, 2002) the demand for pulses will have to be catered by both public and private sector institutions as neither of them alone cannot fulfill the requirements. The weakness can be converted into an opportunity as the country transforms itself from the biggest pulses producing cum importing country to a key pulses producing cum exporting nation.

REFERENCES

Anonymous, (2001). Agricultural Statistics at a Glance, 2001, Directorate of Economics and Statistics, Ministry of Agriculture, Government of India.

Response to Keynote 2: The Current State of Breeding - How Did We Get Here?. Summit Proceedings: Summit on Seeds and Breeds for 21st Century Agriculture. Washington, DC, September 6-8, 2003.

Fowler, C and Lower, R.L. (2005). Politics of plant breeding. Plant Breeding Reviews. Ed Jules Janick, John Wiley & Sons, Inc.Vol. 25: 21-55.

Frey, K.J. (1997). National Plant Breeding Study - II: National Plan for Promoting Breeding Programs for Minor Crops in the U.S. Special Report 100. Iowa Agriculture and Home Economics Experiment Station, Ames, Iowa.

IFPRI. (2002). International Food Policy Institute, Sustainable Food Security for All by 2020: Proceedings of an International Conference, September 4-6, 2001, Bonn, Germany.

Jonathan Knight . (2003). A dying breed. *Nature* 421:568-570.

Punnathara, C.J. (2007). Why is the nation's "pulse" weak? Business Line, May 27, 2007.

Sharma, S.P. (2008). Emerging systems of varietal development and seed production in post Protection of Plant Varieties and Farmers' Rights (PPV&FR) Act regime. In: Proceedings of Winter school on PPV&FR Act and its implementation. Eds. S.K. Chakrabarty and Malavika Dadlani. IARI, New Delhi.

Tracy, W.F. (2004). What is Plant Breeding? In: Proceedings of the Seeds and Breeds Summit, Washington, DC. September 8-9, 2003. Published by the Rural Advancement-USA, Pittsboro.

UPOV, (2005). International Union for Protection of New Varieties of Plants. Available at http://www.UPOV.org.

□□□

Chapter 12

Wide Hybridization in *Vigna species*

PANDIYAN. M , N.RAMAMOORTHI, S.K.GANESH AND S. JEBARAJ
Tamil Nadu Agricultural University, National Pulses Research Centre
Vamban Colony, Pudukkottai, District

Introduction

The genotypes of most crop plants contains only a fraction of the naturally occuring genetic variation of their immediate wild progenitors and more distantly related taxa. Plant breeders are constantly seeking new sources of allelic variation to promote yield and improve its stability, to increase pest and disease resistance and to broaden the eco-geographic adaptation of crop plants. When the desired allelic variation is not available within the cross compatible gene pool , breeders will attempt crosses with more distantly related taxa - distant hybridization. This approach was applied in the past with a number of success stories. However, in most cases it appears that the scope for wide scale introgression of exotic traits from distantly related taxa into crop plants is quite limited, mainly because a universal feature of speciation in the occurrence of crossing barriers. In the case of the legume family, these include hybrid zygote breakdown and chromosomal rearrangements, which in wide crosses, the use of embryo rescue techniques will help the rescue. Wild relatives have always attracted breeders, for these contribute many useful traits. Their genetic background, particularly of species in the secondary or tertiary genepool, makes transfer difficult requiring may generations of backcrossing to acquire proper quality. There have been many attempts and evaluation and maintenance of wild relatives is usually a task taken up by most genebanks. Genetic modification, the modern way of transferring wanted genes, has barely begun for the legume food crops. This paper presents some examples of successful use made of wild relatives of greengram and other wild relatives of *Vigna* for breeding during the past decennia.

The major cultivated species of genus *Vigna* are mung bean or green gram (*Vigna radiata* (L) Wilczek, urd bean or black gram (*V. mungo* (L) Hepper, cowpea (*V. unguiculata* (L) Walp.) and minor cultivated species are moth bean (*V. aconitifolia* (Jacq.) Marechal), rice bean (*V. umbellata* (Thunb.) Ohwi and Ohashi) and adzuki bean (V. angularis (Wild) Ohawi and Ohashi).

The wild species are *V. radiata* var. *sublobata, V. mungo* var. *silvestris, V. trilobata, V. acconotifolia, V. glabrescense, V. hainiana, V. settulosa, V. pilosa, V. bournea, V. marina, V. minima, V. khandalensis, V. dalzalliana* and *V. vexillata*. The wild species, in general are resistant to biotic and/or abiotic stresses and hence have been used to transfer the resistance/tolerance to cultivated types. *V. radiata* var sublobata and *V. mungo* var. silvestris have been used successfully (as male) to transfer gene(s) for resistance to diseases to cultivated species; *V. radiata* and *V. mungo*, respectively. The crosses involving some other wild species with mungbean are reported to be successful but have not been used as an effective tool for improvement of mungbean. The wide hybridization involving cultivated species, mungbean x black gram, blackgram x mungbean, mungbean x ricebean have been reported to be successful.

The introgressed materials developed through wide crosses can also contribute as genetic reservoirs for novel genes apart from contributing to the improvement of yield and yield components. The present study was taken up keeping the objectives in mind in order to generate variability through interspecific hybridization involving *Vigna radiata* with species in secondary and tertiary gene pools and to compare the variability created for yield and yield components among segregants.

Materials and Methods

The genotypes of greengram and other wild *Vigna accessions* used as parents for the synthesis of interspecific hybrids are furnished in Table 1.

Table 1. Name of the *accessions* selected for hybridization and their special features

S.No	Species	Accession No	Special features	Origin
1.	*Vigna radiata*	VRMG(g) 1	Short duration cultivated species	Tamil Nadu
2.	*Vigna radiata* var *sublobata*	*Vigna radiata* var *sublobata 1*	Moderate resistance to MYMV & Bruchids	Tamil Nadu
3	*Vigna radiata* var *sublobata*	*Vigna radiata* var *sublobata 2*	Moderate resistance MYMV & bruchids	Tamil Nadu
4	*Vigna radiata* var *sublobata*	*Vigna radiata* var *sublobata 3*	High number of pods per plant	Tamil Nadu
5	*Vigna radiata* var *sublobata*	*Vigna radiata* var *sublobata 4*	Immune to MYMV and resistance to bruchids	Kerala
6	*Vigna radiata* var *sublobata*	*Vigna radiata* var *sublobata 5*	Lengthy pod, high number of seeds per pod & bruchids resistance	Kerala
7	*Vigna radiata* var *sublobata*	*Vigna radiata* var *sublobata 6*	High number of pods per plant, earlyness.	Kerala

Vigna radiata and six wild *Vigna radiata* accessions were raised during Rabi 2002-2003 in a crossing block for hybridization. To investigate the effect of stigmatic fluids/ exudates on pollen germination in the interspecific crosses, experiments were carried out by supplementing the standardized pollen germination media (James et *al.,* 1987) with the extracts of stigma from the corresponding female parent representing each cross combination. The extract was prepared by grinding 50 mg of stigma in one ml of distilled water and added to 25 ml of the media. The data were analyzed in completely randomized design (CRD). Wherever the percentages were used, the values were transformed into Arcsine values the transformed treatment means were compared with Duncan's Multiple Range Test (DMRT). The slides were photographed using built in camera of the OLYMPUS microscope.

The number of plants survived over germinated seeds were taken to assess the lethality of F_1 hybrids.

$$\text{Hybrid lethality} = \frac{\text{No. of plants died}}{\text{No. of seeds germinated}} \times 100$$

The set seeds from the above mentioned crosses were sown in 2 rows along with one row of male and female parents with a spacing of 45 x 30 cm during summer 2003. The biometrical observations were made in the following characters viz., 1. Plant height (cm), 2. Number of branches per plant, 3. Length of branches (cm), 4. Days to 50 per cent flowering , 5. Number of clusters per branch, 6. Number of clusters per plant, 7. Number of pods per plant , 8. Pod length (cm), 9. Number of seeds per pod,10. Hundred seed weight (g),11. Grain yield per plant (g),12. Dry matter production,13. Days to maturity

The pollen fertility analysis was carried out in the parents and their hybrids by acetocarmine staining technique. The swollen and well-stained pollens were counted as fertile, while shrunken and unstained as sterile

$$\text{Pollen fertility} = \frac{\text{No. of viable pollen}}{\text{Total no. of pollen observed}} \times 100$$

The seeds from individual F_1 plants were collected separately and were sown as progeny rows during kharif 2003. Observation for all the quantitative traits as that of F_1 generation except the traits 50 per cent flowering and days to full maturity were recorded. The descriptive parameters such as mean, range, SE, SD, skewness and kurtosis were computed. The frequency distribution of the F_2 segregants for six traits that are significantly correlated with single plant yield namely, number of branches per plant, length of branch, number of clusters per branch, number of clusters per plant, number of pods per plant and hundred seed weight was worked out and the results were represented as frequency graph. Thirteen quantitative characters recorded for F_1 generation was also recorded for all F_2 plants except days to 50 per cent flowering and days to full maturity.

Results

The result of crosses pertaining to crossability, pollen fertility, F_1 and F_2 morphology, number of seed set and extent of germination in different interspecific crosses attempted are presented below. In *V. radiata* x *V radiata var. sublobata* cross, 114 pollinations resulted in 25 (21.92%) pods with 95 viable and plumpy seeds (Table 1). Germination was 60 % and 25 plants of the germinated seeds gave rise to mature F_1 plants. The hybrid break down percentage was 43.86 (Table 2)

Table 1. Crossability between *Vigna radiata* and other *Wild Vigna accessions* for both direct and reciprocal crosses

Parents	*V. radiata* var. *sublobata*			*V. mungo* var. *silvestris*			*V. haineana*			*V.umbellata*			*V. vexillata*			*Vigna trilobata*		
	No. of Crosses	Pod set	Cross ability %	No. of Crosses	Pod set	Cross ability %	No. of Crosses	Pod set	Cross ability %	No. of Crosses	Pod set	Cross ability %	No. of Crosses	Pod set	%	Crosses	Pod set	%
V.radiata (female)	114.0	25.0	21.9	115.0	8.0	6.9	230.0	18.0	7.8	280.0	31.0	11.0	205	8.0	3.9	110.0	8.0	7.2
V. radiata (male)	225.0	15.0	6.6	105.0	5.0	4.7	190.0	21.0	11.0	152.0	8.0	5.2	105	3.0	2.8	95.0	5.0	5.2

Table 2. Details of germination, hybrid lethality and break down

Parents and Crosses	No.of crosssed seeds obtained	No. of seeds germinated	Hybrid leathality (%)	% of hybrid break down	% of germination	No. of seedlings attained maturity
V. radiata	50	48	6.25	93.75	96.00	45.00
*V. radiata*var *sublobata*	30	15	20.00	80.00	50.00	12.00
V. radiata x V. radiata var sublobata	95	57	56.14	43.86	60.00	25.00

The percentage of lethality in the F_1 varied from 20 per cent (*V. radiata* x *V.radiata* var. sublobata 3) to 56.14 (*V. radiata* x *V. radiata* x *V.radiata* var. sublobata 4) (Table 2)

Fertility status of F_1 hybrids

Pollen fertility status of parents and F_1 hybrids in the interspecific crosses of wild *Vigna accessions* were studied under compound microscope and the results are furnished in Table 3.

Table 3. Pollen fertility percentage of parents and F_1 hybrids

Parents and Hybrids	Pollen fertility (%)
V. radiata	87.84
V. radiata var. *sublobata*	75.65
V. mungo var *.silvestris*	70.23
V. hainiana	81.75
V. umbellata	82.58
V. vexillata	69.54
V. trilobata	65.85
V.radiata x V. radiata sublobata	69.72
V. radiata x *V.mungo* var. *silvestris*	60.00
V. radiata x *V. hainiana*	59.25
V. radiata x *V.umbellata*	0.00
V. radiata x *V.vexillata*	51.30
V. radiata x *V. trilobata*	43.5
V.radiata var. *sublobata* x *V.radiata*	49.86
V. mungo var. *silvestris* x *V.radiata*	15.00
V.hainiana x *V.radiata*	38.33
V.umbellata x *V.radiata*	40.11
V.vexillata x *V.radiata*	32.06
V.trilobata x *V.radiata*	56.78

The average pollen fertility in the parents was 87.84 in *Vigna radiata* and 75.65 in *Vigna radiata var. sublobata* . The low per cent of 65.85 was observed in *Vigna trilobata*. In the direct cross, the cross between *V. radiata x V.radiata* var. *sublobata* showed fertility of 69.72. while *Vigna radiata* x *Vigna umbellata* recorded zero pollen fertility. In the reciprocal *Vigna trilobata* has recoerded 56.78 per cent while *Vigna mungo* var. *silvestris* x *Vigna radiata* had low fertility percentage of 15.00.

Among parental species *V. umbellata* registered highest value of 67.00 for plant height followed by *V. radiata* (45.50). Among the direct crosses, the hybrid *V. radiata* x *V. radiata* var. *sublobata* was found to be taller (44.00). In the reciprocal crosses the *V. umbellata* x *V. radiata* registered the highest value of 48.50. Among the parental species *V. trilobata* possessed more number of branches (5.00). The hybrid of the direct cross *V. radiata* x *V. mungo var. silvestris* registered more branches (3.00). The hybrid of the cross *V. hainiana* x *V. radiata* and *V. umbellata* x *V. radiata* registered more branches per plant (4.00) among the reciprocal crosses.

V. umbellata possessed lengthiest branch (52.00 cm) among parental species. The highest branch length 29.00 cm was recorded by the hybrid *V. radiata* x *V. radiata* var. *sublobata* among the direct crosses. The hybrids of the cross *V. umbellata* x *V. radiata* registered the maximum branch length of 36.80 cm as compared to other reciprocal crosses. Among the parental species *V.hainiana* was the earliest to flower with 37 days closely followed by *V.radiata* with 38 days. Among the direct crosses *V. radiata* x *V. vexillata* registered early flowering (38 days) followed by *V. radiata* x *V. hainiana* (39 days). The hybrid of the cross *V. hainiana* x *V. radiata* was earliest to flower (38 days) among reciprocal crosses. *V. trilobata* registered the highest value of 8.00 for number of clusters per branch among the seven parental species. *V. radiata* x *V. hainiana* and *V. radiata* x *V. vexillata* possessed highest number of clusters per branch (3.0) among the direct cross hybrids, while *V. radiata* var. *sublobata* x *V. radiata, V.hainiana* x *V. radiata and V. umbellata* x *V. radiata* registered highest value of 3.00 clusters per branch among the reciprocal crosses. The maximum of 12 clusters per plant was observed in the parent *V. trilobata*. The cross *V. radiata* x *V. umbellata* recorded the maximum value of 8.00 for this trait among the direct crosses. Three out of six reciprocal hybrids registered the highest value of 8.00 for this trait. Number of pods per plant, for this trait, *V. trilobata* recorded the highest value of 55.00 followed by *V. radiata* (45.00). In the direct cross *V. radiata* x *V. mungo* var. *silvestris* registered more number of pods (35.00), in case of reciprocal crosses highest value of 35.00 was observed in the hybrid V. *radiata* var. *sublobata* x *V. radiata*.The wild parent *V. vexillata* recorded 16.50 which was highest value for length of pods. *V. radiata* x *V.* vexillata registered the highest value (8.50) among the direct crosses. The reciprocal crosses of the same combination also recorded the highest value of 15.30 . The highest value of 15.50 seeds was observed in the parental species *V. vexillata* followed by *V. radiata* which recorded 12.00 seeds per pod. The hybrid of the cross *V. radiata* x *V mungo* var. *silvestris* recorded (11.00) while the reciprocal hybrid *V. vexillata* x *V. radiata* exhibited highest value of 15.20 for this trait. V. *umbellata* possessed very bold seeds among the parental species registering 5.40 for the trait hundred seed weight. *V. radiata* x *V. vexillata* recorded the highest value of 3.20 for this trait among the direct crosses and the other hybrid *V. umbellata* x *V. radiata* performed better with value of 5.50. For single plant yield trait the value of 5.85 registered by *V. radiata* was highest among parents. In the direct crosses the value of 15.00 recorded by the hybrid *V. radiata* x *V. trilobata* was highest. In the reciprocal direction the cross *V. vexillata* x *V. radiata* registered the highest value of 5.00. For dry matter production *V. radiata* (25.50) registered the highest value among parents. In the direct crosses the value of 20.00 recorded by the hybrids *V. radiata* x *V. vexillata* was highest. In the reciprocal crosses *V. umbellata* x *V. radiata* revealed highest value of 25.60. *V. radiata* was the parent to mature earliest at 65 days followed by *V. hainiana* in 66.00 days. The hybrid of the cross *V. radiata* x *V. vexillata* was shortest in duration with 65 days. The hybrid of the reciprocal cross *V. hainiana* x *V. radiata* was earliest with 65 days.

Mean Performance of Different Characters of Wild *Wild Vigna Accessions* F_2 Segregants

The variation for the plant height was predominant in both direct as well as reciprocal crosses. Among the F_2 segregants of the direct cross *V. radiata* x *V. trilobata* registered the maximum mean value of 33.87 cm followed by the cross *V. radiata* x *V. hainiana* (32.7 cm). In the reciprocal crosses, the F_2 segregants of the cross *V. umbellata* x *V.radiata* registered highest mean value of 45.16 cm. The general mean was found to be 1.71 for number of branches per plant in F_2 segregants. The number of branches per plant ranged from 0.90 (*V.radiata* x *V. vexillata*) to 2.73 (*V .trilobata* x *V. radiata*). The maximum number of branches per plant 2.50 was observed in *V.radiata* x *V. trilobata* among the direct crosses. In the reciprocal crosses, *V.trilobata* x *V. radiata* exhibited highest mean value of 2.73. The mean value for the length of branches per plant varied from 10.54 cm to 25.20 cm with over all mean of 19.21 for F_2 segregants. The cross combination *V. radiata* x *V.hainiana* exhibited highest mean value of 22.68 cm and 25.20 cm in the direct and reciprocal combinations respectively.

The number of clusters per branch ranged from 1.00 (*V.mungo var. silvestris* x *V. radiata*) to 4.14 (*V.radiata* x *V. mungo* var. *silvestris*) with the over all mean of 2.37. The number of clusters per plant ranged from 2.20 (*V. vexillata* x *V. radiata*) to 8.50 (*V. radiata* x *V. mungo* var. *silvestris*) with over all mean of 5.25 among the crosses. The range was found to be between 4.30 (*V.vexillata* x *V.radiata*) to 27.87 (*V.trilobata* x *V.radiata*) with the over all mean value of 17.15. The cross *Vigna radiata* x *V. trilobata* recorded highest mean value of 26.88 for number of pods per plant among direct crosses. The pod length ranged from 4.42 cm (*V.mungo* var. *silvestris* x *V.radiata*) to 15.72 cm (*V.vexillata* x *V radiata)* with over all mean of 7.40 cm for F_2. The highest pod length 7.48 cm observed by the direct cross *V.radiata x V. vexillata* while reciprocal cross possessed highest value of 15.72 cm in the *V.vexillata* x *V. radiata*. The range for number of seeds per pod from 4.00 (*V.mungo var. silvestris* x *V.radiata*) to 14.80 (*V.radiata* x *V. vexillata*) while the over all mean was 8.33. The cross *V.radiata x V.vexillata* possessed values of 12.00 and 14.80 for direct and reciprocal respectively. The hundred seed weight ranged from 1.70 g (*V.trilobata x V.radiata*) to 3.43 g (*V.vexillata* x *V. radiata*) with the over all mean of 2.47 g for F_2 segregants. In the direct crosses, *V. radiata* x *V. mungo* var. *silvestris* possessed highest mean value of 2.94 g while *V. vexillata* x *V. radiata* registered highest mean value of 3.43 g observed in the reciprocal cross. The seed yield ranged from 0.93 g (*V.mungo var. silvestris* x *V.radiata*) to 5.21 g (*V.radiata* x *V. mungo* var. silvestris) with a over all mean of 3.25 g for F_2 segregants. The highest mean value of 5.21 g was observed in the direct cross of *V.radiata x V.mungo* var.*silvestris* while maximum value of 4.99 g was recorded by the reciprocal crosses *V.vexillata x V.radiata*. The dry matter production varied from 26.41 g (*V.vexillata* x *V. radiata*) to 109.36 g (*V.radiata* x *V. mungo* var. *silvestris*) with the over all mean of 67.80 for F_2 crosses. The maximum mean value of 109.36 g was recorded by the direct cross of *V. radiata* x *V. mungo* var. *silvestris* where as in the reciprocal crosses recorded highest mean value of 65.63 in V. *radiata* var. *sublobata* x *V. radiata.*

DISCUSSION

The present investigation was undertaken with an objective to transfer useful traits from the wild relatives of greengram *viz.*, the extent of crossability, morphological features and fertility of hybrids and possibility of obtaining superior recombinants in the F_2 generation through recombination of genomes were studied and the results are discussed in detail.

Utilization of wild species for improving their cultivated counterparts is steadily increasing in various crops where genetic diversity in the primary gene pool is limited. Exploitation of the wild species, such as those which belong to the secondary or tertiary gene pools of the cultigen is not always easy since various types of mechanisms that prevent gene flow between species are commonly encountered. On the other hand, because of genetic distance of these wild species from the cultigen, there is a good possibility that they possess variation in characters of economic value, which is lacking in the cultivated germplasm e.g., pest and disease resistance. The wild relatives of greengram such as *V. umbellata, V. vexillata* and *V. trilobata* possess desirable genes for many yield components coupled with resistance to bruchids and

MYMV. Transferring of these genes into cultivated species, could result in development of high yielding resistant types. The use of wild *Vigna accessions* in greengram breeding programme has been difficult because of problems encountered in obtaining successful F_1 hybrids due to crossability barriers. In spite of these difficulties, wide hybridization between *V. radiata* and its wild relatives was successfully accomplished by many workers Pandae *et al.* (1990), Ganeshram (1993) and Subramanian and Muthiah. (2000). Crossability is a pre-requisite for gene transfer in wide hybridization. An understanding of crossability relationship among the species had been helpful not only in choosing methods for producing F_1 hybrids, but also in tracing phylogenic relationship among species. In the present study, successful pod set was observed in all the 12 interspecific crosses with *Vigna radiata* either as ovule or pollen parent. This result was in agreement with the reports of Ahuja and Singh (1977), Gopinathan *et al.* (1986), Egawa (1990) and Mendioro and Ramirez (1994). The percentage of lethality among interspecific hybrids varied from 0.00 to 56.14 per cent . Similar observations on hybrid lethality and inviability were noticed in interspecific crosses involving different wild *Vigna accessions* by AL- Yasiri and Corijne (1966), Adinarayanamurthy *et al.* (1993) and Ganeshram (1993). Stebbins (1958) had attributed the hybrid weakness, inviability, lethality and sterility as mechanisms of nature for maintaining the integrity of related species. In general, the pollen fertility among the direct crosses were higher as compared to their corresponding reciprocal crosses which indicated that the approach using the cultivated species as a female parent is likely to generate better hybrids and segregants. Similar result was reported by various authors for differential pollen fertility among interspecific crosses of wild *Vigna accessions* Mendioro and Ramirez, 1994, Ravi *et al.*, 1987, Subramanian and Muthiah 2000, Monika *et al.*,2001, and Sidhu and Satija, 2003). Among the crosses the pollen fertility was highest in the cross *V. radiata* x *V. radiata* var. *sublobata* and this observation supports the view of Pandae *et al.*(1990) and Mendioro and Ramirez (1994) that *V. radiata* var. *sublobata* is the probable proginator for *V. radiata*. The range of pollen fertility observed in all the F_1 hybrids except the cross *V. radiata* x *V. umbellata* was high enough to obtain sufficient viable F_2 segregants. In the cross *V. radiata* x *V. umbellata* all the pollens were sterile, hence no viable F_2 segregants could be generated. Considering the importance of this cross for the resistance related traits it is essential to device methods for enhancing fertility in F_1 that could aid in developing breeding materials with resistance, hence with this objectives cytological analysis was carried out for this hybrid.The primary criterion used for the evaluation of hybrids was the *per se* performance for different traits. In the present study, among the crosses, the reciprocal cross *V. umbellata* x *V. radiata* exhibited high mean value for important traits *viz.*, plant height, number of branches per plant, length of branch, number of clusters per branch, number of clusters per plant, hundred seed weight and dry matter production.

The reciprocal cross *V. vexillata* x *V. radiata* exhibited higher mean performance for characters *viz.*, length of pod, number of pods per plant and single plant yield while the hybrid of direct cross of same parents exhibited high value for three characters in the desirable direction *viz.*, length of pod and days to full maturity and 50 per cent flowering. The hybrid of the direct cross *V. radiata* x *V. radiata* var. *sublobata* recorded high *per se* performance for the characters plant height and length of branch while the higher number of clusters per branch and number of pods per plant were recorded in the reciprocal crosses. Hybrid of the cross *V. radiata* x *V. mungo* var. *silvestris* registered higher number of branches per plant, number of cluster per plant and number of pods per plant. For seed yield the hybrid of the cross *V. radiata* x *V. trilobata* registered the highest mean value among the direct crosses. Hence the segregants that could be recovered from these promising interspecific hybrids might serve as better breeding base for improvement of yield and yield components. Such promising interspecific hybrids were also reported by Subramaninan and Muthiah (2001) and Ganeshram (1993). For most of the direct as well as reciprocal crosses, the F_2 progenies revealed high *per se* performance and also exhibited high variability for all the 11 traits. For all the traits in majority of the crosses, the skewness was positive indicating that predominance of dominant alleles and Darbeshwar Roy (2000). In this situation selection for traits in the early generation will not be fixable hence selections in later generations or by adopting modified breeding procedures such as intermating the segregants followed by recurrent selection may shift the gene action towards additive effects. Since sterility factors will gradually reduce over generations in case of interspecific crosses and more recombined populations will be available for selection, the effecting selection in the later generation will be more effective.

REFERENCES

Adinarayanamurty, V.V., Rao, M.V.B., Satyanarayana, A. and Subramanyam, D. (1993). The crossability of *V. mungo* and *V. radiata* with *V. trilobata*. *Intl. J. Trop. Agri.*, 11: 209-213.

Ahuja, M.R., and Singh, B.V. (1977). Induced genetic variability in mungbean through interspecific hybridization. *Indian J.Genet. and Plant breed.*, 3(1): 133-136.

AL-Yasiri, S.A. and D.P. Coryne, D.P. (1966). Interspecific hybridization in the genus *Phaseolus*. *Crop Sci.*, 6: 59-60.

Darbeshwar Roy. (2000). Plant breeding analysis and exploitation of variation. pp. 300- 304.

Egawa, Y. (1990). Phylogenetic relationships in Asian *Wild Vigna accessions*. Proc: Mungbean Meeting, 90, Thailand. pp. 87-94.

Ganeshram, S. (1993). Evaluation of some genotypes interspecific hybrids and derivatives of greengram (*V. radiata* (L.) *Wilczek)* x Black gram (*Vigna mungo* (L.) *Hepper*) crosses. M.Sc. (Ag.) Thesis, Tamil Nadu Agricultural University, Coimbatore.

Gopinathan, M.C., Babu, C.R. and Shivanna, K.R. (1986). Interspecific hybridization between rice bean (*Vigna umbellata*) and its wild relative (*V. minima*): Fertility – Sterility Relationships. *Euphytica*, 35: 1017-1022.

James D, Ariyanayagam RP, Dungan EJ. 1987. Comparative studies of *in vitro* germination of pigeonpea (*Cajanus cajan* (L.) Millsp.) and *Atylosia platycarpa* Benth. *Tropical Agriculture* 64, 313-346.

Mendioro, M. S. and Ramirez, D.A. (1994). Morphological and cytological characterization of F_1 hybrids of *Vigna mungo* (L.) Hepper x *V. glabrescens* (L.) and *Vigna radiata* (L.) Wilczek x *Vigna mungo*. *Phil. Agric.*, 76 (4): 425-433.

Monika, K., Singh, P. and Sareen, P.K. (2001). Cytogenetic studies in mungbean- ricebean hybrids. *J. Cytol. Genet.*, 2: 13-16

Pandae, K, Raghavanshi, S.S. and Prakesh, P. (1990). Induced high yielding amphiploid of *Vigno radiata* x *Vigna mungo*. *Cytologia.*, 55: 249-253.

Ravi, J., Singh, P. and Minocha, J.L. (1987). Meiotic behaviour of interspecific hybrids of *Vigna radiata* x *Vigna mungo*. In: First Symposium on Crop Improvement, Feb.1987, India. pp. 23-27.

Sidhu, N and Satija, C.K. (2003). Cytomorphological characterization of amphidiploids of *Vigna radiata* x *V. umbellata*. *Crop Improv.*, 30 (1): 25 – 32.

Stebbins, G. L. (1958). The inviability, weakness and sterility of interspecific hybrids. *Adv. Genet.*, 9: 147-215.

Subramanian, A. and Muthiah, A. R. (2000). Interspecific hybridization between *V. radiata* (L.) Wilczek and blackgram *V. mungo* (L.) *Hepper*. *Legume Res.*, 24(3): 154-158.

□□□

Chapter 13

Association and Path Analysis for Earliness Breeding in Cow Pea (*Vigna Unguiculata* (L.) Walp).

ANANDAN, A. AND R. ESWARAN
Department of Agricultural Botany, Faculty of Agriculture
Annamalai University, Annamalainagar-608 002

Introduction

Cow pea (*Vigna unguiculata*(L.)Walp) is of world wide importance, but is markedly of greater importance in West Africa, Latin America and South East Asia, where it provides more than half the protein and some essential nutrients in human diets (Oluwatosin,1998).It is an important multipurpose grain legume and serves as an ideal crop for soil and water conservation because of its ability to grow fast and cover the soil surface quickly. However the prevelant crop varieties in the region are low yielding. Thus selection for high yielding lines coupled with earliness is desirable.

In a self fertilizing crop like cow pea, the breeders aims at creating new variability by effecting crosses between carefully chosen parents and subsequent selection. Before embarking on any selection programme to develop high yielding types, it is imperative to have knowledge on the nature and magnitude of association and their inter-relationship among different characters in the material for the characters which we want to improve.

Materials and Methods

Thirty genotypes of cow pea (*Vigna unguiculata* (L.)Walp)were taken for the present study. The experiment was conducted at Plant Breeding Farm, Department of Agricultural Botany, Annamalai University. The seeds were sown in randomized block design, replicated twice. An inter row spacing of 45 cm and interplant spacing of 15 cm was adapted. Care was taken enough to maintain only single plant per hill. Recommended agronomic practices and need based plant protection measures were taken and observations were recorded on five randomly selected competing plants per replication for nine earliness traits viz., days to first flowering, number of leaves at the time of first flowering, height of first flowering node, plant height at the time of first flowering, days to 50 per cent flowering, days to 50 per cent maturity ,number of leaves at the time of 50 per cent maturity, seed yield per plant and total dry mater production. The association between earliness and other component as well as among the component traits were estimated as the method suggested by Goulden,1952. The path coefficient analysis was worked out following the method suggested by Dewey and Lu,1959.

Results and Discussion

The genotypic correlation coefficient among the nine earliness characters are presented in Table 1.Among the earliness traits, days to first flowering showed highly significant positive association with days to 50 per cent flowering followed by days to 50 per cent maturity and number of leaves at the time of first flowering. Similarly days to 50 per cent flowering, number of leaves at the time of 50 per cent maturity and seed yield per plant. Number of leaves at the time first flowering evidenced positive significant association with plant height at the time first flowering and days to 50 per cent maturity. The results are in corroboration with Mathur,1995.Significant positive association of days to 50 per cent flowering with days to first flowering ,number of leaves at the time of 50 per cent maturity, days to 50 per cent maturity implies that selection of these characters would lead to simultaneous improvement for earliness in cow pea.

The genotypic correlation coefficients of earliness with nine characters which showed high positive genotypic correlation were further portioned into direct and indirect effects and the results are presented in Table 2.Among the different effects seed yield per plant showed maximum positive direct effect towards days to 50 per cent flowering by plant height at the time of first flowering and number of leaves at the time of 50 per cent maturity. Among the indirect effects days to first flowering exhibited positive indirect effect through days to 50 per cent maturity and number of leaves at the time of first flowering. Height of first flowering node exhibited highest indirect effects through plant height at the time of first flowering towards days to 50 per cent flowering .Total dry mater production exhibits its indirect effect through days to 50 per cent maturity. Hence selection of these aforementioned characters would be very much useful in breeding for earliness genotypes in cow pea.

Table 1. Genotypic correlation coefficients for earliness traits in cow pea

characters	2	3	4	5	6	7	8	9
1	0.2388**	0.0536	-0.1.15	0.6766**	0.3091*	0.0450	0.0450	-0.0373
2	-	0.1333	0.3604*	0.9236**	0.3813*	0.0870	0.3228*	0.2004
3		-	0.2587	-0.1873	0.0987	0.2596	-0.2045	0.0576
4			-	0.2577	0.2228	-0.2123	-0.2587*	-0.2199
5				-	0.8355**	0.9456**	0.4604**	0.2466
6					-	0.1621	0.3107*	0.0606
7						-	0.2005	0.1799
8							-	0.1369
9								-

** Significant at 1 per cent level , * Significant at 5 per cent level

Legends 1. Days to first flowering
2. Number of leaves at the time of first flowering
3. Height of first flowering node
4. Plant height at the time of first flowering
5. Days to 50 per cent flowering
6. Days to 50 per cent maturity
7. Number of leaves at the time of 50 per cent maturity
8. Seed yield per plant
9. Total dry matter production

Table 2. Path analysis in cowpea

Characters	1	2	3	4	6	7	8	9	Genotypic correlation
1	**-1.8159**	0.5376	-0.1371	-0.1536	0.6173	0.3666	-0.1971	-0.3571	0.6766
2	0.1371	**-2.9029**	-0.3716	0.2531	0.1378	0.1510	0.4731	0.1401	0.9236
3	-0.3421	-0.1631	**1.0201**	0.6042	-0.3718	0.1758	-0.1271	0.0368	-0.1873
4	-0.1571	0.3452	1.7101	**3.8778**	0.2378	-0.5671	-0.3751	-0.9361	0.2577
6	0.5368	0.1723	0.2526	-0.4217	**1.4298**	-0.2371	0.3344	0.1982	0.8355
7	0.3628	-0.1315	0.5653	0.1263	-0.2361	**2.9433**	0.1781	0.8070	0.9456
8	-0.4010	-0.2461	1.1847	0.1378	0.1537	-0.2367	**5.2668**	-0.1320	0.4604
9	0.1208	-0.2651	0.1351	0.1020	0.2061	-0.1371	0.0848	**-3.8651**	0.2466

Residual effect:0.6355 Diagonal values (Bold) indicate direct effects

Legends: 1. Days to first flowering
2. Number of leaves at the time of first flowering
3. Height of first flowering node
4. Plant height at the time of first flowering
5. Days to 50 per cent flowering
6. Days to 50 per cent maturity
7. Number of leaves at the time of 50 per cent maturity
8. Seed yield per plant
9. Total dry matter production

REFERENCES

Dewey, D.R and Lu. L.H. (1959). Correlation and Path analysis of components of Crested wheat grass seed production.*Agron.J.*,51: 515-518.

Goulden, C.H. (1952). Methods of Stastical Analysis.John Wiley and Sons, Inc., New York.

Mathur, R. (1995). Genetic variability and correlation studies in segregating generations of cow pea. *Madras Agric.J.*, 82: 150-152

Oluwatosin, O.B.(1998). Genetic and environmental variability in starch,fatty acids, mineral nutrients composition in cow pea (*Vigna unguiculata* (L.)Walp). *Journal of the science of food and agriculture,* 78:1-11.

□□□

Chapter 14

Spectrum of Induced Chlorophyll Mutations in Cowpea (*Vigna Unguiculata* (L.) Walp.)

KANAGARAJ, G., S. THIRUGNANA KUMAR, S. THIRUMAL RAJ, R. BALAMURUGAN AND P.A. VELUSAMY
Department of Agricultural Botany, Faculty of Agriculture, Annamalai University
Annamalai Nagar - 608 002, Tamil Nadu, India

Introduction

The potentiality of mutation breeding as a tool in crop improvement programme has been well realized in many crop plans. Chlorophyll mutations provide one of the most dependable index for the evaluation of genetic effects of mutagenic treatments and have been studied in various crop plants by several workers. In this study a pioneering attempt was made in cowpea (a tropical pulse crop) with EB (an intercalating agent) along with gamma rays (ionizing radiation), EMS (alkylating agent) and combination of gamma rays with EB and EMS to study the relative effects of these mutagens in inducing chlorophyll mutations in M_2 generation.

Materials and Methods

Dry seeds of cowpea (*Vigna unguiculata* (L.) Walp) genotype Co. Vu. 623 were subjected to 40 to 70 kR gamma irradiation, at Sugarcane Breeding

Institute, Coimbatore, India. The concentration of EB with 0.08, 0.10, 0.13, 0.15, 0.18 and 0.20 mM and EMS 20 and 30 mM were prepared in double distilled water, adjusted to pH 7.00. For each concentration, three samples of sixty seeds were taken along with control. The seeds were first presoaked for 6 h. in double distilled water and then for 6 h. in mutagen solution. The combination treatments were performed by first treating the seeds with gamma rays and then with mutagen solution. The treated seeds were then washed in running water for 30 minutes and were sown in the field along with controls, in a randomized block design, replicated thrice. Ten randomly selected normal looking M_1 plants in each treatment were tagged and the seeds harvested separately. The seeds were shown in M_2 generation on a plant to progeny basis. Seedlings from five to twenty days age were scored for frequency, spectrum and segregation pattern on chlorophyll mutations.

Result and Discussion

Chlorophyll mutations were classified following the method of Gustafsson (1940). The following types were identified in M_2 generation.

i. Albina

White, lethal (6-7 days), no chlorophyll or carotenoids are formed.

ii. Xantha

Yellow to yellowish, white and carotenoids present, but chlorophylls absent, lethal (12-15 days).

iii. Chlorina

Yellowish green and lethal (10-12 days).

iv. Viridis

Creamy white and lethal (7-10 days).

Mutation frequencies were calculated for 200 M_1 progenies and 28,815 M_2 seedlings. Chlorophyll mutation frequency was estimated both as per cent of M_2 families (M_1 plant basis) and also as mutations in per cent of M_2 plants (M_2 seedlings basis).

The frequency of chlorophyll mutants in general increased at the higher dose level. However, a dose dependent relationship was not observed. Gamma rays showed higher mutation frequency than in combination with EB and EMS. Out of the combination treatments, gamma rays plus EMS, induced higher mutation frequency than gamma rays plus EB treatments. EB treatments showed no chlorophyll mutants. EMS in individual treatments showed less chlorophyll mutants, than gamma irradiation. The rate of chlorophyll mutation was as high as 4.33 per cent at 70 kR gamma rays, while with EMS the maximum rate was 1.90 per cent at 30 mM. Gamma rays post-treated with EMS showed a maximum mutation rate of 3.27 per cent while gamma rays with EB post – treatment showed a maximum rate of 2.83 per cent (Table 1).

Table 1. Frequency of chlorophyll mutation in M_2 generation

Treatment Dose / Conc.	M_1 plants Studied	M_1 plants Segregated	M_2 plants Studied	M_2 plants Segregated	Mutation frequency Studied	Mutation frequency Segregated
Gamma rays (kR)						
40	10	5	1300	18	50.00	1.38
45	10	5	1877	25	50.00	1.33
50	10	6	1066	34	60.00	3.19
55	10	6	1478	29	60.00	1.96
60	10	7	1211	38	70.00	3.14
65	10	6	1342	36	60.00	2.68
70	10	6	1016	44	60.00	4.33
EB mM						
0.08	10	0	1297	0	-	-
0.10	10	0	1179	0	-	-
0.13	10	0	1129	0	-	-
0.15	10	0	1668	0	-	-
0.18	10	0	1417	0	-	-
0.20	10	0	1477	0	-	-
EMS mM						
20.00	10	5	1404	16	50.00	1.14
30.00	10	6	1205	23	60.00	1.91
Gamma rays + EB						
45 kR + 0.08 mM	10	5	1467	27	50.00	1.84
55 kR + 0.08 mM	10	6	1132	32	60.00	2.83
65 kR + 0.08 mM	10	6	1238	35	60.00	2.83
Gamma rays + EMS						
45 kR + 20 mM	10	5	1290	26	50.00	2.02
55 kR + 20 mM	10	5	1462	27	50.00	1.85
65 kR + 20 mM	10	5	947	31	60.00	3.27
Control						
Control (dry seeds)	10	-	1213	-	-	-

Albina, Viridis, Chlorina and *Xantha* were the four types of chlorophyll mutations observed in this present study. *Viridis* was the most frequent type that occurred in M_2 generation. While *Albina* showed an increasing trend with increase in the dosage of gamma rays and EMS, *Xantha* expressed a decreasing trend. The combination treatments showed inconsistent pattern (Table 2).

Table 2. Frequency of different types of chlorophyll mutants in M_2

Treatment Dose / Conc.	Total chlorophyll mutant seedlings in M_2	Relative percentage of chlorophyll mutants			
		Albina	*Viridis*	*Chlorina*	*Xantha*
Gamma rays (kR)					
40	18	27.78	33.33	16.67	22.22
45	25	28.00	32.00	16.00	24.00
50	34	23.53	23.53	29.41	23.53
55	29	27.59	34.48	17.24	20.69
60	38	26.32	36.84	15.79	21.05
65	36	36.11	41.67	13.89	13.89
70	44	38.64	29.55	15.91	15.91
EB mM					
0.08	0	-	-	-	-
0.10	0	-	-	-	-
0.13	0	-	-	-	-
0.15	0	-	-	-	-
0.18	0	-	-	-	-
0.20	0	-	-	-	-
EMS mM					
20.00	16	12.50	43.75	06.25	37.50
30.00	23	17.39	52.17	08.70	21.74
Gamma rays + EB					
45 kR + 0.08 mM	27	29.63	29.63	11.11	29.63
55 kR + 0.08 mM	32	21.88	31.25	15.63	31.25
65 kR + 0.08 mM	35	37.14	28.57	22.80	11.29
Gamma rays + EMS					
45 kR + 20 mM	26	30.77	38.46	11.54	19.23
55 kR + 20 mM	27	51.85	14.81	14.81	18.52
65 kR + 20 mM	31	41.94	19.35	19.35	19.35

All the treatments including gamma rays and EMS as individual and in combination treatments, such as gamma rays + EMS and gamma rays in combination with EB segregated for multiple mutations. Single mutations was not observed in any of the treatment at the different dose levels studied (Table 3).

The segregation ratio was maximum in treatments with gamma rays (Table 4) than in combination with EB and EMS. Among the combination treatments, gamma rays plus EMS expressed the maximum segregation ratio. The segregation ratio showed an increasing trend both under individual and combination treatments of gamma rays, EMS and in the combination of gamma rays plus EB. The expected segregation ratio of 25 per cent was not observed at any dose level and in any of the mutagen treatments studied.

Table 3. Estimates of M_1 plants segregating for single and multiple chlorophyll mutations

Treatment Dose/Conc.	Number of M_1 families segregating	Chlorophyll mutants in M_2		Plant progenies segregating for mutations of		
		Total number	Mean No. of M_1 plant progeny	One type	Two types	Three types
Gamma rays (kR)						
40	5	18	3.60	1 (20.00)	3 (60.00)	1 (20.00)
45	5	25	5.00	2 (40.00)	2 (40.00)	1 (20.00)
50	6	34	5.67	1 (16.66)	3 (50.10)	2 (33.33)
55	6	29	4.83	1 (16.66)	3 (50.10)	2 (33.33)
60	7	38	5.29	2 (28.57)	2 (28.57)	3 (42.86)
65	6	36	6.00	1 (16.66)	2 (33.33)	3 (50.10)
70	6	44	8.67	1 (16.66)	3 (50.10)	2 (33.33)
EB mM						
0.08	0	0	-	-	-	-
0.10	0	0	-	-	-	-
0.13	0	0	-	-	-	-
0.15	0	0	-	-	-	-
0.18	0	0	-	-	-	-
0.20	0	0	-	-	-	-
EMS mM						
20.00	5	16	3.20	2 (40.00)	2 (40.00)	1 (20.00)
30.00	6	23	3.83	2 (33.33)	3 (50.10)	1 (16.66)
Gamma rays + EB						
45 kR + 0.08 mM	5	27	5.40	1 (20.00)	3 (60.00)	1 (20.00)
55 kR + 0.08 mM	6	32	5.22	1 (16.67)	3 (50.10)	2 (33.33)
65 kR + 0.08 mM	6	35	5.83	1 (16.67)	3 (50.10)	2 (33.33)
Gamma rays + EMS						
45 kR + 20 mM	5	26	5.20	2 (40.00)	2 (40.00)	1 (20.00)
55 kR + 20 mM	5	27	5.40	1 (20.00)	2 (40.00)	2 (40.00)
65 kR + 20 mM	6	31	5.17	1 (16.67)	3 (50.00)	2 (33.33)

Table 4. Segregation ratio for chlorophyll mutation in M_2 generation

Treatment Dose/Conc.	Total number of chlorophyll mutants	Relative percentage of M_2 plants with recessive traits				Mean segregation ratio
		Below 4	5-10	10-15	15-20	
Gamma rays (kR)						
40	18	5	-	-	-	1.28
45	25	4	1	-	-	1.66
50	34	1	4	1	-	4.14
55	29	2	4	-	-	2.83
60	38	4	3	-	-	3.60
65	36	3	3	-	-	3.33
70	44	2	1	3	-	5.63
EB mM						
0.08	0	-	-	-	-	-
0.10	0	-	-	-	-	-
0.13	0	-	-	-	-	-
0.15	0	-	-	-	-	-
0.18	0	-	-	-	-	-
0.20	0	-	-	-	-	-
EMS mM						
20.00	16	6	-	-	-	0.71
30.00	23	3	2	-	-	2.03
Gamma rays + EB						
45 kR + 0.08 mM	27	4	1	-	-	1.91
55 kR + 0.08 mM	32	2	4	-	-	3.67
65 kR + 0.08 mM	35	2	4	-	-	3.24
Gamma rays + EMS						
45 kR + 20 mM	26	2	3	-	-	2.25
55 kR + 20 mM	27	1	3	1	-	3.33
65 kR + 20 mM	31	1	5	-	-	4.23

Simple correlation between lethality, injury and sterility in M_1 with M_2 chlorophyll mutation frequency revealed a significant and positive association between M_1 lethality and chlorophyll mutation frequency only (Table 5).

Among the three mutagens, gamma rays as a sole mutagen treatment was the highest in inducing chlorophyll mutations. The order of the mutagens and their combination treatments in inducing chlorophyll mutations are:

Gamma rays > Gamma raya > Gamma rays > EB
+ +
EMS EB

Table 5. M_1 lethality, injury and sterility and its correlation with M_2 chlorophyll mutation frequency

Dose / conc.	M_1 generation			M_2 generation chlorophyll mutation (CM) frequency (%)
	Lethality (L)	Injury (I)	Sterility (S)	
Gamma rays (kR)				
40	30.29	0.99	25.58	1.38
45	28.25	0.00	28.94	1.33
50	38.94	0.00	31.58	3.19
55	43.27	2.23	33.52	1.96
60	36.74	11.08	29.28	3.14
65	40.87	0.00	29.68	2.68
70	52.41	18.86	39.17	4.33
EB mM				
0.08	18.22	20.30	33.30	0.00
0.10	38.41	27.17	38.57	0.00
0.13	39.71	24.43	39.87	0.00
0.15	36.34	16.47	30.57	0.00
0.18	36.34	21.94	32.99	0.00
0.20	36.86	28.81	28.79	0.00
EMS mM				
20.00	29.19	35.44	23.46	1.14
30.00	40.68	40.67	32.30	1.91
Gamma rays + EB				
45 kR + 0.08 mM	43.97	29.57	29.40	1.84
55 kR + 0.08 mM	56.52	45.40	30.22	2.83
65 kR + 0.08 mM	65.22	40.22	37.19	2.83
Gamma rays + EMS				
45 kR + 20 mM	77.30	44.24	35.16	2.02
55 kR + 20 mM	82.61	37.63	32.41	1.85
65 kR + 20 mM	88.89	35.61	36.23	3.27
Combination	(L) vs (CM)	(I) vs (CM)	(S) vs (CM)	
'r' value	0.48**	-0.01	-0.06	

** Significant at 1 per cent level

The order of the relative frequency of the different kinds of chlorophyll deficient mutants can be represented as follows:

Gamma rays	: *Viridis*	>	*Xantha*	>	*Albina*	>	*Chlorina*
EMS	: *Viridis*	≥	*Xantha*	>	*Albina*	>	*Chlorina*
Gamma rays + EB	: *Viridis*	≥	*Albina*	>	*Xantha*	>	*Chlorina*
Gamma rays + EMS	: *Albina*	>	*Viridis*	>	*Xantha*	>	*Chlorina*
EB	: No chlorophyll mutation						

On an overall basis, only small differences were observed in the frequencies and spectrum of chlorophyll mutations with all the mutagens and their combinations.

Viridis was the most common type and occurred with higher frequency. However, there was no specificity in the occurrence of chlorophyll deficient types. All the types occurred in all the treatments. Sjodin (1962) observed in *Vicia faba*, that *Viridis* was the most common type and he stated that such differences in specificity may be, by and large governed by the inherent genotypic differences in different mutants. The sensitivity of an organism to the mutagens depends upon the genetic architecture and the mutagen employed (Blixt and Moss Berg, 1967), besides the amount of DNA and its replication time in the initial stages (Varughese and Swaminathan, 1968). It may also be due to the heterogeneity of the organelle genomes with respect to its size, activity and abundance.

The failure of EB to induce chlorophyll mutation may also be due to its genomic specificity as it has been shown that EB is a DNA intercalating agent (Waring, 1966) and is an inhibitor or DNA polymerase (Meyer and Simpson, 1969), binds preferentially to the mitochondrial DNA over nuclear DNA (Mahler and Bastos, 1974). The mutagenic specificity of EB with mitochondrial genome had been well recognized with the induction of cytoplasmic petite mutation in yeast, resulting from deficiency of the respiratory mechanism in mitochondria (Nagley and Linnane, 1972; Slonimski *et al.*, 1968); induction of cytoplasmic male sterility in the reverted Tift 23 DB_1 lines of pearl millet (Burton and Hanna, 1976) and induction of cytoplasmic male sterility in sugar beet and wheat (Kinoshita, 1983). It has been reported that the genetic determinants of CMS are housed in the mitochondrial genome of wheat (Ogihara, 1981; Sasaki *et al.*, 1980), pearl millet (Smith *et al.*, 1987) and plasmid like DNAs of mitochondrial origin in sugar beet (Powling, 1982) wheat (Handa *et al.*, 1984). So, it may be inferred that the mutagenicity of EB may be confined with mitochondrial genome. In fact the observed high plant injury in M_1 generation with EB than in gamma rays treatments and EB enhanced gamma rays mutagenicity as evidenced from the highest M_1 injury in gamma rays post-treated with EB supports the specificity of the mutagen and it seems perhaps it may be related to respiratory inhibition of mitochondria. At the same time, the failure to recover chlorophyll mutants in treatments with EB alone may be either due to failure to induce or to high lethality owing to highly toxic nature of EB.

The significant differences in the spectra of mutants observed in the various mutagenic treatments may be attributed to the differences in the nature of premutational events caused by them. It is quite conceivable that differences in the nature of selection sieves through which genetic lesions are pre-assumed to pass before their fixation, also contribute to the spectral differences (Auerbach, 1967).

The occurrence and isolation of more than one type of mutation from a single M_1 plant is termed as multiple mutations. In this study, there were segregations for more than one type under all the treatments excepting the treatments with EB. Sharma and Orav (1965) observed in peas that the occurrence of multiple mutation to be non-random.

In general, the expected segregation ratio of 25 per cent was not observed. Palanisamy (1975) also reported such an aberrated ratio in cowpea. It may be due to the fact that the induced genetic changes with a low transmission frequency were probably closely associated with certain minute chromosomal aberrations. The preference and selection of plumpy, well filled seeds and avoidance of shriveled, ill-filled seeds, while raising M_2 generation may also be a cause, apart from the associated zygotic inviability and lethality of the selected seeds.

The association between M_1 lethality and M_2 chlorophyll mutation indicated about the relation between the potentiality of the mutagens to induce physiological abnormalities and death in M_1 generation and to induce genetic lesions to cause chlorophyll deficient seedlings in M_2 generation.

REFERENCES

Auerbach, C., (1967). The chemical production of mutations. *Science,* 158: 1141-1147.

Blixt, S. and R. Moss Berg, (1967). Studies of induced mutations in peas. XX. A preliminary test of the mutagenicity of different chemicals. *Agric. Hort. Gen.,* 16: 238-250.

Burton, G.W. and Hanna, W.W. (1976). Ethidium bromide induced cytoplasmic male sterility in pearl millet. *Crop. Sci.,* 16: 731-732.

Handa, H., Tsunewaki, K.. Kunisada, T. and Yamagishi, H. (1984). Small circular DNA molecules in wheat mitochondria. *Mol. Gen. Genet.,* 194: 368-372.

Kinoshita, T., (1983). Cytoplasmic mutation induced by chemical mutagens. *Wheat Information Service,* 56: 44-46.

Mahler, H.H. and Bastos, R.N. (1974). Coupling between mitochondrial mutation and energy transduction. *Proc. Natl. Acad. Sci.,* 71: 2241-2245.

Meyer, R.R. and Simpson, M.V. (1969). DNA bio-synthesis in mitochondria; differential inhibition of mitochondrial and nuclear DNA polymerase by the mutagenic dyes ethidium bromide and acriflavin. *Biochem. Biophys. Res. Commun.,* 34: 238-244.

Nagley, P. and Linnane, A.W. (1972). Bio-genesis of mitochondria XXI. Studies on the nature of the mitochondrial genome in yeast, the degenerative effect of EB on mitochondrial genetic information in a respiratory competent strain. *J. Mol. Biol.*, 66: 181-193.

Ogihara, Y. (1981). Characterization of mitochondrial DNAs in two alloplasmic and a normal line of common wheat by four restriction enzymes. *Wheat Information Service*, 52: 38.

Palanisamy, G.A., (1975). Investigations on the induction of mutations in cowpea (*Vigna sinensis* (L.) Savi). M.Sc. (Ag.) Thesis, Tamil Nadu Agricultural University, Coimbatore, India.

Powling, A., (1982). Restriction endonuclease analysis of mt DNA from sugar beet with normal and male – sterile cytoplasms. *Heredity*, 49: 117-120.

Sasaki, M., Nakata, N. and Yasumuro, Y. (1980). Male-fertility restorer mutation induced by X-rays in wheat. *Gamma Field Symp.*, 19: 49-62.

Sharma, S. and Orav, (1965). Investica Acad. Sci. Estonian SSR. *Ses. Biol.*, 4: 471.

Sjodin, J., (1962). Some observations in X_1 and X_2 of *Vicia faba* L. after treatment with different mutagen. *Hereditas*, 48: 565-586.

Slonimski, P.P., Perrodin, G. and Croft, J.H. (1968). Ethidium bromide induced mutation of yeast mitochondria. Complete transformation of cells into respiratory deficient non-chromosomal "Petites", *Biochem. Biophys. Res. Commun.*, 30: 232-239.

Smith, R.C., Chowdhury, M.K.U. and Pring, D.R.. (1987). Mitochondrial DNA rearrangements in *Pennisetum* associated with reversion from CMS to fertility. *Plant Mol. Biol.*, 9: 227-286.

Varughese, G. and Swaminathan, M.S. (1968). A comparison of the frequency and spectrum of mutations induced by gamma rays and EMS in wheat. *Indian J. Genet.*, 28: 158-165.

Waring, M.J., (1966). Structural requirements for the binding of Ethidium to nucleic acids. *Biochem. Biophys. Acta.*, 114:234-244.

□□□

SECTION - II
CROP PRODUCTION

Chapter 15

Pulse Production Technology

RAVICHANDRAN, M.
Professor and Head, Department of Soil Science & Agricultural Chemistry
Faculty of Agriculture, Annamalai University, Annamalainagar-608002

Introduction

Pulses are vital and integral component of Indian agriculture since time immemorial and are considered excellent crops for natural resource management, environmental security, crop diversification and consequently for viable agriculture. India grows a variety of pulse crops in a wide range of macroclimatic conditions and it was recognized as a major player in pulses contributing 25% to the global production (Fig.1). In spite of being the largest producer in the world, India has to import pulses (Fig.2) to the tune of two million tonnes every year to meet its domestic requirement. During 2006 - 2007, pulses production in the country was 14.2 m t from 23.76 m ha (Fig.3) including 12 m ha of area in rice fallow. The projection for 2007-08 is 15.19 m t which is all time high. In spite of having largest area for chick pea, pigeon pea, lentil, dry bean and total pulses in the world, India's position in average productivity of these pulses has never been decent that is 25th position in chick pea, 19th in pigeon pea, 23rd in lentil, 104th in dry bean, 50th in pea and 98th in total pulses . This indicates that except chick pea and pigeon pea, average productivity of other pulses in the country is significantly lower (Fig.4) than the average yield in the world. In chick pea and pigeon pea, India contributes 67% and 75% of global production (Fig.5). The per hectare productivity of pulses in India is very low (598 kg ha^{-1}) when compared to the average productivity of 1494 and 937 kg ha^{-1} in other developed and developing countries respectively, as well as global average productivity of 797 kg ha^{-1}. To meet the domestic requirement, India imported pulses to the tune of 2.82 m t during 2006–07. To make the nation pulse sufficient, productivity level of pulses has to increase substantially to 1200 kg ha^{-1} by 2020. This gigantic task needs vigorous development and promotional efforts to transfer the improved agro - technology packages viz., crop improvement and crop management which are capable of doubling the present level of productivity.

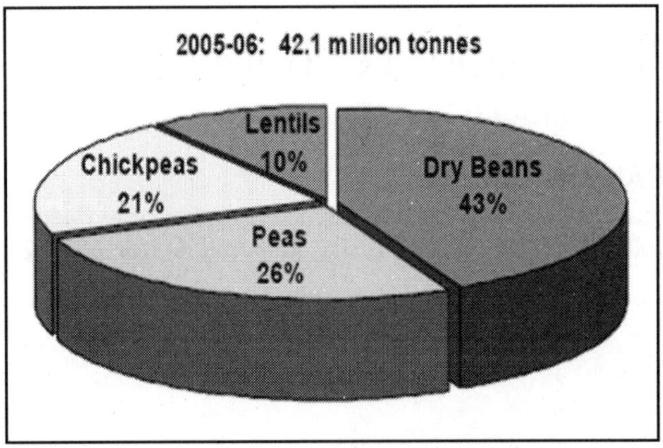

Figure 1 : Global pulse production of selected crops

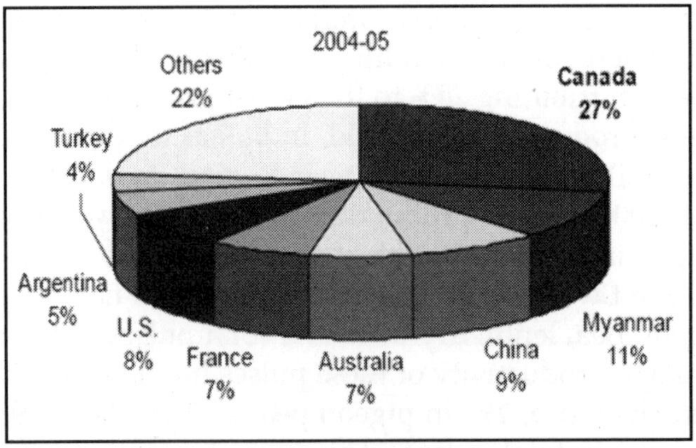

Source : FAO

Figure 2 : World pulse exporters

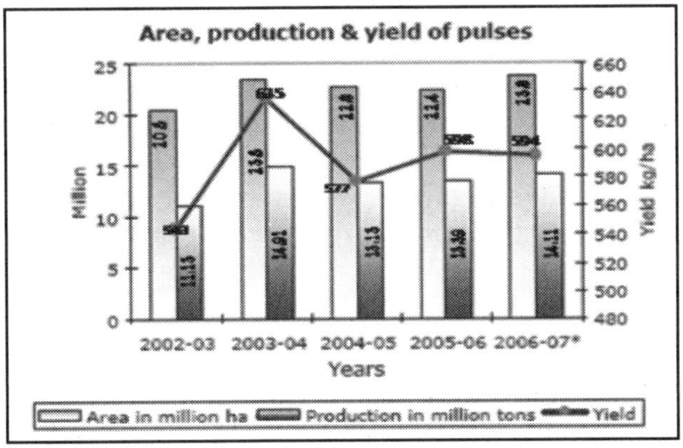

Source : Directorate of Economics and Statistics, Ministry of Agriculture, GOI

Figure 3 : Area, Production and Yied of Major Pulses in India

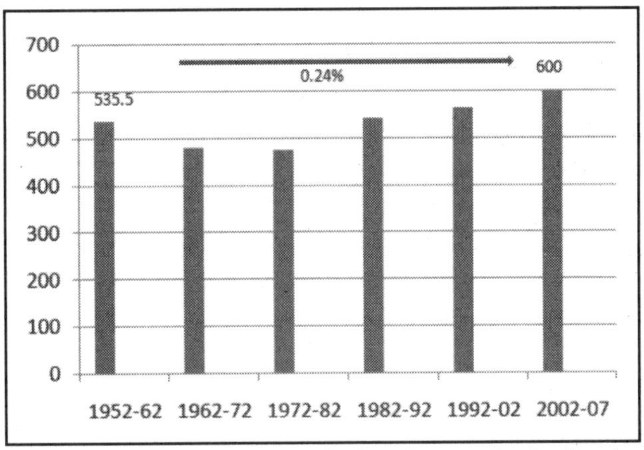

Figure 4 : Productivity trend of pulses in India (kg ha^{-1})

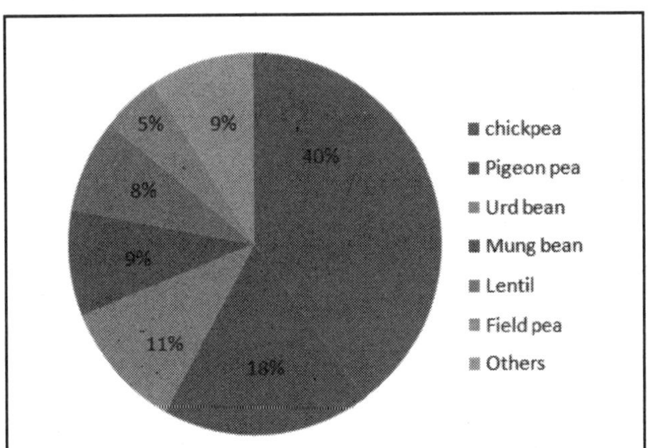

Figure 5 : Contribution of total productivity in India

The yield potential of different pulses in research plots observed during 1999-2002 are 20-22; 15-17, 20-25,11-12,10-12,20-22,15-18 q ha^{-1} compared to the national average yield of 8.06,7.97, 11.35,3.81,4.40,10.34,7.32 q ha^{-1} for chickpea, pigeon pea (early),pigeon pea (late), mung bean, urd bean, field bean and lentil respectively .Thus we can see the critical yield gap in major pulse crops. The wide yield gap between optimal and current yields shows that there is an ample scope for improvement in increasing productivity by improvement in pulse crop management at farmers' fields. Narrowing the yield gap aims not only at increasing yields but also at improving the efficiency of inputs and farmers' profitability. While large number of production technologies are generated, their adoption become meager. Transfer of improved pulse production technologies remains the most neglected component in the past, and consequently the benefit of improved varieties and production technology could not be harnessed. Existing technologies developed so far have the potential to bring at least 30 - 35 per cent improvement in pulse production (Table.1). A number of constraints viz., edaphic, environmental, technical and socio-economical etc are responsible for their low productivity. Edaphic constraints comprise rainfed cultivation of pulses on marginal and sub marginal soils having low inherent productivity. The soils of arid and semi arid tropics support 90% of pulses production in India. Soil related constraints such as physical, chemical and biological in different parts of country is responsible for low yields. The productivity of pulses remain stagnant for long time. Some of the causes of low yields of pulse crops in India are (Ganeshamurthy *et al.*, 2003)

Table 1 : Yield advantage (%) of improved technologies of pulse crops in on-farm demonstrations

Technology	Chickpea	Pigeon pea	Lentil	Mung bean	Urd bean	Field pea
Improved varieties	18.3	30.2	25.0	22.1	26.8	26.2
Fertilizer management	14.3	14.0	17.1	14.8	19.2	16.8
Rhizobium inoculation	12.5	14.0	18.1	12.4	14.8	20.6
Weed management	26.1	33.0	24.5	29.9	31.0	24.4
IPM module	19.3	32.0	15.9	23.1	18.9	23.2
Package technology	25.8	33.0	33.0	40.7	37.7	39.6

Survey of Indian Agriculture 2007 P41

- Pulses are mostly grown in under rainfed condition
- Most of the pulses are grown on marginal lands
- Most of the pulse growing soils are deficient in available plant nutrients
- Pulses receive least attention in fertilizer schedule of cropping system
- Research activities of pulses are confined mainly to inter and multiple cropping system
- In case of pulses neither intensive nor extensive method of cultivation, sustainability and profitability of the system are in vogue.

Production Technologies

A twin approach will increase the productivity through package of practices and expansion of pulse cultivation through time and space as a result of identification of short duration varieties of pulses for multiple cropping in place of low yielding crops, and in kharif or rabi fallow areas where ever possible. The major contribution, however, can be stipulated by raising the productivity of pulses with the existing technology in terms of improved varieties, fertilizers, rhizobium culture and plant protection measures. The productivity of pulses can be increased substantially, provided these inputs are used judiciously and made available in time of crop growth.

Some of the important strategies to increase the pulse yields are: Selection of varieties, efficient agronomic management (use of quality seeds/seed treatments, tillage methods, time of sowing, spacing/population maintenance), monitoring soil fertility depletion, integrated nutrient management, integrated weed, pest, and disease management, expansion of quality seed production through public/private partnership, intercropping, efficient use of farm implements and aggressive extension efforts to spread the existing improved farm technology.

Area Expansion

Pulses as a group has tremendous scope for expansion in space and time. Sustainable agriculture oriented towards market economy also demands inclusion of pulses in major cereals based cropping system so as to boost up the overall productivity, sustainability and profitability of the system. An estimated 2.5 m ha of additional area can be brought under different pulses through cropping system manipulation, crop diversification and multiple cropping systems. In addition, there is a scope for introduction of pulses in new niches such as waste land, reclaimed soils and rice fallow land by efficient watershed management and as a replacement for less remunerative crop by adopting various development activity.

In uplands, where there is still possibility for giving at least one or two irrigations through harvested rain water, crops like pulses can be successfully grown after harvest of wet season rice. Intercropping of rice with short duration pulses like green gram, black gram and pigeon pea etc also shows a good prospect for improving productivity and farmer's income in rainfed upland. Intercropping of upland rice with pigeon pea (4:1 row ratio) recorded higher net return over sole crop in Orissa. Rice based inter cropping with pigeon pea, black gram recorded much higher (3.3-3.9 kg ha^{-1} mm) rain water use efficiency. Under irrigated medium land situation, there is still scope to introduce a third crop of short duration pulses like cow pea, green gram and black gram where facilities are available to provide 1-2 life saving irrigations to the third crop.

The country has approximately 12 m ha under rice fallow. The non availability of irrigation water and delay in vacating the field after rice does not normally permit double cropping. Rainfed rabi pulses in rice fallow has great potential for income and improving food security and human nutrition. In coastal rice fallows, urd bean and mung bean in rabi season have tremendous scope owing to mild temperature during winter season. Utera (Sowing in standing rice crop) cropping of black gram in rice fallows aim at utilizing residual moisture and nutrients of rice respectively, no fertilizers are applied basally. Due to this, the productivity of rice fallow pulse is very low i.e., 250 kg ha^{-1}. There is a greater scope available for increasing rice fallow pulse production.

Improved Varieties

Genetic manipulation for efficient input use needs to be pursued to make cultivation profitable venture. Development of high yielding varieties insulated against major biotic stresses has increased the scope of pulses in cereal based cropping systems both in rainfed and irrigated areas. But the genetic potential on yield of available varieties in India is comparatively low and stagnant at particular level. The genetic bases of pulses are very narrow i.e. genetic variability is limited. Genetic base can be broadened by extensive exploration and collection of germplasm of economic as well as related wild species. Development of short duration varieties offers greater promise for introduction of pulses in new as well as in the existing cropping system. Fertilizer responsive varieties have to be developed and cultivated.

Cultural Methods

It is assessed that availability of quality seed along with critical input in time can uplift about 30 % production even with the available varieties.

Optimum population stand should be maintained for all pulses (33 no m^{-2}) except red gram, which needs 11 no m^{-2} is the basis for higher yield (Table.2). The gaps are to be filled with pre-sprouted seeds. The sowing of seed at optimum time is so important (between Jan 15th and Feb 15th for rice fallow pulses) (Tables 3&4). Poor drainage/water stagnation during rainy season causes heavy losses to pulses on account of low plant stand and increased incidence of Phytophthora. Ridge planting has been found very effective in ensuring optimal plant stand and consequently higher yield.

Table 2 : Effect of plant population on yield of irrigated blackgram (CO5)

Treatment	Grain yield kg/ha	Haulm yield kg/ha
33 Plants / m^2	1103	2754
40 Plants / m^2	1337	3123
50 Plants / m^2	1176	3817
CD (p = 0.5 %)	15	67

Subramani et al., 2002, MAJ, 89 (4-6): 305-306

Table 3: Seed yield of blackgram (kg/ha) at different dates of sowing 1998

Date of sowing	Yield
January II fortnight	739
February I fortnight	702
February II fortnight	409
March I fortnight	362
March II fortnight	262
CD	32

Table 4: Suitable time of sowing for APK1 redgram at Aruppukkottai (ARS, 2000)

Treatments	Grain yield (kg/ha)
September I fortnight sowing	812
September II fortnight sowing	304
October I fortnight sowing	158
October II fortnight sowing	78
CD (5%)	34

Nutrient Management

Among various constraints to low productivity of pulses, mineral nutrition play key role in exploiting the genetic potential of pulse crop. Pulse crop generally remove 30-50 kg N, 2-7 kg P, 12-30 kg K, 3-10 kg Ca, 1-5 kg Mg, 1-3 kg S, 200-500 g Mn, 5 g B, 1 g Cu, and 0.5 g Mo. Hence unbalanced use of nutrients caused disparity in the rate of nutrient removal and application leading to excessive mining of some nutrients, which affect the long term productivity of soil. Inadequate and imbalance of plant nutrition assumes great significance, as pulses are generally grown on marginal and sub marginal lands which are impoverished if soil nutrients and the fertilizer use is negligible. Further, constraint in availability of soil nutrients which not only brings down the productivity but also deteriorates the quality of pulse grains. Therefore, adequate and balanced nutrition is more important but often ignored by the resource poor farmers. Balanced and judicious use of fertilizers is the key to efficient use of nutrients, higher production and for maintenance of soil productivity. INM modules in cropping system mode need to be developed with major emphasis on balanced and efficient fertilizer application that would mean rational use of inorganic fertilizers, organic manures, biofertilizers and micronutrients for supply of nutrients for pulse production in such a manner that would ensure efficiency of fertilizer use, harnessing best possible positive and synergistic interaction among the various other factors of production (seed, water and agro-chemicals etc.) for realizing higher yield and reducing cost of production in pulses. Deep rooting characteristics, leaf shedding ability, capacity to fix N and solubilizing P in association with VAM make pulse crop the most effective nutrient recycling agents in the nature. Efforts should be made to exploit useful soil microorganisms to promote crop growth and enhance efficiency of applied inputs. Biological efficiency is likely to result when pulses are included in the cropping system. The N requirement is largely met through biological N fixation by rhizobium. About 40% pulse growing regions have low to medium population of native rhizobium. However in early stages of plant growth when rhizobium is not well established, a starter dose of 10-20 kg N ha^{-1} is necessary. Inoculation of seeds with rhizobium (600 g ha^{-1}) and phosphate solubilizing bacteria (600 g ha^{-1}) are known to increase grain yield in the range of 12-23 %. The methods, time and rate of N to be applied without any adverse effect on symbiotic N fixation, plant growth, economy, yield and soil fertility is important. Pulses require less N but need more P. Amongst others, P deficiency in soil is usually the key factor for poor yield of pulses on all types of soil. Phosphorus application to pulses has therefore become the basis of fertilizing pulses. The recommended dose of P varies from 30-80 kg P_2O_5 ha^{-1} and in Tamilnadu 25-50 kg ha^{-1} is recommended. The N and P management in pulses assumes greater importance than other nutrients.

All pulses which respond to P may also respond to K where the soils are deficient or medium in K status. The application of K regulates the utilization of other nutrients in plant system. When soil application of fertilizers are not possible (in rice fallow), foliar fertilization of 2% DAP and 1% KCl is recommended at 30 and 45 DAS.

In the major pulse growing areas, 44 districts have shown 40-60% sulphur deficiency and 82 districts with 50-60% zinc deficiency. Most of the pulses respond to S application and the response is up to 40 kg ha^{-1}. For the supply of both P and S, superphosphate is the best source. For S alone, gypsum is the most economical, easily available and efficient source. Very encouraging response to application of S and Zn has been found with cost benefit ratio of 10-21%. Deficiency of micronutrients and S is a serious limiting factor in obtaining derived response to applied P in pulses. Pulse crops are often exposed to multi-micronutrient deficiencies causing yield reductions to the tune of 10-40 % depending on the severity and stage of occurrence. In light textured, low organic matter and calcarious soils, deficiency of B and Fe also occur. Unless corrective measures through the application of these nutrients are taken, response to applied P cannot be fully realized. By investing small amount in these corrective measures, farmers can get a quantum jump in their profits. The mean response to applied Zn varied from 160-460 kg^{-1} in different pulse crops. Similarly the mean response to B was 60-420 kg ha^{-1}. As B and Mo are required by rhizobia, their application increases the yield of several pulse crops.

Method of fertilizer application is a non-monetary input which influences growth and consequently the crop yields. Pelleting of seeds with superphosphate, sodium molybdate and soaking of seeds with 8% $MnSO_4$ solution results in higher yield of black gram and green gram in deltaic region. The time gap between the maximum vegetative and reproductive stage in pulses is small and has low partitioning efficiency. To meet the increasing demand of nutrients at this stage, foliar application of macro and micro nutrients had been proved beneficial. The subject of feeding plants through foliage has gained much importance in recent years. The uptake of necessary elements becomes difficult for the plants when fertilizers are applied to the soil due to formation of certain less soluble complexes.

Foliar fertilization is attracting renewal interest as an efficient and economical way of supplementing the supply of macro and micro nutrients, because it avoids loss by fixation in the soil. It is advantageous over soil application in specific occasions in certain crops. For instance, in calcareous soil, deficiencies of Fe, Zn and B are inevitable due to precipitation of Zn and

Fe as insoluble carbonates and Ca –B antagonism. In soils with high indigenous P supply, Zn and Fe deficiency are common. Some of the other nutrient antagonisms commonly encountered are Cu-Mo, Cu-Fe, K-Mg etc. On such occasion, soil application of zinc sulphate or borax may not be useful. Thus, specificity of crops, soil conditions and the specific occasions in certain crops (rice fallow) warrant foliar application of nutrients. The foliar fertilization can be adopted wherever possible except for unavoidable circumstances where soil application is only feasible and advantageous over foliar application. Foliar nutrients usually penetrate the leaf cuticle or stomata and enter the cells facilitating easy and rapid utilization of nutrients. So resorting to foliar fertilization is the best option. Numerous evidences have indicated the beneficial effect of foliar application of nutrients by influencing the crop both in quantity and quality. However, to cope with the increasing demand for higher food production, the need for judicious use of foliar application is inevitable.

In case of black gram, normally it produces large number of flowers but only a few are retained and developed in to pods. This crop suffers from excessive vegetative growth due to application of N fertilizers. Although the inflorescence is profuse the yield is low due to poor pod setting and harvest index. Among the various cultural factors, time of sowing and fertilizer management appeared to have major role in determining the high grain yield of pulse crop. To avert all these problems foliar nutrition is a better tool in nutrient management of black gram (Table.5).

Table 5: Supplementary foliar spray of 250 liters of solution / ha with the following chemical will boost the yield of pulses.

Chemicals	Black gram	Green gram
Urea	7.5kg	7.5-10 kg
DAP	1.95kg	1.9-2.6 kg
Muriate of potash	1.31 kg	1.31 - 1.75kg
Potassium sulphate	1.05 kg	1.50 Kg
Succenic acid	40 g	50 g
Teepal	125 ml	125 ml
First spray	25 DAS	25 DAS
Second spray	40 DAS	40 DAS

A number of growth regulators (NAA, salicylic acid etc) have been developed to maximize the yield in many crops. Application of plant growth regulators (PGRS) can play an important role in regulating excess vegetative growth and increasing the yield of crops by making the plant photosynthetically more effective (Sinha and Ghildyal, 1973), enhancing the flower buds and checking their abscission (Murthy and Singh, 1983) and transporting photosynthates to sink. Similarly, several physiological alterations have been reported due to growth retardants such as cycocel (CCC) for optimizing yield in various crops (Oshio and Izumi, 1986). The role of mepiquat chloride (MC) in arresting vegetative growth and increasing assimilate translocation towards reproductive parts is proven fact. The use of mepiquat chloride at later period of crop growth definitely may help in this aspect which can result in enhanced grain yield of black gram. Hormonal imbalance, inadequate nutrient uptake and some environmental factors are the main cause for flower drop in pulses. In maintaining proper hormonal balance, to increase the uptake of nutrients and enhance assimilate partitioning targeting towards higher pod set and yield, homobrassinolides are found to contribute significantly. Brassinosteroids have significant growth promoting activity and pleiotrophic effect, as they influence varied development processes like growth, germination of seeds, rhizogeneses, flowering and senescence. Brassinosteroids induced plant growth was reported to be associated with increased metabolic processes and synthesis. Brassinosteroids are now receiving a great deal of international attention. A greater role of brassinosteroids realizing better crop yield during 21st century is predicted.

Irrigation Management

About 87% area under pulses is rainfed and consequently pulses face severe moisture stress with low productivity. Quantum jump in productivity can be achieved by applying come up/life saving irrigation especially in rabi pulses grown on residual moisture. Pulses in general will respond to one or two irrigations under moisture stress conditions. Pre sowing irrigation is essential followed by one or two irrigations depending upon the state of soil moisture. Sprinkler irrigation can be used for most efficient use of scarce irrigation water. The moisture use efficiency of black gram is 0.6 kg/ha/mm, while it was 0.12 for red gram, 0.84 for green gram, and 0.53 for cow pea. It is inferred that pulses have lower moisture use efficiency. Green gram possess higher WUE followed by black gram. This indicates that in low rainfall area, green gram would be the choice either as sole crop or as inter crop.

Integrated Pest/Disease Management

Pulse crops suffer heavy loss due to a wide variety of diseases, insect pests and plant parasitic nematodes and the annual yield loss is estimated to be 20 per cent in pigeon pea, 15 per cent in chickpea and 30 per cent in urd bean and mung bean. On an average, 2.5 to 3.0 million tonnes of pulses are lost annually due to pests (Ali, 1998). Integrated management strategies involving use of resistant varieties, use of disease free seeds, manipulation of cultural practices, management of vectors, and biological and chemical control methods (Vidhyasekaran and Muthamilan, 1995) will help in overcoming the loss. To prevent seed borne diseases pre-seed treatment with fungicides (Bavestin @ 2 g / kg seeds, carbendazim or thiram @ 2 gm / kg of seeds) / bio control agents (Trichoderma viridi @ 4 gm / kg of seeds) is to be resorted. Further, resource conservation technologies in planting, irrigation and manuring offer greater promise for improving pulses production and reduce the cost of production. These technologies need to be popularized extensively in pulse growing areas. This coupled with timely weed control by pre-emergence application of weedicide (pendimethaline @ 1.5 kg ai. ha^{-1} etc.) and one hand weeding after 25 DAS can enhance pulse productivity in the range of 15-32% in different pulse crops.

CONCLUSION

From the foregoing discussion, it can be inferred that the increase in productivity could be achieved when constraints are properly addressed. Thus, the pulses productivity can very easily be increased and sustained provided integrated approach is handled. Each component of improved technology is equally important. With the introduction of high potential pest and disease resistant genotypes, rhizobium inoculation, proper seed rate, executing optimum time of sowing, practicing INM, IPM, IDM and IWM techniques, the productivity of pulses could certainly be doubled in the coming years.

REFERENCES

Sinha, S.K. and Ghildyal, M.C. (1973). Increase in yield of Bengal gram by 2, 3, 5 tri - tri - iodobenzoic acid. *Crop Science,* 13:283.

Ali, M. (1998). Research, Develoment and Management for production of pulses. In: IPM System in Agriculture. Volume 4. Pulses, R.K.Upadhyay, K.G.Mukerji and R.L. Rajak (Eds). Aditya Books Private Limited, New Delhi, Pp 1-40.

Ganeshamurthy, A.N., Srinivasa Rao, C.H., Singh K.K. and Masood Ali. (2003). Management of phosphourus for high pulse productivity in different Agroclimatic Regions of India. *Fertilizer News,* 48(5): 23-42.

Murthy, Y.S and Singh, K. (1983). Flower abscission response of Vicia faba to treatments with growth substances. *Acta Botanica Indica.* 11: 65-68.

Oshio, H and Izumi, K. (1986). S-3307, a new plant growth retardant, its biological activities, mechanism and mode of action. In: Plant Growth Regulators in Agriculture, pp 198-208. Food and Fertilizer Technolog Centre, Taipei,Taiwan.

Vidhyasekaran, P and Muthamilan, M. (1995). Development of formulation of Pseudomonas fluoorescens for the control of chick pea wilt. *Plant Dis.* 79:782-786.

□□□

Chapter 16

Legumes in Sustainable Cotton Farming

SANKARANARAYANAN, K.*, C.S. PRAHARAJ* AND N.GOPALAKRISHNAN**
Senior Scientist (Agronomy)* and Project Coordinator and Head**
Central Institute for Cotton Research, Regional Station
Coimbatore - 641 003 Tamil Nadu, India

Introduction

Cotton inherits a tremendous significance in Indian economy as nearly one-third of India's export earnings in the form of yarn, cloth and ready-made garments are from the textile sector. India ranks second in global cotton production scene after China and the country accounts 26% of world's total cotton area and 12% of global cotton production. The share of cotton in world textile manufacturing is around 45% as against 75% in India. Moreover, as the most important natural fibre crop, it plays a dominant role both in agrarian and industrial economy. An estimated 60 million Indians get employment directly or indirectly through its cultivation, marketing, value addition and export.

Despite the global importance & need for the commodity and highest acreage in India (9.2 million ha), cotton productivity revolves only around 502 kg as against the world average of 733 kg per hectare (AICCIP 2006-07). In addition, the spate of cotton farmers in the last decade traverses through agonies like the pest resistances to insecticides, higher cultivation cost, ever decreasing & unstable returns and above all unfavorable climate. Monocropping, erratic rains, increasing pest attacks, degraded soils and indiscriminate use of pesticides have added to the farmers' woe and taken their toll in reducing the crop's performance in terms of productivity and remunerability. Moreover, the cotton soils have been rendered lesser productive because of continuous monocropping, use of high analysis fertilizers and non addition of organics. For e.g., a study conducted by the Centre for Research in Rural and Industrial Development (CRRID) in seven villages of Punjab's Bhatinda district reveals that 85% of cotton farmers are under debt trap. Hundreds of cotton farmers in the Vidarbha of Maharashtra are caught in the vicious cycle of debt and poverty. The problem continues throughout the country because of its low productivity and high cultivation cost that does not support the rural livelihoods. Hence, the sustainability issue is becoming great concern for all of us.

Legumes - an essential component in sustainable agro-ecosystems - greatly contribute by rendering its Yeoman Services like carbon sequestration, organic N addition, soil stabilization, and nutrient recycling. Legumes have long been recognized as an essential part of crop rotation even during ancient Roman times (Rogers, 1976). With increasing concern about the environmental and socio-economic sustainability of agricultural production system, there has been a renewed interest in diversifying agricultural production system through legumes alone and in combination/mixture with other crops.

As a companion crop, pulses or grain legumes - a major component of legumes play a vital role in the cotton based cropping systems. Similar to cotton, the present scenario in pulses describes India as the largest producer, importer and consumer of pulses in the world. Accounting nearly 25 per cent of global production, 15 per cent trade and 27 per cent of consumption, a sizeable population in the country still depends on pulses for vegetarian diets in meeting its protein need. As against the estimated requirement of around 17 million tones, the country produced only 14.1 million tones during 2006-07 and the shortfall is increasing with each passing year due to ever growing population and less priority on its production. As a result of mismatch in demand and supply position, price stability has been a concern and is soaring during the past 2-3 years. Therefore, keeping in view the potentials of both higher yields & margins in growing pulses in various combinations with cotton farming systems, addition of pulses may prove handy in rescuing cotton farmers from the debt trap besides ensuring sustainability of cotton production systems and helping the state in bridging the gap made by the mismatch in its demand and supply.

Legume Systems in Sustainable Cotton Production

1. Intercropping

Intercropping is growing of two or more crops simultaneously with distinct row arrangement in the same piece of land every year. Here the plant interference is minimal. Plant interference is the response of an individual plant or species (pulses) to its environment as modified by the presence of another individual plant (cotton) or another species. In contrast to intercropping, mixed cropping is also in practice in areas of likelihood for crop failures where simply seed mixture of 2 or more crops was planted.

Cotton with the relatively longer life cycle (175-210 days) and its slow growth in the initial stages offers required suitability for raising intercrops. Besides yield addition, these intercrops serve as an insurance against the crop failure, menace to pests, diseases and vagaries of weather to which the cotton monocropping is subjected to, especially in the vast rainfed tracts in the country. In addition, under Indian condition where a major chunk of cultivated area is rainfed, intercropping systems are preferred against the crop failure. Since suitability of an intercrop mainly depends upon branching habit, leaf area development and growth characteristics of the base cotton crop, thus, more compact, early mature, compatible and less competitive legume genotypes are ideally taken up as intercrop in cotton. In addition, the intercrops that can reduce the menace of sucking pests and bollworms-the most obnoxious one- are also most preferred in cotton. Dominant intercropping systems are given in Table-1.

Table 1. Dominant cotton based legume intercropping system

States	Cropping system
M.P., Maharashtra and Gujarat	Cotton intercropped with black gram, green gram, soybean, groundnut and red gram
Andhra Pradesh	Cotton intercropped with black gram, green gram, soybean and red gram
Tamil Nadu	Cotton intercropping with groundnut and black gram
Karnataka	Cotton intercropping with groundnut, black gram and green gram

Adopted from Chittapur (2004)

2. Multi-tier Cropping

An extrapolation of different intercropping combination is the multi-tier cropping where multiple crops provide an efficient mean of harvesting solar energy. Here, similar to the above ground canopy, the root systems of the component crops are also located at distinct soil zones for optimal exploration of moisture and nutrients (Palaniappan, 1985). Performance study under rainfed conditions revealed that introduction of tall third-storey crop was not advantageous due to competition for resources especially for moisture as was reflected in plant water relations. The major contributing factors in suitability of cotton + red gram intercropping system (with the highest yield in terms of cotton equivalents) are the negative temperature differential, reduced transpiration rate and high temperature and relative water content (Potiraj and Srinivasan, 1992).

3. Crop Rotation

Cotton when grown successively in the same plot without any other crops in the sequence led to failure of cotton over a period of time. Even a fallow period intervening the two successive cotton crops did not sustain yields (Praharaj and Rajendran 2007, Praharaj *et al.* 2004). Thus, more productive, more efficient yet remunerative cropping systems are to be taken up for attaining desired sustainability. Within broad limits, the choice of crops by a farmer is set by the principles of ecology and the laws of economics. Amongst the crops well adapted to a farm and favoured by the economy of the region, there appears always to be one crop which is especially favoured and is called most efficient crop (or cropping systems).

Study on one-year rotation also shows that the cotton based system having cotton (Aug-Feb), Radish (Feb- Mar), Black gram (Apr-Jun) & Sorghum (June-Aug) was more remunerative (with 11.2 q seed cotton, 150 q of radish tubers, 5.30 q of black gram and 38.8 q per hectare of sorghum grain). Trial at Central Institute for Cotton Research, Coimbatore reveals that Cotton -Maize - Tomato - Wheat - Sunflower - Cowpea (6 Crops on a 2 year rotation) produced the maximum net income. Even monocropping of cotton grown once in 2 years performed significantly better over than that grown every year in the same land. Here in these crop rotations, half the recommended dose of fertilizers performed similar with full dose in all crops except in straw yield of cereals only (higher with full dose).

A long-term study on cotton at Auburn, Alabama, showed that winter annual legumes rotated with cotton, produced cotton yields equivalent to those grown using fertilizer N. Here, yield increase to the tune of 11% was recorded in a 2-year cotton-legume-corn rotation as compared to continuous cotton grown without legumes each year. A three year rotation involving cotton-vetch, corn-rye (fertilized with 60 kg of conventional N per hectare), followed by soybeans, produced about the same cotton yields as of two year rotation (Fyre *et al.*, 1988).

Advantages of Legumes in Cotton Systems

1. Higher Productivity

In natural ecosystems, functional diversity leads to higher productivity and stability (Petchey and Gaston, 2002). Here, crop mixtures provide the necessary functional differences in agronomic traits for better crop performance and realization of potential yield (Wolfe, 1985).

Intercropping of green gram (mung) in cotton in 1:1, 1:2 or 2:1 row ratios was beneficial both in yield and economics as reported from Punjab and Haryana. The practice of raising green gram as an intercrop in cotton was advantageous in terms of higher net returns in rainfed areas, and was recommended at several locations in Andhra Pradesh, Karnataka, and Tamil Nadu. Short duration legumes like green gram and black gram appeared ideal as these can additionally contribute 5-6 q of grain per hectare with minimal impact on seed cotton yield (Mudholkar and Basu, 1995). At Coimbatore, intercropping of three rows of green gram in the interspaces of cotton and application of 45 Kg N recorded higher seed cotton yield of 14.8 q as compared to sole cotton crop of 13.2 q per hectare (Kunesekaran and Iruthayaraj, 1980). Certain green gram genotypes viz, T 44 and NPRG 3 are even more suitable in cotton systems while TCH 665 cotton cultivar is more adopted to intercropping system over MCU 9. Yet varieties MCU 9 and MCU 11 behaved similarly to a green gram intercropping system (Jaganathan *et al.*, 1994).

Modern statistical tools such as multiple regression analysis shows that soil moisture in 30-60 cm layer on 30th and 45th day greatly determined the fate of intercrop black gram (Balasubramanian and Subramanian, 1994). Black gram cultivar, APK 1 intercropped in cotton produced 28 per cent higher yield over that in check (CO 5) (Rajarathinam *et al.*, 1996).

At Nanded (Maharashtra), red gram as an intercrop in desi cotton did not differentiate the seed cotton yields obtained under it *versus* sole cotton. However, in Marathwada region of the state, cotton + pigeon pea was far superior to sole crop of Tur, and however, intercropping of cowpea with cotton produced the mixed response. Although no beneficial interaction was realized in Punjab, Haryana and Delhi, intercropping trials at Guntur recorded consistently higher monetary returns with reduced pest incidence and weed menace.

2. Stability in Yields

Biological diversity is more important in yield stability (Willey 1979). Perhaps the most interesting biological and economic aspect of intercropping is the potential for compensation among the components of the system that leads to greater stability in total yield of component crops. Cotton + Black gram raised in paired row system (2:1) produced highest mean seed cotton yield equivalent (1815 kg/ha).

In another trial, although cowpea suppressed cotton yields but the reduction in yield was compensated by cowpea grain yield (Rusinamhodzi *et al.,* 2006). Intercropping cotton with cowpea increased the productivity with land equivalence ratios (LER) of 1.4 and 1.3 for 1:1 and 2:1 intercrop respectively.

3. Higher Profitability

Variation in gross or net return in a cotton production system is a function of yields, prices of the commodities and of course, cultivation cost. What distinguishes intercropping system from monoculture is the introduction of several components into the income calculation such as relative prices of two or more commodities, efficiency of production that results in lower production cost, and compensation between the two or more intercrop components. Thus, some of the profitability related questions can be solved by appropriate intercropping. As it is less likely that prices of several crops will go down between planting and harvest than that might occur with a single crop, thus the diverse mix of enterprises would provide higher income than any single crop.

With the advent in demand of black gram, intercropping of black gram in cotton is remunerative at many locations in Tamil Nadu, Karnataka, Andhra Pradesh and Madhya Pradesh and was most profitable over cotton + soybean and sole cotton (Giri *et al.* 2006). Higher monetary return was also associated with cotton + black gram (Tomar *et al.* 1994 and Wankhede *et al.* 2000). At Lam (A.P), cotton planted in paired rows and intercropped with three rows of black gram was most suitable for realizing higher monetary return.

In north western Rajasthan, intercropping of green gram in paired rows of cotton brought about higher net monetary returns over sole cotton crop. In the rainfed zones of Central India, the practice of raising green gram as an remunerative intercrop in cotton was recommended at many locations like Akola, Nagpur, Parbhani Nanded, Achalpur, Indore, Badnawar and Khandwa. Green gram was also found to be ideal intercrop in cotton at Surat and Banswara in Gujarat and at many places in Andhra Pradesh, Karnataka, and Tamil Nadu.

At Dharwad (Karnataka) and at Lam in Guntur, soybean intercropping was remunerative. In Nandyal region of A.P., soybean intercropping in cotton gave a gross monetary return of Rs.11,247 per hectare. Field trial conducted at Banswara, Rajasthan revealed that intercropping of black gram, green gram or cowpea in rainfed crop increased the net profit compared with cotton in pure stand (Singh *et al.* 1981). Cotton intercropped with soybean, black gram and green gram also gave an additional per hectare profit of Rs.1057, Rs. 748 and Rs. 708 respectively (Patel *et al.* 1995).

A compilation on intercropping system reveals that all these recorded higher net returns per hectare than sole crop of cotton and groundnut (Koraddi *et al.* 1991). Intercropping of groundnut and cotton ir 1:1 row proportion with normal fertilization to cotton and 25% of that to groundnut recorded the highest net income of Rs. 9532/- per hectare as compared to sole crop of cotton which recorded a net income of only Rs. 3974.

At Guntur, moderate net returns of Rs.11,855 per hectare was obtained from cotton intercropped with pigeon pea compared to Rs.10,400 from sole cotton crop. The practice of cotton-pigeon pea intercrop in the row ratios of 6:1 or 8:1 is popular among cotton farmers of Marathwada region of Maharashtra and Warangal, Khammam and Adilabad districts of Andhra Pradesh.

4. Income Stability

Dimension of economic buffering is related to the nature of intercropping or mixture of species. The first factor is having an yield advantages, thus, if higher total yields result from intercropping, there will be an economic advantage as the most valuable species are well represented in the harvest. The second factor relates to the biological buffering described as that if two crops are planted throughout a field, rather than in half of field, and if a devastating insect, disease, or drought situation selectively eliminates or strongly suppresses one component, the remaining crop is likely to take advantage of the available resources and produce a relatively high yield over the entire area.

Cotton + black gram intercropping both in paired row or skip row was more profitable with a net profit of Rs.1005/- over that in pure cotton (Rs. 825/-) in dry tracts of Kovilpatti in Tamil Nadu. Substantial higher net return from cotton + black gram system in coastal districts of A.P. (Rs.17,208) along with cost benefit ratio of 3.38 were recorded in two rows of black gram as an intercrop in paired cotton rows (Krishnasamy *et al.*, 1995).

Intercropping studies with groundnut, black gram and pearl millet under normal and paired row planting under two cotton genotypes revealed that

Varalaxmi gave 23.5 per cent higher seed cotton yield and Rs. 603/- more monetary returns from a hectare over MCU 5. Both planting methods recorded almost equal yields and gross income. Groundnut and black gram did not reduce the seed cotton yield while pearl millet reduced the yield. Highest gross monetary returns were obtained with Varalaxmi cotton + groundnut followed by Varalaxmi cotton + black gram. A maximum return of Rs. 14,124/ - resulted in cotton intercropped with cluster bean followed by cotton intercropped with black gram (Rs.9511/-), green gram (Rs.8458/-) and mustard (Rs.5257/-) when compared to pure cotton crop (Rs.5007/-) (Sanandchari *et al.,* 1980). The cost benefit ratio was also maximum in cotton intercropped with cluster bean.

5. Soil fertility Management

Since crop nutrition through fertilizers incurs a substantial cost and thus, it is essential to find economy in fertilizer use. In a trial with fast growing leguminous green manures viz., sunnhemp (Praharaj *et al.,* 2004), lucerne, cowpea and clitoria with three N levels viz., 40, 60 and 80 Kg/ha in cotton under irrigated conditions, it revealed that growing and incorporation of all the green manures on 40 days after sowing increased the seed cotton yield (Subramanian *et al.* 1995 , Janardhanan 1989). In addition, 25 per cent of N fertilizer can be reduced if the green manures are grown and incorporated in the field. Incorporation of sunnhemp as green manure increased 9.85 per cent seed cotton yield and resulted in maximum nutrient uptake than without green manuring (Raskar, 2004). Several workers were also opined that growing of legumes, as an intercrop was beneficial to soil health and soil fertility (Basu, 1992).

There was an increase in percentage of N fixation by cowpea intercrops. Both sole cowpea and the intercrops showed positive N balances of 92 kg/ha for sole cowpea and 1:1 intercrop, and 48 kg/ha for 2:1 intercrop. Yet, cowpea fixed N transferred to the companion cotton crop was very low with 1:1 intercrop (3.5 kg N/ha) and 2:1 intercrop (0.5 kg N/ha) (Rusinamhodzi *et al.* 2006). Nevertheless, crop residues from intercrops and sole cowpea increased succeeding maize yields more than residues from sole cotton.

The practice of green manuring through cover cropping is widely adopted in the subtropical regions where winter fallow is common. At Sirsa, Haryana, *Melilotus indica* was observed to be promising. Intercropping of legumes helped to reduce the 'N' requirements of cotton to the extent of 25 per cent. AT Siruguppa, Karnataka, *in situ* green manuring of green gram (Praharaj *et al.* 2004) and cowpea was found to substitute 25 per cent of recommended dose of nitrogen.

Studies carried out under rainfed conditions with multicut legumes used as long year covers in chilli - *desi* cotton and hybrid cotton at Dharwad (Karnataka) revealed that chilli - *desi* cotton and *Stylosanthes hamata* as cover crop at 1:2 row proportion with a cutting interval of 45 days saved the recommended NPK nutrients (100:50:50 Kg N, P_2O_5 and K_2O per hectare) by about 25-50% in the subsequent year under continuous cropping (Anonymous, 2003). Cover cropping accounted to organic N addition to the extent of 144 Kg/ha and increased soil organic carbon from 0.58 to 0.73% (average of two years) leading to overall improvement in soil fertility.

In case of hybrid cotton *cv* DHH - 11, of the four legume covers (viz., lucerne, *Stylosanthes hamata, S. scabra* and *Centrocema sp*) screened, Lucerne was found to be most suitable at 1:2 row proportions in cotton (120cm x 60 cm) at 30 days cutting frequency. Cotton yield was increased by 13.62 per cent and the system reduced weed intensity to the extent of 91.0 and 59.8 % with Lucerne and *Stylosanthes hamata* respectively over sole cotton. Besides soil moisture conservation, cover cropping promised a fertilizer reduction to the extent of 25-50% in subsequent seasons. Lucerne green manuring with 50 % N recorded higher seed cotton yield comparable to sole cotton with NPK (Khamb, 2003). Thus these studies reveal the possibility of sustained production even under mono-cropping systems as intercropped green manure could help maintain soil fertility although without much economic advantages.

6. Organic Farming

Growing cotton organically entails using cultural practices, organic manures (with fertilizers in rock form), and biological controls rather than synthetic fertilizers and pesticides. A systems approach to organic production involves the integration of suitable practices viz., cover crops, strip cropping, crop rotation, etc. into a larger system. Through good soil and biodiversity management, farms can become increasingly self-sufficient in fertility, while pest problems are diminished, and some pests are even controlled outright. Thus, a diverse rotation, using legumes and other cover crops, is at the heart of good humus and biodiversity management.

Nutrient management in organic farming solely depends on manures and legumes. Since availability and economical feasibility of using manure is less practicable, *in situ* green manuring with fodder cowpea and its burying at 40 days after sowing will ensure a steady N supply during the grand growth and flowering periods, when the N demand peaks up in the crop. It hastens microbial activity of soil, reduce the weed growth and enhance the natural enemy build up. This provides around 400-500 kg dry matter per hectare with 2.5 per cent N and contribute 10-12 Kg N/ha during squaring. Other benefits include smothering of weeds, controlling soil erosion and nurturing natural enemies of cotton pest (Rajendran *et al.* 2000). *In situ* incorporation of legumes in biomass production, cotton yield & nutrient status of soil are given in Table-3.

Table 2. Effect of intercropping against cotton leaf hopper and bollworms

Treatments	Leafhopper (No./5plants)				Boll worm incidence (%)				Yield (Kg/ha)
Days after sowing	30	45	60	75	45	60	75	90	
Inter crop									
Black gram	3.3	10.1	17.3	20.9	24.0	21.3	21.1	21.4	455
Green gram	3.6	12.3	18.1	25.8	24.6	24.2	21.1	23.6	440
Cluster bean	2.9	8.9	14.4	20.4	21.8	14.7	18.4	20.9	662
Mustard	5.4	17.6	25.3	26.4	32.8	25.6	23.9	23.6	349
Pure cotton	7.2	20.0	30.3	29.4	36.1	27.8	27.2	26.2	343
SEd	0.7	1.3	2.02	3.7	4.7	2.7	3.4	1.9	21.6
CD (P=0.05)	1.5	2.6	4.13	7.6	9.5	5.6	6.8	3.9	44.4

(Adopted from Suresh and Dasan 1996)

Table 3. *In situ* incorporated legumes biomass production (45th DAS), cotton yield & nutrient status of soil

Intercropping system	Seed cotton yield (q/ha)	Organic carbon (%)	Total N (%)	Av.P_2O_5 (Kg/ha)	Av.K_2O (Kg/ha)
Cotton sole crop	11.0	0.53	0.06	54.3	238.0
Cotton +sunhemp	14.4	0.54	0.08	57.5	269.0
Cotton+cowpea	12.1	0.61	0.09	61.8	266.0
Cotton+blackgram	12.7	0.59	0.07	58.6	254.0
Cotton+Lucerne	13.9	0.50	0.07	45.6	253.5
Cotton+soybean	11.6	0.63	0.07	40.5	257.5
Cotton+soybean(black)	12.8	0.62	0.07	60.3	263.0
Cotton+Horsegram	14.7	0.61	0.09	59.9	296.5
Cotton+FYM 10t/ha	12.8	0.63	0.1	45.8	294.5
S.Em.	0.44	0.01	0.003	1.7	6.29
C.D.at 5%	1.28	0.03	0.007	4.97	18.08
Initial soil analysis		0.54	0.06	47.0	212.4

(Adopted from Gidnavar et al., (1992)

Dense stand of Dhaincha (*Sesbania aculeate*), a legume, can be raised around cotton field at a width of 2 m; its lopping cut and spread between cotton rows at 65-70 DAS. Its fast decomposing leaves provide N during early boll development period and stalks act as temporary mulch, preventing soil moisture evaporation (Rajendran *et al.*, 2000). At Darwad (Karnataka), three rows of few leguminous crops were intercropped in normally spaced cotton (120 x 60 cm) and incorporated after 45 days. Horse gram, sunhemp and lucerne resulted in higher lint yield of cotton. At Coimbatore, cotton yield increased by 16-20% with the application of 12-18 t/ha of green manure. Green manuring of intercropped cowpea/sun hemp/horse gram was also beneficial in cotton and reduced N requirements by 25%.

Many a time, a combination of organics is suitable for higher yield realization. At Coimbatore, combined application of FYM @ 5 t/ha and sunhemp seeded @ 15 kg/ha in inter-rows as green manure (buried at 45 DAP) with or without cotton residues @ 2.5 t/ha produced highest seed cotton yield and was significantly higher over control and NPK (Praharaj and Rajendran 2007). Thus, application of cost effective organics available locally could play as an effective substitute for inorganic fertilization and sustaining the yield.

7. Weed Management

One of the best approaches for reducing problems caused by weeds is increasing the crop density either of sole crop or intercrop. Shading the top soil and competition for water and nutrients will certainly suppress weed germination and growth (Altieri and Liebman, 1986). Legumes are often used for cover because they are less competitive with cotton and have the capacity to yield N. Inclusion of forage legumes in cropping system can also be an effective approach to weed control (Entz *et al.* 1995).

Since mechanism of weed control varies with the cropping system in practice, yet, crop diversification prevents the development of population of specific weeds adapted to cultural practices and resource competitiveness associated with a mono-culture of a single crop or row crops (Liebman and Dyck, 1993). Thus, use of crop specific herbicide- tillage practices results in the proliferation of specific weed species associated with continuous cropping system.

8. Soil Water Conservation

Besides protecting soil and associated erosion, cover crop provides N to the subsequent cotton when tilled into the soil, thereby improving soil tilth by adding organic matter. Sometimes cover crop serves as a catch crop when planted to reduce nutrient leaching following a main crop. Dense and quick growing cover crops are also used to suppress the problem caused by weeds as a smother crop or allelopathic cover. Even, mere presence of most cover crops reduce the competition from weeds. When crops are planted no till into such covers, and cover crop is not killed, then it is referred to as live mulch. Thus, live mulch involves planting of suppressed legume so that can recover following harvest of the grain crop. Legume live mulches can provide ground cover for erosion control and suppress weed during crop growth (Worshan and White, 1987).

Living plants and their residue on the soil surface protect the soil from impact of rain drops, reduce soil surface pore blockage, and reduce the velocity of running water (Hargrove and Fyre, 1987). Crop residues from the root and shoots increase soil organic carbon, water stable aggregation, and macro porosity thereby hastening the rate of water infiltration and soil water retention (Rasse *et al.* 2000).

Besides supply of BNF-N, cover crops often provide excellent habitat for predatory and parasitic insects and spiders. Some good insectary plants often used as cover crops include alfalfa, sweet clover, vetch, red clover, white clover and cowpeas. Migration benefits from the cover crop to the main crop is sometimes associated with the post-bloom period of cover crop. Here, mowing the cover crops in alternate strips may facilitate their movement, while the remaining strips continue to provide refuge for other beneficial species. Long-term cotton cover crop studies have also been tried in Louisiana (Millhollon and Melville, 1991). In the long-term Louisiana study, cotton yields declined for the first 9 years when cover crops were used, but increased steadily thereafter. In the final 4 years of the study, cotton yields were 360 kg higher following vetch, compared to fallow + 60 kg of fertilizer N per hectare. Averaged over the 30-year study period, the highest cotton yields followed wheat + 60 kg of fertilizer N, hairy vetch alone, common vetch alone, or vetch + 40 kg of N per hectare.

9. Pest Control

The agro-ecology approach in understanding insect population dynamics and pest management is proposed by Altieri (1983). Considering a larger biographic region, it includes studying natural ecosystem to find sustainable models that can be emulated in cropping system. Multiple cropping systems provide one portion in the array of possible methods that helps to implement this approach.

Sometimes a number of multiple cropping systems are favoured as insects were less prevalent than in sole crops (Altieri and Liebman, 1986). Insects may have more difficulty in finding host plants in an intercrop due to presence of non host plants, to camouflage of the preferred host, to changes in the texture or color of the total background, to masking of a chemical attractant, or to presence of a repellent from a non host plant. Even if insects successfully find the host, there may be interference with reproduction and survival. Mechanical barriers may be present in the form of non host plants, insects may leave the field more quickly if it is not a pure crop stand, or there may be differences in either the microclimate or the natural enemy population in the intercrop compared to a sole crop environment.

Among the various causes responsible for the low yield, the losses caused by the insect pest are of prime importance in cotton. Lowest pink boll worm (*Pectinophora gossypiella)* incidence was observed with cotton intercropped with soybean (Vennila, 2002). Intercropping also influenced the growth, development and maturity of cotton that in turn affected *P. gossypiella* damage. Red gram served as an attractant crop for population of *H. armigera* emerging from cotton implying the impact of chronology of cropping system determining host availability and population dynamics of the pest. Cotton + cowpea was most effective in reducing boll worm infestation and comparable with cotton intercropped with soybean system (Rajput and Daware 2002). The lowest population of leafhopper (2.89 to 20.44/5 plants) and bollworm (14.66 and 20.94 %) was observed in cotton intercropped with cluster bean over different periods and this was on par with black gram intercropped cotton, when compared to 7.22 to 36.05/5 plants and 26.22 and 36.05 per cent of leafhopper and bollworm incidence respectively in pure cotton crop (Suresh and Dasan 1996). Hopper & boll worm management under intercropping is given in Table-2.

Similarly, a trap crop is planted specifically to attract certain insects. It is then sprayed with some type of insecticide, in conventional management, or left to detain the pests from the cotton crop, or the entire trap crop is tilled under to kill the pest insects. Early sown cotton has been used as a boll weevil trap crop. Using fall-planted-cotton trap crops to reduce the number of over-wintering boll weevils was proposed (Javaid and Joshi, 1995). In a Mississippi study, Laster and Furr (1972) showed sesame (*Sesamum indicum)* to be more attractive than cotton to the cotton bollworm. Lygus bug may also be kept out of cotton by using nearby alfalfa as a trap crop. Unmowed or strip-mowed alfalfa is preferred by that pest over cotton (Grossman, 1988). Strip cropping of alfalfa prevents the immigration of pest at harvest time and keeps one of the main cotton pests, *Lygus Hesperus* under control.

Strip cropping takes place when harvest-width strips of two or three crops are planted in the same field, and thereby creating space biodiversity. Strip cropped cotton fields with alfalfa increases beneficial arthropod populations. Among the most notable are carabid beetles that prey on cutworms and armyworms (Grossman, 1989). Planting cotton into strip-killed crimson clover allows him to grow cotton without any insecticides.

Thus, in term of productivity, profitability, maintenance of soil health and fertility, pest reduction and organic cotton production, legumes are suitable as a best companion crop for cotton.

Future Lines of Work

- Breeding of short duration, early, determinate, compact and high yielding legumes with suitability to intercropping.
- Appropriate implements and machinery for working with legume intercropping need to be developed. Post emergence herbicides selective to cotton based cropping systems are to be identified.
- Round up ready legumes (e.g. soybean) for round up ready cotton for sustainability of cotton based intercropping system.
- Awareness in non adopted areas to popularize the manifold advantage of legumes. in cotton production system.
- Identification of location specific cotton based legume cropping system for further fine tuning to meet the multiple needs of the farmers.

REFERENCES

AICCIP (2007). Annual Report. All India Coordinated Cotton Improvement Project. Central Institute for Cotton Research, Regional Station, Coimbatore.

Aliteri, M.A. (1983). Agroecology: The scientific basis of alternative agriculture. Div. Biol.Control.Univ.California, Berkeley.

Altieri, M.A, and Liebman, M. (1986). In: "Multiple Cropping System" (C.A. Francis Eds.),, Macmillan,New York. pp.183-218

Anonymous (2003). AICCIP Annual report, Cotton Agronomy, Physiology and Biochemistry 2002 - 03 UAS, Dharwad, India.

Balasubramanian, T.N, and Subramanian, S. (1994).Correlation and multiple regression studies in rainfed cotton+blackgram intercropping system. *Madras Agric.J.*, 81 (3):155-157.

Basu, A.K. (1992). Integrated nutrient supply system in cotton based cropping system. *Fert.News*, 37(4):47-54.

Chittapur, B.M. (2004) Cotton based cropping System for sustained Productivity. In: *Proceedings of International Symposium on strategies for sustainable cotton Production-A Global vision.* 23-25 Nov.UAS,Dharwad, Karnataka (India).

Entz., M.H., Bullied, W.J. and Katepa-Mupondwa, F. (1995). Rational benefits of forage crops in Canadian Prairie Cropping system. *Journal of Production Agriculture,* 8: 521-529.

Fyre, W.W., Blevins, R.I., Smith, M.S., Corak, S.J. and Varco, J.J. (1988). Role of annual legumes cover crops in efficient use of water and nitrogen. In: Cropping strategies for efficient use of water and nitrogen, (Ed.) W.L Hargrove, Madison, WI: American Society of Agronomy, pp.129-154.

Giri, A.N., Deshmukh, M.N. and Gore, S.B. (2006). Nutrient management in cotton-based cropping system. *Indian Journal of Agronomy* 51(2):116-118.

Gidanavar, V.S., Shashidhara, G.B., and Manchunathaiah, H.M. (1992). Soil fertility management in monocrop cotton through legume incorporation. *Fmg. Systems* 8 (1&2): 53-55.

Graham, P.H.(1998). Biological dinitrogen fixation: symbiotic. In: Principles and application of soil microbiology, (Ed.) D.M.Sylvia, P.Hartel, J. Fuhrmann and D.Zuberer, Upper saddle River, NJ: Prentice Hall, pp.322-345.

Grossman, Joel. (1988). *Lygus. IPM Practitioner.* April. pp. 11-12.

Grossman, Joel. (1989). Enhancing natural enemies. *IPM Practitioner.* November-December. p. 10.

Hargrove, W.L. and Fyre, W.W. (1987). The need for legume cover crop in conservation tillage production. In: The role of legumes in conservation tillage system, (Ed)., j.F.Power,ankney, IA: Soil Conservation Society of America, pp.1-4.

Jaganathan, N.T., Pothiraj, P., Venkitaswamy, R., Premsekhar, M. and Prusothaman, S. (1994). Identification of cotton and green gram varieties for cotton-green gram intercropping system. *Madras Agric.J.,*81 (8):449-451.

Janardhanan, K.V. (1989). Intercropping in cotton. *Cotton Dev.* 12 (1-2):29-32.

Javaid, I., and Joshi, J.M. (1995). Trap cropping in insect pest management. *Journal of Sustainable Agriculture.* 5 (1-2)117-136.

Khamble, A.S. (2003). Effect of insitu green manuring and nitrogen levels on hybrid cotton DHH-11 under transitional tract of Dharwad. M.Sc (Agri) Thesis, UAS,Dharwad (India).

Koraddi, V.R., Channal, S.K., Guggari, A.K. and Kamath, K.S. (1991). Studies on planting pattern and fertilizer requirements for intercropping of cotton and groundnut under assured rainfall conditions. *Karnataka J. Agric. Sci.,* 4 (3&4): 126-128.

Krishnaswamy, S., Mohammed Ali, A. and Manoharan, S. (1995). Productivity and Profitability of intercropping in summer cotton and its nitrogen management. *Madras Agric.* J., 82 (11): 584-587.

Kunesekaran, V. and Iruthayaraj, M.R (1980). It pays to intercrop in cotton. *Curr.Res.*9 (2):27-29.

Larue, T.A. and Patterson, T.G. (1981). How much nitrogen do legumes fix? *Adv.Agron.*, 7: 25-47.

Laster, M.L., and Furr, R.E. (1972). *Heliothis* population in cotton-sesame intercroppings. *Journal of Economic Entomology* 65 :1524-1525.

Leung, H. Y., Zhu, Y., Molina, I.R., Fan, J.X., Chen, H., Pangga, I., Cruz, C.V. and Mew, T.W. (2003). Using genetic diversity to achieve sustainable rice disease management. *Plant Dis.* 87: 1156-1159.

Liebman, M. and Dyck, E. (1993). Crop rotation and intercropping strategies for weed management. *Ecological Applications* 3:92-122.

Mahler, R.L., Bezdicek, D.F. and Witters, R.E. (1979). *Agron.J.*, 71:348-351.

Millhollon, E.P., and Melville, D.R. (1991). The long-term effects of Winter Cover Crops on Cotton Production in Northwest Louisiana. Louisiana Experiment Station Bulletin No. 830. 35 p.

Mudholkar, N.J. and Basu, A.K. (1995). Intercropping in cotton :a review of the experiments conducted under AICCIP. *J. Indian Soc. Cotton Improv.* 20:105-11.

Palaniappan, S.P. (1985). Cropping systems in tropics. Principles and Management. Wiley Eastern Limited,. New Delhi p.215.

Patel, P.G, Patel, O.M. and Patel, U.G. (1995). Intercropping in irrigated cotton G.cot.Hyb.6.*Journal of Research*, Gujarat Agricultural University 20:1-6.

Petchey, O.L, and Gaston. K.L. (2002). Functional diversity (FD), Species richness and community composition. *Ecol. Lett.* 5: 402-411.

Potiraj, P. and Srinivasan, G. (1992). Evaluation of cotton based multitier cropping system under rainfed conditions. *Madras Agric.J.*, 79 (7): 633-369.

Praharaj, C.S. and Rajendran, T.P. (2007). Long term quantitative and qualitative changes in cotton and soil parameters under cultivars, cropping systems and nutrient management options. *Indian Journal of Agricultural Science* 77 (5): 280-85.

Praharaj, C.S., Sankaranarayanan, K. and Rajendran, T.P. (2004). Studies on fibre yield and quality as influenced by cropping systems and integrated nutrient scheduling the cotton. Contributory paper published in the International Symposium on "Strategies for Sustainable Cotton Production - A Global vision", on 23-25th Nov. 04, UAS, Dharwad, p.69-74.

Rajarathinam, S., Renganayaki, K., Shanmugasundaram, P. and Pothiraj (1996). APK-1. A high yielding black gram variety for intercropping with cotton. *Madras Agric.* J. 83(1): 1-2.

Rajendran, T.P., Venugopalan, M.V. and Tarhalkar, P.P. (2000). Organic cotton farming in India. CICR Technical bulletin No1/2000, Pub. by Director, Central Institute for Cotton Research, Nagpur, p.13-14.

Rajput, K.P and Daware, D.G. (2002). Effect of different intercrops on incidence of boll-worm complex of cotton. *J Cotton Res. Dev.*16 (2): 217-220.

Raskar, B.S. (2004). Effect of irrigation methods, fertilizer levels and green manuring on yield and nutrient balance in summer cotton. *J. Cotton Res. Dev.*18 (2):180-183.

Rasse, D.P., Smucker, A.J.M.and Santos, D. (2000). Alfalfa root and shoot mulching effect on soil hydraulic properties and aggregation. Soil Science Society of American Journal, 64:725-731.

Rogers, H.H. (1976). Forage legumes (with particular reference to lucerne and red clover).In: Report of the plant breeding Institute for 1975. Cambridge, UK: Plant Breeding Institute, p. 22-57.

Rusinamhodzi, L., Murwira, H. K. and Nyamangara, J. (2006). Cotton–cowpea inter-cropping and its N_2 fixation capacity improves yield of a subsequent maize crop under Zimbabwean rain-fed conditions. *Plant and Soil,* 287 (1-2): 327-336.

Sanandchari, A., Padaki, G.R and Apparao, P. (1980). Intercropping in rice fallow cotton with pulses. *Cotton Dev.*12 (2):5-6.

Singh, V.D., Chauhan, P.S. and Singh, V. (1981). Companion cropping in rainfed hybrid cotton. Indian J. *Agron.* 26(4):450-51.

Subramaniam. V, Jaganathan, N.T., Venkitaswamy, R., Premsekhar, P. and Purushotahaman, S. (1995). Effect of fast growing legumination intercrops and nitrogen levels on cotton. *Madras Agric.* J. 82(1): 40-41.

Suresh, S. and Dason, A.A. (1996). Effect of intercropping and time of sowing on cotton leaf hopper and bollworm. *Madras Agric.* J. 83(1): 56-57.

Tomar, R.S.S., Sharma, R.K., Patidar,G.L. and Julka, R. (1994). Performance of American cotton in relation to planting pattern and intercropping with legumes. *Indian Journal of Agronomy* 39(3):397-402.

Vennila, S. (2002). Rainfed cotton based cropping system vis-à-vis insect pest infestation. *Journal of Cotton Research and Development.*16 (2):149-152.

Wankhede, S.T., Tukhedo, A.B., Solanke, V.M., Malvi, S.D. and Katkar, R.N. (2000). Effect of intercropping on yield of cotton. *Crop Res.,* 19 (3):409-413.

Willey, R. W. (1979). Intercropping- its importance and research needs. Part I. Competition and yield advantages. *Field Crops Abstr.* 32:1-10.

Wolfe, M.S. (1985). The current status and prospects of Multiline cultivars and variety mixtures for disease resistance. *Ann. Rev Phytopathol.* 23: 251-273.

Worshan, A.D. and White, R.H. (1987). Legumes effects of weed control in conservation tillage. In: The role of legume in conservation tillage System, (ed.) ,J.F.Power, Ankeny, IA:The Soil Conservation Society of America. pp.113-118.

□□□

Chapter 17

Effect of Grain Cowpea Inter Cropping in Castor

THANUNATHAN, K., K.MUTHULINGAM, M.THIRUPPATHI AND M.RAVICHANDRAN
Faculty of Agriculture, Annamalai University, Annamalainagar - 608 002

Introduction

Cowpea (*Vigna unguiculata* (L.) Walp), a versatile and an important food legume is cultivated mostly in Africa, Asia and the Mediterranean. Being a drought tolerant crop with better growth in warm climates, cowpea is the most popular crop in the semi-arid regions of the tropics, where other food legumes do not perform as well. In Tamil Nadu, the area of cowpea, which is grouped under other pulses is estimated to be around 1 lakh hectares (Anon, 2001). Intercropping is one of the potential cropping systems to use the natural resources more efficiently than a single crop (Srilatha *et al.*, 2001). Intercropping in castor significantly increased the production and net profit per unit area per unit time (Rajput and Mishra, 1995). In order to assess the effect of grain cowpea intercropping in castor, the present study was undertaken.

Materials and Methods

Field experiment was conducted at Experimental Farm, Annamalai University, Annamalainagar, Tamil Nadu during summer season of 2007. The type of soil is clayey loam, low in available N (53 kg ha^{-1}), medium in available P_2O_5 (12 kg ha^{-1}) and high in available K_2O (155 kg ha^{-1}) with a pH of 7.9. The experiment consisted of castor as main crop and cowpea as intercrop at their sole population and under additive series and paired row system. The treatments were sole castor (T_1), sole grain cowpea (T_2), castor + grain cowpea under normal intercropping (T_3), castor + grain cowpea under paired row intercropping (T_4), sole castor + panchagavya spray (T_5), sole grain cowpea + panchagavya spray (T_6), castor + grain cowpea + panchagavya spray under normal intercropping (T_7) and castor + grain cowpea + panchagavya spray under paired row intercropping (T_8). The experiment was laid out in Randomized Block Design with three replications. Castor Cv. TMV5 was grown under normal and paired row system. Grain cowpea Cv. CO (CP)7 was intercropped in between castor rows and pairs adopting a spacing of 10 cm between cowpea plants in each row. A fertilizer dose of 30:15:15 kg N, P_2O_5 and K_2O ha^{-1} was applied to sole as well as all the inter crop plots and the recommended dose of 25:50:0 kg N, P_2O_5 and K_2O ha^{-1} was applied to sole crop of grain cowpea plots. Panchagavya (3%) was sprayed on 30 and 45 days after sowing of main and intercrops.

Results and Discussion

Intercropping grain cowpea in castor reduced the production potential of castor when compared to sole cropping (Table 1). But the castor yield equivalent was improved by intercropping cowpea. Higher castor yield equivalent of 1319 kg ha^{-1} was recorded with castor + grain cowpea under normal intercropping (T_3). Similar findings of higher castor yield equivalent due to intercropping castor with pulses was reported by Padmavathi and Raghavaiah (2004).

Foliar application of panchakavya at 3% improved the production potential of both main and intercrop. Cultivation of grain cowpea in castor field under normal intercropping and panchakavya foliar spray (T_7) recorded higher castor seed equivalent of 1426 kg ha^{-1} followed by castor + grain cowpea + panchakavya spray under paired row intercropping (T_8-1324 kg ha^{-1}). The present finding was in line with the reports of Somasundaram *et al* (2003).

Based on the above results, it can be concluded that when compared to sole cropping of castor, intercropping of cowpea in castor was found to be highly remunerative to farmers.

Table 1. Effect of castor + grain cowpea intercropping system on yield and castor seed yield equivalent

	Treatment	Castor yield (kg ha^{-1})	Grain cowpea yield (kg ha^{-1})	Castor yield equivalent (kg ha^{-1})
T_1	Sole castor	826	-	-
T_2	Sole grain cowpea	-	681	-
T_3	Castor + grain cowpea (normal)	553	639	1319
T_4	Castor + grain cowpea (paired row)	588	590	1296
T_5	Castor + panchagavya foliar spray	958	-	-
T_6	Sole grain cowpea + panchagavya foliar spray	-	723	-
T_7	Castor + grain cowpea + panchagavya foliar spray (normal)	618	674	1426
T_8	Castor + grain cowpea + panchagavya foliar spray (paired row)	590	611	1324

REFERENCES

Anon, (2001). Agrostat. Commissioner of Agriculture, Government of Tamil Nadu, Chennai.

Padmavathi, P. and Raghavaiah, C.V. (2004). Productivity and returns of castor (*Ricinus communis*) based intercropping systems with pulses and vegetables under rainfed conditions. *Indian J. Agrl. Sci.*, 74(5): 235-8.

Rajput, R.L. and Mishra, M.K. (1995). Studies on intercropping in castor (*Ricinus communis*) under rainfed condition. *Haryana J. Agron.*, 11(2): 141-144.

Somasundram, E., Sankaran, N., Meena, S., Thiyagarajan, T.M., Chandaragiri, K.K. and Panneerselvan, S. (2003). Response of greengram to varied concentrations of panchakavya (organic nutrition) foliar application. *Madras Agric. J.* 90(1-3): 169-172.

Srilatha, A.N., Masthan, S.C. and Shaik Mohammed. (2001). Evaluation of biological and economic efficiency in castor - legume intercropping systems. *Res. Crops*, 2(3): 445-448.

□□□

Chapter 18

The Response of Seed Enhancement Techniques to Mid Storage Correction in Groundnut (*Arachis Hypogaea* (L.)) Cv Vri 2 Kernels

SATHIYA NARAYANAN, G., M. PRAKASH, J. GOKULAKRISHNAN AND B. SUNIL KUMAR
Department of Agricultural Botany, Annamalai Univerisity

Introduction

Groundnut (*Arachis hypogaea* L.) is an important oilseed crop in India and commonly called as poor man's nut. It is also known as peanut, monkey nut and goober nut. Seed being a living entity, deterioration beyond physiological maturity is inevitable and will be pronounced when seeds are stored under hostile conditions. Rapid loss of vigour and viability of oilseeds during storage is one of the major constraints faced by the Indian seed industry amounting to loss of valuable seeds and the corresponding financial implications. Although ageing of seeds cannot be arrested completely when they are stored under ambient condition, it can however, be controlled to an appreciable extent by adoption of suitable storage technologies. Developing controlled storage condition might solve this problem but it is often not feasible for storing bulk quantity of seeds at least for the present. Hence, with this background the present study was attempted to find out suitable mid storage treatment for groundnut

Materials and Methods

Genetically pure seeds of groundnut cv. VRI 2 obtained from the Oilseed Research Station, Virudhachalam formed the base material for this study. The bulk seeds were graded for uniformity using appropriate round perforated metal sieves of sizes 18/64". In this experiment five months old seeds of groundnut kernel were imposed with various physical seed enhancement techniques such as treatment with thiram (@ 2 g kg^{-1} as slurry/dry treatment) (T_1 and T_2), halogen mixture (@4 g kg^{-1}) as dry/slurry treatment (T_3 and T_4), pelleting with leaf powders (arappu, neem, pungam) (T_5, T_6 and T_7),@ 200 g kg^{-1} and pelleting with vasambu rhizome powder @ 200 g kg^{-1} (T_8) of seed along with control. Seeds were also given to the mid storage correction such as hydration - dehydration, halogen mixture (@ 4 g kg^{-1} for) as dry/slurry treatment along with untreated control. The treated seeds were evaluated for germination percentage and dry matter production.

Results and Discussion

The germination percentage of the invigorated seeds varied significantly due to treatments and their interactions (Table 1). Among the invigoration treatments, I_3 recorded the highest germination (80 per cent) followed by I_2 (78 per cent), while the control I_0, recorded the lowest germination (72). Between the interaction effects, I_3T_4 seeds recorded the highest germination (86 per cent) while the lowest was with control, I_0T_0 (69 per cent). The dry matter production of the invigorated seeds varied significantly due to treatments and their interactions (Table 2). Among the invigoration treatments, I_3 recorded the highest drymatter production (2.056 g.10 $seedlings^{-1}$) followed by I_2 (2.019 g.10 $seedlings^{-1}$), while the control, I_0 recorded the lowest drymatter production (1.901 g.10 $seedlings^{-1}$). Between the interaction effects, I_3T_4 seeds recorded the highest drymatter production (2.283 g.10 $seedlings^{-1}$) while the lowest was with control I_0T_0 (1.713 g.10 $seedlings^{-1}$).

The study revealed that the germination potential of groundnut kernels increased irrespective of the treatment, when they were subjected to mid storage correction. In the untreated control, germination reduced drastically. The halogen slurry treated seeds of groundnut kernels imposed with mid storage correction of halogen slurry treatment maintained higher germination. The enhancement of germination over control was 24% in groundnut kernel. These findings are in agreement with the findings of Basu and Pal (1979) in rice, Basu *et al.* (1985) in mustard and Anbazhagi (1997) in groundnut. It might be due to enlargement of embryo within the seed as reported by Austin *et al.* (1969).

Table 1. Response of physical seed management techniques to mid storage correction in groundnut kernel cv VRI 2 – Germination percentage

Treatments	I_0	I_1	I_2	I_3	Mean
T_0	69 (56.17)	71 (57.42)	73 (58.92)	75 (60.00)	72 (58.12)
T_1	70 (57.00)	73 (58.69)	75 (60.00)	77 (61.34)	74 (59.26)
T_2	71 (57.42)	74 (59.34)	76 (60.67)	78 (62.27)	75 (59.92)
T_3	75 (60.00)	77 (61.34)	80 (63.44)	84 (66.43)	79 (62.80)
T_4	76 (60.67)	80 (63.44)	83 (65.92)	86 (68.32)	81 (64.59)
T_5	73 (58.69)	76 (60.89)	79 (62.73)	81 (64.41)	77 (61.68)
T_6	71 (57.42)	74 (59.34)	76 (60.89)	78 (62.03)	75 (59.92)
T_7	72 (58.05)	75 (60.23)	77 (61.57)	79 (62.97)	76 (60.70)
T_8	73 (58.69)	76 (60.68)	78 (62.27)	80 (63.68)	77 (61.33)
Mean	72 (58.24)	75 (60.16)	78 (61.83)	80 (63.49)	76 (60.93)
		I	T		I x T
CD (P = 0.05)		0.48	0.72		1.44

Figures in parenthesis are Arcsine Transformed value

Table 2. Response of physical seed management technique to mid storage correction in groundnut kernel cv VRI 2 – dry matter production (g.10 seedlings^{-1})

Treatments	I_0	I_1	I_2	I_3	Mean
T_0	1.713	1.720	1.783	1.810	1.757
T_1	1.847	1.867	1.917	1.960	1.898
T_2	1.853	1.870	1.927	1.973	1.906
T_3	1.983	2.077	2.160	2.180	2.100
T_4	2.080	2.183	2.257	2.283	2.201
T_5	1.910	2.047	2.080	2.100	2.034
T_6	1.917	1.960	1.967	2.043	1.972
T_7	1.877	1.977	2.047	2.070	1.993
T_8	1.930	1.973	2.030	2.083	2.004
Mean	1.901	1.964	2.019	2.056	1.985
		I	T		I x T
CD (P = 0.05)		0.01	0.01		0.02

The halogen slurry treated seeds of groundnut kernel subjected to the mid storage correction with halogen slurry treatment recorded higher values of dry matter production. The increase was to the tune of 33 per cent over control in groundnut kernel. The results are in conformity with the findings of Anbazhagi (1997) in groundnut. Activation of protein synthesis by the RNA through mid storage correction might be the reason for the increased dry matter production in the present study. The increase might be due to higher mitochondrial activity, formation of high energy compound and vital bimolecular and preservation of ultra structures (Henckel, 1964), control to free radical reaction in lyophilized bacterial cells (Heckly and Dimmick, 1967) and removal of toxic by-products (Dharmalingam and Basu, 1990).

The results of the studies clearly indicated that five months aged seeds of groundnut kernel imposed with various physical enhancement techniques can be subjected to mid-storage correction with halogen slurry treatment @4 g kg^{-1} for the improvement of germination and dry matter production.

REFERENCES

Anbazhagi, G. (1997). Nutrient management on seed yield, quality and storability of bunch Groundnut (*Arachis hypogaea* L.) cv CO 2. M.Sc.(Ag.) Thesis, Tamil Nadu Agricultural University, Coimbatore.

Austin,R.B., Longden, P.C. and Hutchinson,J. (1969). Some effects of "hardening" carrot seed. *Ann.Bot.,* 33: 883-893.

Basu, R.N. and Pal, P. (1979). Physico-chemical control of seed deterioration in rice. *Indian J. Agric. Sci.,* 49(1): 1-6.

Basu, R.N., Rudrapal, A.B. and Pathak, G. (1985). Physico-chemical control of mustard seed deterioration. *Seed Res.,* 13(2): 34-46.

Dharmalingam, C. and Basu, R.N. (1990). Maintenance of viability and vigour in sunflower (*Helianthus annuus* L.). *Seed Res.,* 81(1): 15-24.

Heckly, R. T. and Dimmick, R.L. (1967). Electron paramagnetic resonance in frozed and dried biological materials. *Nature* (Lond), 216:1003-1004.

Henckel, P.A. (1964). Physiology of plants under drought. *Ann. Rev. Pl. Physiol.,* 15: 363-386.

□□□

Chapter 19

Pattern of Pea Seed (*Pisum sativum* (L.)) Deterioration

PARAMASIVAM,V., S.SUNDARESWARAN,* R.VAIDYANATHAN AND D.ANANDA NAYAKI
Vegetable Research Station, Tamil Nadu Agricultural University, Palur 607 102
* Department of Seed Science and Technology, TNAU, Coimbatore

Introduction

Peas are rich in protein and being cultivated in temperate regions like Udhagamandalam, Coonoor, Kodaikanal and Yercaud in Tamil Nadu. The pea seeds are very poor in storability. Hence, the present investigation was carried out to assess the pattern of seed deterioration in peas.

Materials and Methods

Pea (cv. Bonneville) seeds were subjected to accelerated ageing (Delouche and Baskin, 1973) for different durations viz, 0,2,4,6,8,10 and 12h for creating different quality levels and following observations viz., germination (ISTA, 1999), speed of germination (Maguire, 1962), root and shoot length, dry matter production, vigour index(Abdul -Baki and Anderson,1973), dehydrogenase activity (Kittock and Law, 1968), electrical conductivity (Presley, 1958), free sugars (Somogyi, 1952) and free amino acids (Ching and Ching, 1964) were recorded. The data were statistically scrutinized by the method described by Panse and Sukhatme (1967).

Results and Discussion

The results of the present investigation revealed that the speed of germination, dehydrogenase activity and the integrity of the seed coat measured through electrical conductivity, free sugars and free amino acids in seed leachate were the first parameters to deteriorate within two days of ageing (Table 1 & 2). These results are in conformity with the findings of Delouche and Baskin (1973) in which they stated that first was the degradation of membrane followed by impairment of energy and synthesizing mechanism, decline in the rate of respiration and biosynthesis, decline in the rate of germination etc., were in the process of deterioration. The deterioration alters the semipermeability properties of the membrane integrity (Berjak and Villiers, 1972). Berjak (1968) suggested that loss of viability may be the result of loss of integrity of cell membranes.

Table 1: Pattern of deterioration of pea seeds under accelerated ageing condition

Ageing periods (days)	Germination (%)	Speed of Germination	Root Length (cm)	Shoot length (cm)	Dry matter production (mg/seedling)
0	81.0 (64.43)	7.0	12.3	9.6	117.0
2	80.0 (63.51)	6.5	12.2	9.1	115.0
4	56.0 (48.58)	4.6	11.3	8.6	103.0
6	48.0 (43.84)	3.6	11.1	8.3	97.0
8	14.0 (22.23)	1.9	10.4	8.0	87.0
10	9.0 (18.00)	0.9	10.1	7.1	85.0
12	7.0 (14.93)	0.8	6.8	6.5	74.0
CD (0.05)	7.89*	0.41*	0.91*	0.77*	13.74*

(Figures in parentheses are transformed values)

Subsequently, loss in germination was the next visible parameter affected within 4 days of accelerated ageing (Table 1). Ghose *et al.* (1981) suggested that the complete loss of activity of dehydrogenase enzyme prevent the germination in rice seeds. According to Chauhan *et al.* (1984), the percentage decrease of dehydrogenase activity was 40 in 3 days aged soybean seeds and it was 60 in 15 days aged barely seeds. In the present study, the dehydrogenase enzyme activity declined significantly with 7per cent after 2 days of ageing (Table 2).

Table 2: Pattern of deterioration of pea seeds under accelerated ageing condition

Ageing periods (days)	Vigour index	Dehydrogenase activity (O.D)	Electrical conductivity (μmhos/cm)	Free sugars (μg)	Free amino acids(μg)
0	1779	0.44	354.0	108.0	13.0
2	1703	0.41	482.0	178.0	24.0
4	1138	0.35	530.0	242.0	44.0
6	934	0.32	724.0	332.0	83.0
8	257	0.30	969.0	477.0	141.0
10	160	0.27	1118.0	520.0	185.0
12	88	0.24	1223.0	620.0	240.0
CD(0.05)	85.55*	0.008*	31.38*	23.51*	12.42*

Rudrapal and Basu (1979) in wheat, Kalavathi (1985) in soybean evidently established a positive association between dehydrogenase enzyme activity and seed vigour. Another interesting finding was after 6 days ageing, the seeds produced twin seedlings. This might be due to the chromosomal aberrations and point mutations occurred during ageing (Harrison, 1966 and Purkar and Banerjee, 1979).

The third parameter was drymatter production which was reduced to a detectable extent within 6 days of ageing. Detectable reduction in shoot length was measured only 8 days after accelerated ageing. Root length was the last parameter so for studied in this investigation which took 10 days to have significant reduction. (Table1).

REFERENCES

Abdul - Baki, A.A. and Anderson, J.D. (1973). Vigour determination of soybean seeds by multiple criteria. *Crop Sci.*, 13: 630-633.

Berjak, P. (1968). A study of some aspects of senescence in embryos of Zea mays L. Ph.D., Thesis, Univ. of Natal.

Berjak, P. and Villiers, T.A. (1972). Ageing in plant embryos II. Ageing induced damage and its repair during early germination. *New Phytol.,* 71:135-144.

Chauhan, K.P.S., Purkar, J.K. and Banerjee, S.K. (1984). Ageing induced changes in seeds. *Seed Res.,* 12(1): 53-71.

Ching, T. and Ching, K.K. (1964). Freeze drying of pine pollen. *Plant Physiol.,* 39: 705-709.

Delouche, J.C. and Baskin, C.C. (1973). Accelerated ageing techniques for predicting the relative storability of seed lots. *Seed Sci. & Technol.,* 1: 427-452.

Ghose, B., Adikary, J. and Banerjee, N.C. (1981). Changes of some metabolite in rice seed during ageing with special reference to seedling vigour. *Seed Sci. and Technol.,*9: 469-473.

Harrison, B.J. (1966). Seed deterioration in relation to storage condition and its influence upon germination, chromosomal damage and plant performance. *Natn. Inst. Agric. Bot.,* 10: 644-663.

ISTA, (1999). International Rules for Seed Testing. Supplement Rules. *Seed Sci.& Technol.,* 27: 25-30.

Kalavathi, D. (1985). Studies on seed viability and vigour in a few soybean genotypes. M.Sc. (Ag.) Thesis, Tamil Nadu Agrl. Univ., Coimbatore.

Kittock, D.L. and Law, A.G. (1968). Relationship of seedling vigour to respiration and tetrazolium chloride reduction by germinating wheat seeds. *Agron. J.,* 60: 286-288.

Maguire, J.D. (1962). Speed of germination – Aid in selection and evaluation of seedling emergence and vigour. *Crop Sci.,* 2: 176-177.

Panse, V.G. and Sukhatme. P.V. (1967). Statistical Methods for Agricultural Workers, ICAR Publication, New Delhi, pp. 327-340.

Presley,J.T. (1958). Relation of protoplast permeability to cotton seed viability and predisposition to seedling disease. *Pl. Dis. Reptr.,* 42: 852.

Purkar, J.K. and Banerjee, S.K. (1979). Genetic changes in relation to seed ageing under tropical storage conditions. *Seed Res.,* 7(2): 190-196.

Rudrapal, A.B. and Basu, R.B. (1979). Physiology of hydration dehydration treatment in the maintenance of seed viability in wheat. Indian J. Expt. Biol., 17: 768-771.

Somogyi. (1952). Note on Sugar determination. *J.Biol. Chem.,* 195: 19-23.

□□□

Chapter 20

Effect of Integrated Nutrient Management on the Performance of *Dolichos Lablab Var. Typicus* in Lignite Mine Spoil

KARUPPAIAH, P. AND K. MANIVANNAN
Department of Horticulture, Faculty of Agriculture, Annamalai University
Annamalainagar - 608 002

Introduction

Nowadays, the increasing pace of industrialization have not only increased the global warming but also brought more of vegetative area under threat. Among the various categories of industries, mining is considered as the biggest destroyer of vegetative area by means of mining and dumping (Trivedi and Gurdeepraj, 1983). In India, the total area under leasing is 27,346 ha for mining and dumping the mine spoil is in more than 4082 places (Dadhwal *et al.*, 1988). In Tamilnadu, the major destruction due to mining is attributed by NLC in Cuddalore district covering an area of 2000 ha where they go for expansion upto 200 acres every year under open cast mine system. In the actual process of mining, the area is completely stripped off vegetation and subsoil is removed in order to reach the lignite bed. This over burden soil is dumped on barren land. Hence, both the mined and barren lands are affected. Researches revealed that minespoil is different from normal soil and as such it is a stressful environment for the plant growth because it lacks the life supporting parameters such as organic matter, nutrients, soil texture, structure, beneficial microorganisms besides the deleterious effects of soil compaction, leading the area not suitable for cultivation and vegetation and causing environmental problems (Gardiner, 1993 and Karuppaiah *et al.*, 1998). Therefore, it is necessary to overtone the ecosystem of the minespoil area. Application of suitable amendments to the spoil has been found to increase the yields of legumes, vegetable crops and fruit crops (Pitchai *et al.*, 1994, Karuppaiah *et al.*, 1998 and Karuppaiah and Manivannan, 2003).

Integrated Nutrient Management (INM) as defined by IFOAM include all agricultural systems that promote environmentally, socially and economically sound production of food. These systems take soil fertility as a key to successful production. By respecting the natural capacity of plants, animals and soil, it aims to optimize quantity and quality in all aspects of agriculture and the environment. INM dramatically reduces the use of chemo-synthetic fertilizers by reframing the quantity and quality of organic and inorganic nutrients supplied to the soil. Since FYM, pressmud and humic acid have been proved to be the richest source of organic matter (Schritzer and Khan, 1972) and are cheaply available in the lignite mine area, an experiment have been formulated to determine the appropriate combination of organic amendments with inorganic nutrients for the effective cultivation of *Dolichos lablab var typicus* in the lignite mine spoil.

Materials and Methods

The experiment was carried out at Annamalai University in collaboration with NLC, Neyveli with *Dolichos lablab var typicus* as a test crop. First the representative mine spoil samples were collected and analysed for its physico-chemical characters and compared with the normal soil of the experimental area. To study the effect of humic acid and other amendments such as FYM, pressmud and inorganic N, P and K, seventeen treatments have been formulated in Randomised Block Design with 3 replications. The field situated at Mine-I afforestation site at neyveli was thoroughly ploughed and levelled. Fifty one plots of 10 m^2 area with 1 m distance between each plot was formed with a bund size of 45 cm. The treatments imposed are presented in the Tables 2, 3 & 4. The treatments were randomly arranged in each replication. In all the treatments, except control, the fixed quantity of FYM, pressmud, humic acid in the form of potassium humate, phosphorus, potassium and one third of nitrogen were applied at the time of sowing. Remaining N was applied in two splits at 20 days interval. The biometrical characters observed were germination percentage, plant height, number of branches per plant, vigour index, leaf area index, chlorophyll content, photosynthetic rate, dry matter production, number of days taken for first flowering, number of pods plant^{-1}, harvest index, yield and uptake of NPK. At the conclusion of each experiment, the plants were collected at a height of 1 cm above the spoil surface and washed successively in acidified and deionised water and dry weights were taken after drying at 60°C for 48 hours. Plant samples were digested and analysed for total nitrogen (Humphries, 1956) and phosphorus and potassium (Jackson, 1973). The chlorophyll content at first flowering stage was determined as described by Arnon (1949). The photosynthetic rate was measured at first flowering stage using the photosynthetic system LCA-2 model.

The physico-chemical properties of the representative mine spoil samples are presented in table 1. The data revealed that the spoil is structureless, sandy clay loam in texture having high bulk density, low exchangeable bases and neutral pH with high spoil compaction. The over burden spoil is also very low in organic carbon content, nitrogen, phosphorus and medium in available K_2O. The contents of available manganese, copper, zinc and iron are normal. But, as a whole, the physico-chemical characters of mine spoil revealed that it is not a medium suitable for establishment of vegetation.

Results and Discussion

Growth and physiological attributes are the essential parameters for the attribution of yield. The biometric characters like germination percentage, seedling vigour, plant height, number of branches per plant, leaf area index and dry matter production are known to have direct influence on the yield (Kannan and Ramani, 1988). In the present study, integrated nutrient management in different levels of FYM, Pressmud, N, P and K fertilizers along with humic acid showed better performance in improving the growth attributes in different magnitudes in the lignite mine spoil compared to the control (Table 5).

Application of FYM 18.5 t ha^{-1} along with pressmud 18.5 t ha^{-1}, NPK 45:75:75 kg ha^{-1} and humic acid 30 kg ha^{-1} (T_{16}) recorded the highest value for germination percentage (83.06%), vigour index (1872.42), plant height (65.25 cm), number of branches per plant (8.40), leaf area index (51.86) and dry matter production (6.25 t ha^{-1}). The next best result was observed in the treatment combination of FYM 18.5 t ha^{-1} along with pressmud 18.5 t ha^{-1} and NPK 45:75:75 kg ha^{-1}. The poor performance of germination and other growth and physiological attributes were observed with control. From the results, it is clearly understandable that the minespoil as such is not encouraging the growth of the plants even in the nodule forming leguminous garden bean. The variation in the growth performance of the garden bean in different treatments could be attributed due to the variation in the superiority of the amendment combinations. Goldsworthy and Fischer (1984) explained that the growth of any plant could be determined by the availability of nutrients, water, light and air. Though the minespoil is very poor in available nutrients, air and water along with spoil compaction, the appropriate combination of organic amendments along with inorganic nutrients and humic acid improved the growth of garden bean in the mine spoil as in the normal soil due to the supply of adequate nutrients and release of soil compaction. Similar results were also reported by Gardiner (1993), Pitchai et al. (1994), Karuppaiah *et al.* (1998) and Karuppaiah and Manivannan (2003). With regard to the chlorophyll content and photosynthetic rate (Table.3), the treatment T_{16} was found to be the best, followed by T_{14}. Goldsworthy and Fisher (1984) explained that the chlorophyll content and photosynthetic rate of a particular plant were directly correlated with the yield. The results of the present study are in accordance with the above findings.

With regard to the flowering characters such as number of days taken for first flowering, the very early flowering was noticed with the control followed by the treatments supplied with low amount of organic and inorganic nutrients supplied. Under the optimum level of organic and inorganic nutrients combination along with humic acid, the first flowering was taken at 28 to 29 days after sowing. Levitt (1973) explained that any stress to the plant could induce early flowering. Here also, the stressful environment of lignite mine spoil induced early flowering. While amendments were applied, the stresses were released and normal flowering was ensured. This result confirms the scope of minespoil reclamation and effective cultivation of garden bean in the lignite mine spoil.

Regarding the yield attributing characters viz, number of pods per plant, yield per hectare and harvest index (Tables 3 & 4), the treatment combination of FYM 18.5 t ha^{-1} + press mud 18.5 t ha^{-1} + NPK 45:75:75 kg ha^{-1} along with humic acid 30 kg ha^{-1} recorded the maximum values of 39.33 numbers of pods per plant , yield (11.7 t ha^{-1}) and harvest index (62.79). This was followed by the treatment combination of FYM 18.5 t ha^{-1} + pressmud 18.5 t ha^{-1} and NPK 45:75:75 kg ha^{-1}. This is due to the fact that integration of different organic and inorganic nutrients along with humic acid offers a balanced supply of major and micronutrients for plant uptake and the integration of FYM 18.5 t ha^{-1} along with pressmud 18.5 t ha^{-1} are useful for the release of compaction and the addition of organic matter in the spoil. In addition humic acid contains many plant growth factors. These favourable physical and chemical conditions created in the spoil due to the right integration of nutrients increased the

Table 1. Physical and exchangeable properties and mineral status comparison of representative mine spoil with normal soil

Elements	Minespoil	Normal soil	Comments on mine spoil
Physical properties			
1. Colour	Very pale brown to pink (10 YR 7/4 to 10 YR 7/3 M)	Grey to brown (10 YR 5/1 M to 7.5 YR 5/M)	Iron containing minerals are abundant
2. Structure	Massive	Blocky to crumb	Very poor agreggates
3.Textural class	Sandy clay loam to sandy loam	Sandy loam	Moderately fine textured
4. Bulk density (mg/m^3)	1.64	1.33	high
Exchangeable properties			
Cation exchange capacity (c mol p $^{(+)}$ kg^{-1})	9.90	20.72	Low
Exchangeable calcium (c mol p $^{(+)}$ kg^{-1})	5.20	13.00	Low
Exchangeable magnesium (c mol p $^{(+)}$ kg^{-1})	2.0	6.00	Low
Exchangeable potassium (c mol p $^{(+)}$ kg^{-1})	0.1	1.00	Low
pH (1:25)	6.52	7.00	Neutral
EC (dSm^{-1}) 1:25)	0.37	0.97	normal
Mineral status			
1. Sesquioxides (R_2O_2)% (iron and aluminium oxides)	14.00	25.00	Medium
2. Iron oxides (Fe_2O_3%)	8.20	18.00	Medium
3. Organic carbon (%)	0.02	0.60	Very low
4. Available N (kg ha^{-1})	81.28	250.00	Low
5. Available P_2O_5 (kg ha^{-1})	3.05	15.00	Low
6. Available K_2O (kg ha^{-1})	178.68	300.00	Medim
7. Total iron (ppm)	9341.52	25000.00	Low
8. Total manganese (ppm)	125.00	1000.00	Low
9. Total zinc (ppm)	85.00	50.00	High
10. Total copper (ppm)	69.44	20.00	High
11. Available iron (ppm)	11.25	24.10	Medium
12. Available manganese (ppm)	5.64	4.23	High
13. Available zinc (ppm)	1.08	1.31	Medium
14. Available copper (ppm)	2.22	2.00	High

Table.2. Effect of integrated nutrient management (INM) practices on growth attributes of *Dolichos lablab* var. *typicus* in lignite mine spoil

Treatments	Germination %	Plant height (cm)	Number of branches per plant	DMP (t ha^{-1})
T_1 – FYM 12.5 t ha^{-1} + PM 12.5 t ha^{-1} + NPK 35:45:45 kg ha^{-1}	72.23 (57.82)	43.21	3.40	3.78
T_2 – FYM 12.5 t ha^{-1} + PM 12.5 t ha^{-1} + NPK 45:75:75 kg ha^{-1}	73.18 (58.61)	44.37	3.42	3.98
T_3 – T_1 + HA 30 kg ha^{-1}	73.81 (58.62)	45.38	3.86	4.02
T_4 – T_2 + HA 30 kg ha^{-1}	75.12 (59.81)	47.74	4.10	4.21
T_5 – FYM 18.5 t ha^{-1} + PM 12.5 t ha^{-1} + NPK 35:45:45 kg ha^{-1}	75.42 (59.81)	49.86	4.82	4.48
T_6 – FYM 18.5 t ha^{-1} + PM 12.5 t ha^{-1} + NPK 45:75:75 kg ha^{-1}	79.94 (63.39)	55.58	4.87	4.69
T_7 – T_5 + HA 30 kg ha^{-1}	78.12 (62.67)	54.41	4.20	4.56
T_8 – T_6 + HA 30 kg ha^{-1}	77.90 (61.95)	56.71	5.06	4.98
T_9 – FYM 12.5 t ha^{-1} + PM 18.5 t ha^{-1} + NPK 35:45:45 kg ha^{-1}	78.86 (62.67)	58.36	5.12	5.07
T_{10} – FYM 12.5 t ha^{-1} + PM 18.5 t ha^{-1} + NPK 45:75:75 kg ha^{-1}	81.22 (61.95)	60.04	6.12	5.16
T_{11} – T_9 + HA 30 kg ha^{-1}	79.62 (62.67)	58.96	5.84	5.62
T_{12} – T_{10} + HA 30 kg ha^{-1}	80.92 (64.10)	60.82	6.20	5.26
T_{13} – FYM 18.5 t ha^{-1} + PM 18.5 t ha^{-1} + NPK 35:45:45 kg ha^{-1}	81.12 (64.91)	59.36	6.00	5.90
T_{14} – FYM 18.5 t ha^{-1} + PM 18.5 t ha^{-1} + NPK 45:75:75 kg ha^{-1}	82.02 (64.91)	63.13	7.93	5.76
T_{15} – T_{13} + HA 30 kg ha^{-1}	81.78 (64.91)	62.25	7.14	6.09
T_{16} – T_{14} + HA 30 kg ha^{-1}	83.06 (65.70)	65.25	8.40	6.25
T_{17} – Control	40.21 (32.26)	30.78	2.21	1.98
S.Ed	0.55	0.33	0.03	0.06
C.D	1.14	0.68	0.07	0.13

HA – Humic acid, PM – Pressmud, FYM – Farmyard Manure

Table.3. Effect of INM practices on physiological characters of *Dolichos lablab* var *typicus* in lignite mine spoil

Treatments	Vigour index	Leaf area index	Chlorophyll content (mg g^{-1})	Photosynthetic rate (μ mol Co_2 $m^{-2}s^{-1}$)	Harvest index
T_1 – FYM 12.5 t ha^{-1} + PM 12.5 t ha^{-1} + NPK 35:45:45 kg ha^{-1}	1284.17	37.02 (36.68)	0.316	60.67	32.18 (27.65)
T_2 – FYM 12.5 t ha^{-1} + PM 12.5 t ha^{-1} + NPK 45:75:75 kg ha^{-1}	1343.41	38.72 (37.16)	0.326	63.14	34.26 (29.41)
T_3 – T_1 + HA 30 kg ha^{-1}	1437.04	39.19 (38.76)	0.354	62.87	37.07 (32.51)
T_4 – T_2 + HA 30 kg ha^{-1}	1503.70	40.23 (39.37)	0.363	67.23	41.32 (39.65)
T_5 – FYM 18.5 t ha^{-1} + PM 12.5 t ha^{-1} + NPK 35:45:45 kg ha^{-1}	1520.19	40.88 (39.75)	0.386	69.14	42.68 (40.79)
T_6 – FYM 18.5 t ha^{-1} + PM 12.5 t ha^{-1} + NPK 45:75:75 kg ha^{-1}	1605.72	42.52 (40.70)	0.401	70.68	47.07 (42.15)
T_7 – T_5 + HA 30 kg ha^{-1}	1585.80	41.51 (40.11)	0.410	70.29	45.87 (40.83)
T_8 – T_6 + HA 30 kg ha^{-1}	1645.69	43.93 (41.76)	0.436	71.02	48.13 (43.50)
T_9 – FYM 12.5 t ha^{-1} + PM 18.5 t ha^{-1} + NPK 35:45:45 kg ha^{-1}	1661.21	45.80 (42.59)	0.449	72.21	49.02 (44.44)
T_{10} – FYM 12.5 t ha^{-1} + PM 18.5 t ha^{-1} + NPK 45:75:75 kg ha^{-1}	1796.11	49.14 (44.63)	0.441	73.51	53.66 (47.10)
T_{11} – T_9 + HA 30 kg ha^{-1}	1782.72	48.18 (44.03)	0.438	72.67	51.26 (45.15)
T_{12} – T_{10} + HA 30 kg ha^{-1}	1817.71	47.11 (43.34)	0.451	74.69	55.96 (48.65)
T_{13} – FYM 18.5 t ha^{-1} + PM 18.5 t ha^{-1} + NPK 35:45:45 kg ha^{-1}	1794.52	48.86 (44.23)	0.448	73.42	54.18 (47.61)
T_{14} – FYM 18.5 t ha^{-1} + PM 18.5 t ha^{-1} + NPK 45:75:75 kg ha^{-1}	1854.18	50.14 (45.84)	0.458	77.23	59.76 (50.65)
T_{15} – T_{13} + HA 30 kg ha^{-1}	1842.24	49.64 (44.83)	0.453	76.05	57.18 (48.56)
T_{16} – T_{14} + HA 30 kg ha^{-1}	1872.42	51.86 (46.06)	0.463	78.39	62.79 (52.41)
T_{17} – Control	807.17	28.02 (26.84)	0.258	46.37	19.26 (13.82)
S.Ed	10.95	0.30	0.01	0.57	0.75
C.D	22.52	0.62	0.02	1.14	1.55

HA – Humic acid, PM – Pressmud, FYM – Farmyard Manure

Table 4. Effect of INM practices on yield characters and NPK uptake of *Dolichos lablab* var. *typicus* in lignite mine spoil

Treatments	Number of days taken for first flowering	Number of pods plant^{-1}	Yield (t ha^{-1})	Uptake of nutrients (kg ha^{-1}) N	P_2O_5	K_2O
T_1 – FYM 12.5 t ha^{-1} + PM 12.5 t ha^{-1} + NPK 35:45:45 kg ha^{-1}	25.04	19.94	3.92	78.61	14.13	82.65
T_2 – FYM 12.5 t ha^{-1} + PM 12.5 t ha^{-1} + NPK 45:75:75 kg ha^{-1}	26.46	21.48	4.44	81.23	15.42	84.54
T_3 – T1 + HA 30 kg ha^{-1}	26.19	22.98	4.98	87.88	16.10	86.15
T_4 – T2 + HA 30 kg ha^{-1}	27.55	24.47	5.75	91.61	16.68	89.25
T_5 – FYM 18.5 t ha^{-1} + PM 12.5 t ha^{-1} + NPK 35:45:45 kg ha^{-1}	28.04	25.17	5.85	94.69	18.21	90.22
T_6 – FYM 18.5 t ha^{-1} + PM 12.5 t ha^{-1} + NPK 45:75:75 kg ha^{-1}	28.56	27.07	6.50	98.79	18.01	91.14
T_7 – T5 + HA 30 kg ha^{-1}	28.94	26.47	6.14	97.17	19.13	93.04
T_8 – T6 + HA 30 kg ha^{-1}	29.04	29.64	6.86	102.23	20.61	92.21
T_9 – FYM 12.5 t ha^{-1} + PM 18.5 t ha^{-1} + NPK 35:45:45 kg ha^{-1}	29.45	30.36	7.13	106.97	21.78	96.73
T_{10} – FYM 12.5 t ha^{-1} + PM 18.5 t ha^{-1} + NPK 45:75:75 kg ha^{-1}	29.61	33.14	9.03	112.12	22.01	98.52
T_{11} – T9 + HA 30 kg ha^{-1}	29.41	31.84	8.43	108.10	23.16	99.87
T_{12} – T10 + HA 30 kg ha^{-1}	29.68	34.98	9.86	114.00	21.89	101.65
T_{13} – FYM 18.5 t ha^{-1} + PM 18.5 t ha^{-1} + NPK 35:45:45 kg ha^{-1}	29.58	33.28	9.14	110.00	23.86	100.04
T_{14} – FYM 18.5 t ha^{-1} + PM 18.5 t ha^{-1} + NPK 45:75:75 kg ha^{-1}	30.04	37.13	10.48	120.86	25.94	104.365
T_{15} – T13 + HA 30 kg ha^{-1}	29.82	36.02	9.84	118.62	24.86	102.17
T_{16} – T14 + HA 30 kg ha^{-1}	29.88	39.33	11.70	126.00	26.32	108.63
T_{17} – Control	19.14	8.96	0.92	58.94	10.78	56.82
S.Ed	0.02	0.11	0.02	0.30	0.14	0.22
C.D	0.05	0.24	0.05	0.61	0.29	0.45

HA – Humic acid, PM – Pressmud, FYM – Farmyard Manure

Table 5. Production potentiality of Blackgram varieties

Treatments	Number of Pods plant^{-1}		Number of clusterplant^{-1}		Number of seeds pod^{-1}		100 seed weight (g)		Seed yield (kg ha^{-1})		Haulm yield kg ha^{-1}	
T_1 - ADT-2	15.2	15.0	99.2	9.8	11.2	11.0	3.70	3.75	752.40	760.50	1815.02	1845.50
T_2 - ADT-3	17.2	17.4	13.5	13.7	15.1	15.3	3.90	3.92	975.21	987.72	21.05.01	2232.53
T_3 - Chidambaram local	15.9	15.8	10.3	11.5	13.0	12.8	3.82	3.85	865.30	867.40	1957.01	19650.20
T_4 - KM-2	13.0	12.9	7.7	7.9	12.5	12.7	3.54	3.58	530.70	537.30	1685.02	17020.00
T_5 - VGG-8	13.5	14.8	8.9	8.6	8.9	8.9	3.65	3.62	627.50	633.14	1782.60	1793.21
SED	0.80	0.83	1.61	1.63	0.85	0.95	0.03	0.03	46.28	48.28	60.40	61.50
CD (p = 0.05)	1.71	1.77	2.7	3.0	1.7	1.9	0.08	0.08	107.56	108.60	130.00	135.20

nutrient availability. This would have helped to establish a strong vegetative phase which is a source for better synthesis of carbohydrate due to the potential utilization of available micro and macro nutrients, which chiefly influence the photosynthesis and partitioning of photosynthates from source to sink. This might have ultimately increased the fruit yield as reported by Ram Prasad *et al.* (1992). Similar reports were also reported by Gardiner (1993), Pitchai *et al.* (1994), Karuppaiah *et al.* (1998) and Karuppaiah and Manivannan (2003).

In case of plant nutrient uptake, all the treatments significantly performed better than the control. But these were maximum in the treatment combination of FYM 18.5 t ha^{-1} + pressmud 18.5 t ha^{-1} + NPK 45:75:75 kg ha^{-1} along with humic acid 30 kg ha^{-1}. These results give an optimistic view about the use of above said integrated nutrient management for increasing the growth and yield of gardean bean and for the reclamation of lignite mine spoil.

REFERENCES

Arnon, D.I. (1949). Copper enzymes in isolated chloroplasts, polyperoxidase in *Beta vulgaris. Pl. Physiol.,* **24:** 1-45.

Dadhwal, K.S., Bijendra singh, Pratap Narainan and Kaityar, V.S. (1988). Techniques for the rehabilitation of degraded mined lands. *Indian Soil conser,* 18(2): 51-62.

Gardiner, D.T. (1993). Revegetation status of reclaimed abandoned mined land in Western Northern Douta. *Arid Soil Research and Rehabilitation,* 7(1): 79-84.

Goldsworthy, A. and Fischer, N.M. (1984). Crop growth and development physiology phase. In: The physiology of tropical field crops. Pp. 163-205.

Humphries, E.C. (1956). Mineral componentrs and ash analysis. In: Modern methods of plant analysis. Springer Verlag., Berlin, 1: 468-502.

Jackson, M.L. (1973). Soil Chemical analysis. Prentice Hall of India Pvt. Ltd, New Delhi. pp. 498.

Kannan, R. and Ramani, S. (1988). Effect of salinity on plant growth. Ann. Rev. Phytopathol., 13: 99-102.

Karuppaiah, P., Kuppuswamy, G and Manivannan, K. (1998). Environmental management of the Neyveli lignite open cast minespoil. In: Proc. Seminar on the exposure to environmental status in Tamil Nadu. Pp. 30-34.

Levitt, J. (1973). Responses of plants to environmental stresses. Academic Press, London, New York.

Karuppaiah, P. and Manivannan, K. (2003). Growth and physiology of cucumber grown under stress in lignite mine spoil. *Ad.Plant Sci.,* 16(II): 591-596.

Pitchai, J.R., Dick, W.A. and Sutton, P. (1994). Comparison of amendments and management practices for long term reclamation of abandoned mine lands. *J.environ. Qual.,* 23(4): 766-772.

Ramprasad, R., Awasthi, R. and Prasad, R. (1992). Nutritional status of afforested coal mine over burden dumps of Dhanpuri in Madhya Pradesh. *J. Trop. Forestry.* 8(2): 109-118.

Schritzer, M. and Khan, S.U. (1972). In: Humic substances in the environment. Marcel Deuuer Inc., New York.

Trivedi, R.P. and Gurdeepraj. (1983). In: Encyclopaedia of environmental sciences. Environmental Agricultural pollution, Vol.9, Akashdeep Publishing House, New Delhi.

□□□

Chapter 21

Chemical Impregnation Treatment for Greengram (*Vigna radiata* (L.) Wilczek.)

GNANASEKARAN, J., S. PADMAVATHI, G. SATHIYANARAYANAN AND M. PRAKASH
Department of Agricultural Botany, Faculty of Agriculture, Annamalai University Annamalainagar - 608 002, Tamilnadu, India

Introduction

Among the popular pulses, greengram (*Vigna radiata* L.) plays an important role in food value and dietary protein. Greengram is graded superior because of its well balanced amino acid and rich amount of vitamins and phosphorus. The increase in seed yield indicates the scope for further rationalizing the nutrients for efficient production pattern at various stages of growth and flower production for obtaining higher seed yield. The production of greengram has been hampered severely due to lack of quality seeds and proper soil management and prevalence of disease infestation. Thus, an attempt was made to improve the seedling vigour through chemical impregnation treatment with micronutrients which may result in a good field emergence and pave way for harvesting higher yield.

Materials and Methods

Fresh seeds of greengram, AU 1 were impregnated using optimum quantity of water as well as nutrient solution at 1:1 ratio on weight/volume basis for 1 hr, 2 hr, 6 hr, 12 hr and 24 hr under room temperature.

The chemicals used were magnesium sulphate (0.5%) and zinc sulphate (0.5%) in 1:1 proportion on volume by volume basis. The impregnated seeds were air-dried prior to sowing. The treated seeds were evaluated for germination and vigour under laboratory conditions in four replications. Seedling evaluation was carried out on 12th day of sowing in laboratory condition.

Results and Discussion

Evaluation of germination potential and seedling growth revealed significant differences due to treatments. Invariably of the chemical solution, 2 hrs. soaking treatment recorded high germination percentage (90%) (Tables 2 & 3) and vigour index of 2565 for $MgSO_4$ and 2485 for $ZnSO_4$ when compared with water (2008) (Table 1). Rahman *et al.* (1996) observed increased germination percentage and thousand seed weight in tomato due to nutrient management. Manjunatha Reddy and Kulkarni (1986) also observed increased seed setting percentage and reduced number of ungerminated seeds in alfalfa due to micro nutrient application. The growth measurements and dry matter production showed significant differences in 1 hour soaking treatments.

The seeds impregnated with zinc sulphate solution registered more root length (4.06 cm) and shoot length (25.92 cm) in 1 hr soaking duration and magnesium sulphate recorded high dry matter production of seedling (0.255 g per 10 seedlings) at 12 hrs soaking.

Table 1. Effect of seed impregnation with water in seed vigour of greengram

Hour	Germination percentage	Root length (cm)	Shoot length (cm)	Dry matter production (g)	Vigour index
1	75	3.2	25.0	0.130	2115
2	80	3.0	22.1	0.168	2008
6	75	3.0	19.6	0.163	1696
12	70	3.1	25.4	0.182	1995
24	70	2.8	20.3	0.171	1618
SED	0.516	0.030	0.392	0.0007	0.695
CD (1%)	1.580	0.092	1.202	0.0022	2.127

Table 2. Effect of seed impregnation with $MgSO_4$ in seed vigour of greengram

Hour	Germination percentage	Root length (cm)	Shoot length (cm)	Dry matter production (g)	Vigour index
1	70	3.28	25.5	0.184	2014.6
2	90	3.34	25.16	0.125	2565.0
6	85	2.8	21.5	0.174	2065.5
12	45	3.3	16.94	0.255	910.8
24	25	2.17	7.5	0.126	241.87
SED	0.632	0.063	0.063	0.0005	8.199
CD (1%)	1.935	0.193	0.193	0.0016	25.09

Table 3. Effect of seed impregnation with $ZnSO_4$ in seed vigour of greengram

Hour	Germination percentage	Root length (cm)	Shoot length (cm)	Dry matter production (g)	Vigour index
1	75	4.06	25.92	0.128	2248.5
2	90	3.32	24.3	0.139	2485.8
6	30	2.9	9.26	0.107	364.8
12	24	4.2	18.36	0.122	564.0
24	10	2.2	12.4	0.090	146.0
SED	0.903	0.632	0.063	0.0005	6.791
CD (1%)	2.765	0.193	0.193	0.0016	20.781

CONCLUSION

Invariably of the chemical solution, 2 hrs soaking treatment recorded high germination percentage (90%) and increased vigour index of 2565 for $MgSO_4$ and 2485 for $ZnSO_4$. Hence these treatments can be recommended to attain higher productivity in greengram.

REFERENCES

Manjunatha Reddy, B.P. and Kulkarni, G.N. (1986). Influence of foliar spray of zinc and iron on pollen affecting seed yield in alfalfa. *Seed Res.*, 14: 185-188.

Rahman, M., Farid, A.T.M., Shahidullah, M. and Sultana, W. (1996). Improving seed yield and quality of tomato through integrated nutrient management and liming. *Seed Res.*, 24: 34-37.

Chapter 22

Influence of Mother Crop Nutrition on Seed Yield and Quality of Blackgram (*Vigna ungiculata* (L.))

MANONMANI, V[1] AND P.SRIMATHI [2]
Associate professor[1] and Professor[2] (Seed Sci. and Technology)
Dept.of Seed Science and Technology,Tamil Nadu Agricultural University
Coimbatore-641 003

Introduction

Pulses occupy strategic position in agricultural economy of India. Pulse crops are grown on an area of 23.84 lakh ha accounting for nearly 33 percent of world acreage and consumed by 2.2 percent of world population. Among the pulses, black gram is widely considered as an excellent source of high quality protein with good digestibility. Pulses normally produce a large number of flowers but only a few are retained and developed into pods. Although the inflorescence is profuse, the yield is low due to poor pod setting and harvest index.

Foliar application of nutrients constitute one of the important milestones in the progress of agricultural production. All the fertilizers applied to soil at the time of sowing are not fully available to the plants before the crop attains maturity. Supplemental foliar application is one of the many techniques in seed agronomy that caters to seed nutrition at the most vulnerable stage i.e.

seed filling (Shibles and Anderson, 1975). Care of seed during crucial stages of filling and maturation will lead to realization of final quality that losts even during storage (Krishnasamy, 1982). The techniques of supplemental foliar application and pre-harvest sanitation spray are simple, cost effective and can be advocated to any seed crop. Foliar application is credited with the advantage of quick and efficient utilization of nutrients, elimination of losses through leaching and fixation and helped in regulating the uptake of nutrients by plants (Barik and Rout, 1990). The present investigation was undertaken to study the effect of different nutrients on seed yield attributes and storage potential of carry over seeds of black gram.

Material and Methods

The experimental material consists of seeds of black gram variety APK 1 and the nutrients namely, $ZnSO_4$ 1% (T_1), Borax1% (T_2), $FeSO_4$ 1% (T_3), $MnSO_4$ 1% (T_4), Na_2MO_4 1% (T_5), DAP 2.0% (T_6), Urea 1% (T_7), and KCl 1% (T_8) along with control (T_9). The trial was conducted at Agricultural Research Station, Bhavanisagar, and Tamil Nadu Agricultural University,Coimbatore. Foliar application of nutrients was done at initial and fifty percent flowering stages of the crop. The observations on 100 seed weight, seed yield, germination and vigor index were recorded. Seeds were subjected to germination test by sowing 25 seeds in paper medium in four replications. Germination count was made eight days after sowing and germination was expressed as percentage of seeds which produced normal seedlings. Ten random seedlings were then measured for root and shoot length. The vigor index was derived from the formula by Abdul Baki and Anderson (1973).

VI=Percentage germination x seedling length (cm).

The resultant seeds from foliar spraying treatments were kept in storage under ambient condition (28+2 C temperature; 70 + 5% RH) for a period of 12 months. The samples were drawn in bimonthly intervals and analyzed for seed quality attributes like germination per cent, root length, shoot length and vigor index and assessing for its storage potential. The experiments were carried out in Randomized Block Design and the results were subjected to analysis of variance and tested for significant differences (p = 0.05) as per the method of Panse and Sukhatme (1967).

Results and Discussion

The results of present investigation revealed that black gram sprayed with DAP 2.0% followed by urea spray (1.0%) showed superior performance by recording maximum 100 seed weight (5.6, 5.5 gm), seed yield (1240,

1040 kg/ha), germination (92, 90 %) and vigor index (3690, 3256) respectively (Table 1). Similarly increase in plant height, number of branches and leaves with concomitant increase in seed yield (51%) in sesame was reported with DAP application (Ravichandran, 1989, Chaplot *et al.*, 1992). Improving the performance of plants derived from aged seeds of cotton with foliar fertilization has been reported by Sastri (1997).

Table 1. Effect of foliar spraying of nutrients on seed yield of black gram APK 1

Treatments		No. of pods/plant	100 seed weight (g)	Seed yield (kg/ha)	Germination (%)
T_1 - Spraying with $ZnSO_4$	1.0%	40	5.2	710	86
T_2 - Spraying with Borax	1.0%	44	5.3	890	86
T_3 - Spraying with $FeSO_4$	1.0%	49	5.3	895	87
T_4 - Spraying with $MnSO_4$	1.0%	52	5.4	920	87
T_5 - Spraying with Na_2MO_4	1.0%	48	5.4	870	87
T_6 - Spraying with DAP	2.0%	72	5.6	1240	92
T_7 - Spraying with Urea	1.0%	58	5.5	1040	90
T_8 - Spraying with KCl_2	1.0%	40	5.0	740	88
T_9 -Control (No spray)		38	4.8	680	80
CD (P=0.05%)		5.0	0.08	15.5	2.0

Supplemental foliar application of nutrients has been reported to offset the deleterious effect of ageing in cotton (Sastri, 1997). The initial low vigor could be made-up by spraying DAP, borax and magnesium sulphate and thereby achieve faster growth, earlier flowering, increased pollen viability and increase in seed set to an extent of 38% over plants not sprayed. Similar studies were also reported in peas using DAP spray (Vakeswaran, 1998). DAP spray was found to increase seed germination in soybean (Sabir Ahamed, 1989).

Foliar application of DAP and $MnSO_4$ contributed to increase in number of seeds per capsule in soybean (Terman, 1977). The supplementary foliar application was also suggested to green gram (Rajendran, 1990) and black gram (Subramanian and Palaniappan, 1980). Pothiraj *et al.*, (1995) recommended the mixture of NAA (40 ppm) and DAP in cotton to improve boll set.

The seeds fed with foliar nutrients *viz.*, DAP (2.0%) and Urea (1%) maintained shelf life in storage. The seed crops fed with DAP 2%, Urea 1% maintained their storability with good germination (72, 70 %) and vigour index (2088, 1820) respectively upto a period of ten months of storage whereas control seeds maintained their viability only upto eight months of storage (Table 2).

Table 2. Foliar spraying of nutrients on germination (%) of black gram APK 1 in storage

Treatments		Germination %		Root length (cm)		Shoot length (cm)		Vigour index	
		2	12	2	12	2	12	2	12
T_1 - Spraying with $ZnSO_4$	1.0%	86	72	13	10	16	12	2494	1584
T_2 - Spraying with Borax	1.0%	88	74	12	10	15	11	2376	1554
T_3 - Spraying with $FeSO_4$	1.0%	85	68	14	11	16	11	2550	1496
T_4 - Spraying with $MnSO_4$	1.0%	85	65	15	10	17	10	2720	1360
T_5 - Spraying with Na_2MO_4	1.0%	88	65	13	9	16	10	2552	1235
T_6 - Spraying with DAP	2.0%	90	72	19	14	22	15	3690	2088
T_7 - Spraying with Urea	1.0%	88	70	17	13	20	13	3256	1820
T_8 - Spraying with KCl	1.0%	88	65	14	10	15	12	2552	1430
T_9 -Control		84	60	11	9	14	10	2100	1140
CD (P=0.05%)		2.51	2.88	2.4	2.5	2.2	2.26	25.8	25.6

Hence, spraying of nutrient solutions (DAP 2%, Urea 1%) is a viable technique in providing quick short term solutions to mother crop needs. Care must be taken to assess the nutrient needs of different crops during seed filling stage.

REFERENCES

Abdul-Baki, A. A. and Anderson, J.D. (1973). Vigour determination in soybean seed by multiple criteria. *Crop Sci.*, 13 : 360-363.

Barik, T. and Rout, D. (1990). Effect of foliar spray of commercial micro nutrient mixtures on growth, yield and quality of Urd bean. *Legume Res.*, 13 (1) : 50-52.

Chaplot. P.C., Jain, G.C. and Bansal, K.N. (1992). Effect of sulphur and phosphorus in sesame and its residual effect on wheat. *Indian J.Agron.*, 37 (2) : 340-341.

Krishnasamy, V. (1982). Studies on certain aspects of production and storage of sorghum seed. Ph.D. Thesis, Tamil Nadu Agricultural University, Coimbatore.

Panse, V.G. and Sukhatme, P.V. (1967). Statistical methods for Agricultural workers, Indian Council of Agricultural Research Publications. New Delhi, pp.166-175.

Pothiraj, P.N., Jaganathan, T., Venkataswamy, R., Premsekar, M. and Prushothaman, S. (1995). Effect of growth regulators in cotton MCU 9. *Madras Agric.J.* 82 (4) : 283-284.

Rajendran. R. (1990). Response of greengram CO 4 to soil and foliar nutrition. *Madras Agric. J.* 78 (1): 9-12.

Ravichandran, A.K. (1989). Studies on the use of growth regulants and potassium in sesame. M.Sc (Ag), Thesis, Tamil Nadu Agricultural University, Coimbatore.

Sabir Ahmed. A., (1989). Studies on production of quality seed and storage in soybean. M.Sc (Ag), Thesis, Tamil Nadu Agricultural University, Coimbatore.

Sastri. G., (1997). Certain aspects of seed production technology in cotton. M.Sc (Ag), Thesis, Tamil Nadu Agricultural University, Coimbatore.

Shibles Gibson, A. and Anderson, C. (1975). Soybean. In Crop Physiology (Evans, L.T.Ed) Cambridge University Press, Cambridge, London, Pp. 151-189.

Subramanian, A. and Palaiappan, S.P. (1980). Effect of foliar fertilization of nitrogen and phosphorus on protein yield of blackgram varieties. *Madras Agric. J.* 67 (12) : 774 – 776.

Terman, G.L. (1977). Yield and nutrient accumulation determining soybean as affected by applied nutrient. *Agron.J.*, 69 : 234-238.

Vakeswaran, V. (1998). Seed technological studies in pea. M.Sc (Ag), Thesis, Tamil Nadu Agricultural University, Coimbatore.

❑❑❑

Chapter 23

Studies on the Production Potentiality of Blackgram Varieties Under Rice Fallow Condition

SHOBANA, R., P. THANGAVEL AND V. IMAYAVARAMBAN
Faculty of Agriculture, Annamalai University, Annamalainagar - 608 002

Introduction

Pulses are the major source of vegetable protein in Indian vegetarian diet. Protein provides amino acids and others biological important compounds essential for build up of the body tissues. Hence, nutritionists recommend pulses based diets to correct malnutrition. On average, pulses contain 20-25 per cent protein in dry seeds basis, which is almost 2.5 to 3.0 times more than cereal protein. The ICMR has recommended per capita intake of 50 g of pulses per day for balanced nutritional supply (Chaturvedi and Masood Ali, 2002). There was drastic reduction in per capita availability of pulses from 60.7 g in 1951 to 35.9 g in 2004. Hence, the present study was undertaken to increase the production potentiality of blackgram.

Materials and Methods

Field experiments were conducted at Annamalai University Experimental Farm, Annamalainagar, during January-March 2006 and 2007. The experiments were conducted in clay loam soil with a pH of 7.62. The soil was low in available nitrogen, medium in available phosphorus, high in available potassium. Five treatments were laid out in Randomized Block design with five replications. The treatments are ADT 2 (T_1), ADT 3 (T_2), Chidambaram local (T_3), KM 2 (T_4) and VGG 8 (T_5).

Results and Discussion

Yield Attributes

Yield attributes like number of branches per plant at 45 DAS and at harvest, number of pods per plant, number of seeds per pod, hundred seed weight and seed yield are significantly increased in ADT 3 than the other varieties.

The increased number of branches per plant might be responsible for the higher yield of this variety (Nag *et al.,* 2000). The number of pods per plant is one of the major yield components in blackgram. The higher number of pods per plant was registered by ADT 3 in both the years. This might be due to prolonged assimilation activity of leaves, increased number of flower production and lesser flower drop (Patra *et al.,* 2001).

Maximum number of seeds per pod was also higher in ADT 3, which could be due to more photosynthetic activity and translocation of photosynthates rapidly to sink(Inderjith Singh *et al.,* 1999). The highest hundred seed weight was also recorded by ADT 3 which was due to greater assimilate supply to the sink (Patel and Munda, 2001).

Seed Yield

The seed yield was significantly higher in ADT 3 in both the years. This might be due to increased number of branches per plant, number of pods per plant, number of seeds per plant and hundred seed weight. The same findings were reported by Nag *et al.* (2000).

Haulm Yield

Haulm yield was also higher in the ADT 3 due to increased growth and yield parameters.

Table 1. Production potentiality of Blackgram varieties

Treatments	Number of Pods plant^{-1} 2006-2007		Number of clusterplant^{-1} 2006-2007		Number of seeds pod^{-1} 2006-2007		100 seed weight (g) 2006-2007		Seed yield (kg ha^{-1}) 2006-2007		Haulm yield kg ha^{-1} 2006-2007	
T_1 - ADT-2	15.2	15.0	9.2	9.8	11.2	11.0	3.70	3.75	752.40	760.50	1815.02	1845.50
T_2 - ADT-3	17.2	17.4	13.5	13.7	15.1	15.3	3.90	3.92	975.21	987.72	2105.01	2232.53
T_3 - Chidambaram local	15.9	15.8	10.3	11.5	13.0	12.8	3.82	3.85	865.30	867.40	1957.01	1965.20
T_4 - KM-2	13.0	12.9	7.7	7.9	12.5	12.7	3.54	3.58	530.70	537.30	1685.02	1702.00
T_5 - VGG-8	13.5	14.8	8.9	8.6	8.9	8.9	3.65	3.62	627.50	633.14	1782.60	1793.21
SED	0.80	0.83	1.61	1.63	0.85	0.95	0.03	0.03	46.28	48.28	60.40	61.50
CD (p = 0.05)	1.71	1.77	2.7	3.0	1.7	1.9	0.08	0.08	107.56	108.60	130.00	135.20

REFERENCES

Chaturvedi, S.K. and Masood Ali, (2002). Poorman's meat needs fresh fillip. *The Hindu Survey of Indian Agriculture*, 63-69.

Nag, B.L., Alimar Rahman and Rahman, M.A. (2000). Growth analysis and yield performance of blackgram varieties. *Legume Res.*, 23(3): 146-150.

Patra, A.K., Karmakar, S.K. and Mukherjee,S.K. (2001). Response of blackgram (*Phaseolus mungo*) varieties to dates of sowing during winter. *Indian J. Agric.*, 71(6): 405-407.

Patel, D.P. and Munda, D.C. (2001). Growth pattern and yield potential of some promising cultivars of blackgram (*Vigna mungo* L.) under rainfed mid hills of meghalaya. *Crop Res.*, 21(2): 130-133.

Inderjith Singh, Verander Sardana and Sekhon, H.S. (1999). Performance of some promosing cultivars of summer blackgram (*Vigna mungo* L.) and varying seed rates. *Legume Res.*, 22(2): 127-129.

□□□

Chapter 24

Effect of Different Levels of Phosphorus and Potassium on Growth, Flowering and Yield of Vegetable Cowpea Cv Tux. 944

ANUJA, S., K.ILAVARASI AND S. ARCHANA
Department of Horticulture, Annamalai University

Introduction

Cowpea is an important food legume in the tropics and sub-tropics of Asia, Central and South America and part of Southern Europe. Cowpea is consumed in many forms. In India, it is grown for its long green pods as vegetable, seed as pulse and foliage as vegetable and fodder. Regarding vegetable production, fertilizer offers the best means of increasing the yield and to maintain soil fertility. It has been shown in many fertilizer trials that all the three major nutrients viz. nitrogen, phosphorous and potassium are essential for higher yield of vegetable cowpea.

Materials and Methods

The experiment was carried out in the vegetable field unit, Department of Horticulture, Faculty of Agriculture, Annamalai University to study the response of vegetable cowpea to combined application of phosphorus and potassium during season-I [January-March] 2005 and season II

[July-September] 2005. The field trial was laid out in randomized block design with three replications. The recommended dose of nitrogen [20 kg ha^{-1}] was applied to all the treatment plots. The treatment combinations were $P_{50}K_{50}$ 50+50 (kg ha^{-1}) (T_1), $P_{50}K_{60}$ 50+60 (kg ha^{-1})(T_2), $P_{50}K_{70}$ 50+70 (kg ha^{-1})(T_3), $P_{60}K_{50}$ 60+50 (kg ha^{-1})(T_4), $P_{60}K_{60}$ 60+60 (kg ha^{-1})(T_5), $P_{60}K_{70}$ 60+70 (kg ha^{-1})(T_6), $P_{70}K_{50}$ 70+50 (kg ha^{-1})(T_7), $P_{70}K_{60}$ 70+60 (kg ha^{-1})(T_8), $P_{70}K_{70}$ 70+70 (kg ha^{-1})(T_9) and P_0K_0 0+0 (kg ha^{-1})(T_{10}).

The observations were recorded on ten tagged-plants in each replication for plant height, number of branches per plant, days to 50 percent flowering and pod yield per plant. The plant height was measured from the ground level to the tip of the topmost leaf in ten plants of each treatment at the time of harvest and the mean was worked out and expressed in cm. The total number of branches arising from the main stem at the time of last harvest was recorded and expressed in number. Visual observations were made to record the number of days taken under each treatment for first flowering and expressed in days. The pod weight per plant was recorded during each picking and finally total yield per plant was arrived by adding the weight of pod per plant from all the pickings and expressed in grams.

Results and Discussion

The data relating to the effect of phosphorus ad potassium on plant height in season-I and II is given in table-I. The mean plant height was considerably higher in season I (40.64 cm) than in season II (37.47 cm). The range of plant height was 28.70-50.58 cm in Season I, while it was 26.50-46.85cm in Season II.

Maximum plant height (59.58 cm) was recorded in the treatment T_9($P_{70}K_{70}$), followed by T_8 with $P_{70}K_{60}$ (49.78 cm)in Season I. In Season II, the treatment T_9 recorded the maximum plant height of 46.85 cm followed by T_8 (45.85 cm). The minimum plant height was recorded in T_{10} with P_0K_0 (28.70 cm) in Season I. Also in Season II, the treatment T_{10} (26.50 cm) possessed minimum plant height. The results of the present study are in close agreement with the results of Swaroop et al (2001) in vegetable cowpea. The increase in plant height may be due to the presence of sufficient amount of phosphorus which played good role towards division and cell elongation in the meristematic regions. (Bahadhur and Singh, 1990).

The data recorded on number of branches per plant in Season I and II are furnished in (Table 1). The mean branches per plant was nearly the same for Season I and II. The range for number of branches per plant was higher (2.00-4.53) during Season I and it was 2.00-3.94 during Season II. The treatment

T_9 (4.53), followed by T_8 (4.36) had the maximum number of branches per plant in Season I and minimum number of branches per plant (2.41 and 2.00) were recorded in the treatments $T_1(P_{50}K_{50})$ and T_{10} (P_0K_0) respectively. Five treatments exceeded the general mean (3.37). This might be due to enhanced growth capacity consequent to fertilizer application. Abbas et al (1994) in soybean and Swaroop *et at.* (2001) in vegetable cowpea recorded higher number of branches with an increase in phosphorus and potassium application.

Days to 50 percent flowering was earlier (37.83) in season-I and it was 42.96 in season –II in the treatment T_9 $(P_{70}K_{70})$ (Table 2). This indicated that higher level of P and K hastened the flower differentiation. Similar results was also reported by Rajput and Desai (1985) in Dolichos bean.

Table 1. Effect of phosphorus and potassium on plant height (cm) and number of branches of vegetable cowpea.

Treatments	Plant height (cm)		Number of Branches	
	Season 1	Season II	Season I	Season II
T_1-$P_{50}K_{50}$	32.95	31.30	2.41	2.30
T_2-$P_{50}K_{60}$	34.85	32.72	2.69	2.51
T_3-$P_{50}K_{70}$	37.15	34.26	2.99	2.75
T_4-$P_{60}K_{50}$	39.70	36.19	3.28	2.98
T_5-$P_{60}K_{60}$	42.45	38.70	3.58	3.23
T_6-$P_{60}K_{70}$	43.65	39.75	3.76	3.37
T_7-$P_{70}K_{50}$	46.63	42.60	4.05	3.58
T_8-$P_{70}K_{60}$	49.78	45.85	4.36	3.81
T_9-$P_{70}K_{70}$	50.58	46.85	4.53	3.94
T_{10} (Control)	28.70	26.50	2.00	2.00
General mean	40.64	37.47	3.37	3.05
SED	0.84	0.78	0.14	0.09
CD (p=0.05)	1.48	1.28	0.25	0.18

The data pertaining to the pod yield per plant is presented in Table -2. A wider range was observed for this character in Season I (42.60-87.60g) and Season II (34.40 -71.30g). The general mean recorded for Season I was 66.71 g and for Season II it was 54.12g.

Out of ten treatments observed, the treatment $T_9(P_{70}K_{60})$ showed highest yield (87.60g) followed by T_8 (85.00g), which was on par with T_9, whereas T_{10} (42.60g) recorded lowest yield in Season I. During Season II, the treatment T_9 $(P_{70}K_{70})$ recorded the maximum pod yield of 71.30, followed by T_8 (69.90 g) which was on par with T_9. The treatment T_{10} (34.40 g) recorded minimum pod yield. Five treatments exceeded the general mean of 66.71g and 54.12 g in Season I and II respectively. Pod yield was found to be increased with an increase in P and K levels during season -I and II. The results of the present study are in conformity with those of Geetha and Verghese [2001] in cowpea.

Table 2. Effect of Phosphorus and Potassium on days to 50 per cent flowering and pod yield (g plant^{-1}) of vegetable cowpea.

Treatments	Days to 50 per cent flowering		Pod yield (g plant^{-1})	
	Season 1	Season II	Season I	Season II
T_1-$P_{50}K_{50}$	42.50	47.85	51.00	41.00
T_2-$P_{50}K_{60}$	42.00	47.05	55.20	44.50
T_3-$P_{50}K_{70}$	41.58	46.13	59.50	48.20
T_4-$P_{60}K_{50}$	41.18	45.25	64.40	51.30
T_5-$P_{60}K_{60}$	40.68	44.45	70.10	57.70
T_6-$P_{60}K_{70}$	40.61	44.26	72.60	59.80
T_7-$P_{70}K_{50}$	39.11	43.71	79.10	63.10
T_8-$P_{70}K_{60}$	37.91	43.14	85.00	69.90
T_9-$P_{70}K_{70}$	37.83	42.96	87.60	71.30
T_{10} (Control)	45.25	51.05	42.60	34.40
General mean	40.86	45.58	66.71	54.12
SED	0.05	0.26	1.56	1.34
CD (p=0.05)	0.10	0.54	3.01	2.62

REFERENCES

Abbas, M, M.P.Singh, K.B Nigam and V.S. Kandalkas. 1994. Effect of phosphorus, plant densities and plant types on different growth and physiological parameters of soybean [*Glycine max* L.]. *Indian J. Agron*, 39;246-248.

Bahadhur, V.and T. Singh. 1990. Yield and growth response of garden pea [*Pisum sativum*] to nitrogen and Phosphorus Application. *Vegetable Sci*, 17; 205-209.

Geetha, V. and K.Verghese. 2001. Response of vegetable cowpea to nitrogen and potassium under varying methods of irrigation. *J. Tropic Agriculture*, 39; 111-113.

Rajput, J.L and A.G.Desai. 1985. Effect of nitrogen, phosphorus and plant growth regulators on yield and quality of dolichos bean. *Maharastra J Hort.*, 10;108-109.

Swaroop K., A.N. Ganeshmurthy and S.V.S. Rathore. 2001. Response of vegetable cowpea to P,K, and rhizobium inoculation under Andaman condition. *J. Hort.*, 58;254-259.

□□□

SECTION - III
PHYSIOLOGICAL BIO-CHEMICAL AND BIO-TECHNOLOGICAL ADVANCES

Chapter 25

Studies on the Management of Differentially Aged Seed and their Progenies in Cowpea (*Vigna sinensis* (L.) Savi) Cv.Co 6 by Foliar Spray of Nutrients and Hormones

SASIKALA, K
Agricultural Research Station, Tamil Nadu Agricultural University, Pattukkottai

Introduction

High quality seed improves the field stand and maximize the yield up to 15-20%. No agricultural practice can improve the yield beyond the limits set by the seed. High quality seed should be high in germination, vigor and purity. If farmers want to obtain maximum profit from their crop, seed vigor is an important factor of seed quality, which must be taken into account. The production and maintenance of quality in crop seeds is essential, if seeds are to be used to establish crops reliably.

Ageing both before harvest and in storage after harvest, is the major determinant of the germination level and the physiological quality. Sometimes referred to as vigour of a seed population or seed lot, which is the term used to describe a uniform population of crop seeds that is produced. The germination level of a lot and the vigour of the germinable seeds interact with soil conditions eventually to determine whether or not a crop will be established successfully. Ageing can be detected at the level of the population, in individual seeds, in tissues and in cells (Mathews, 1985).

As seed deterioration increases, germination, field emergence, speed of emergence and seedling growth rate will also be gradually reduced. Finally the plant stand, growth, development and productivity will be decreased as seed deterioration increases (Delouche and Baskin, 1973).

To study the influence of environment or season of production on the physiological stamina of the seed, some stress test is essential among which accelerated ageing test has been employed to predict the quality (Thiyagarajan, 1992)

The present study was undertaken with a view to evolve suitable foliar nutrients for better management of aged seed and their progenies in cowpea.

Materials and Methods

Seed lots of varying viability and vigour were obtained by subjecting the seeds to accelerated ageing for 0,2,4,6,8,10 and 12 days at regimes of 40±2° c temperature and 96±1 % relative humidity (Delouche and Baskin, 1973). At the end of each duration, seeds were removed from accelerated ageing cabinet and were equiberrated under ambient condition. The seeds which were not exposed to ageing condition were termed as '0' day aged (or) control. After 8,10 & 12 days of accelerated ageing, seeds were tested for pattern of senescence.

Physiological parameters like germination (%), root length(cm), shoot length (cm) hypocotyl length (cm), epicotyl length (cm), vigour index and seedling dry matter production (mg) were taken.

Foliar spray was given with water, DAP 2%, cytozyme 1 % and vitamin E 1%. (Preparation of vitamin E 1 % solution - A few drops of acetone was added till 400 mg capsule gets completely dissolved. Then 40 ml of water was added to prepare one percent of aqueous solution of vitamin E). First spray was given at 50% flowering and 2nd spray at 15 days after first spray. Biometric observations were taken before nutrients spray on field emergence (%), plant height (cm), No. of leaves/plant, leaf area(cm^2/plant), total chlorophyll content (mg/g), root dry weight (g), stem dry weight (g), leaf dry weight (g) and crop growth rate (mg/cm^2/day). Biometric observations were recorded before and after nutrients spray on No. of fertile seeds pod, No. of sterile seeds/pod, seed yield (kg/ac.) and seed recovery (%).

Results and Discussion

The results were statistically significant for ageing, treatment and their interaction for almost all the parameters for which results were interpreted here.

Deterioration is an inevitable process in any biological life leading finally to the death of an organism. However, there may be differences in the pattern of deterioration between different biological units. In seeds according to the kind, constitution and genetic make up, the deterioration differs. Even among the seeds in a variety, there may be individual variations due to individual seed vigour exhibited in the form of seed weight and also history of seed development and maturation may have a role on pattern of deterioration of seed.

Ageing in all organisms is the sum total of deteriorative process eventually lead to death (Mathews, 1985). Accelerated ageing accelerates the senescence of seed and brings deterioration in viability and vigour (Heydecker, 1972). Accelerated ageing is a reliable test to predict the physiological stamina of the seed and was proven as an useful index of seed vigour (Delouche and Baskin, 1973).

In the present investigation, generally the deterioration in cowpea was slower. It might be due to more protein content. The process of ageing in seed is accompanied by gradual deterioration and loss of vigour. In the present investigation, cowpea seed recorded 98,72,66 and 64% germination in 0,8,10 and 12 days of accelerated aged seed. In most of the seeds, the germination percentage decreased with the duration of ageing (Chauhan *et al.*, 1985) (Table-1). The decline in germinability could be due to ageing phenomena associated with irreversible physical, physiological and biochemical changes occurring inside the seeds (Heydecker, 1972).

Root length is an excellent parameter to predict in severity of ageing stress in any crop seed. The root length deteriorated to an extent of 34 and 61% after 8 and 12 days of ageing respectively indicating the deterioration of root under accelerated ageing was found faster in cowpea.The results indicated that there was a progressive fall in the physiological stamina of deteriorating seed to produce reduced root and shoot length. Cowpea, being a legume crop the shoot length can be conveniently divided into hypocotyl and epicotyl length. The deterioration of hypocotyl and epicotyl lengths were 34 and 16% after 8 days and 37 and 20% after 12 days of ageing respectively indicating that deterioration of hypocotyl length was more than epicotyl length. Longer period of ageing resulted in higher decline in seedling length (Wu, 1977). The present investigation clearly indicated the reliability of root length to predict the level of deterioration of seed lots.

The decline in vigour index was preceded by the decline in germination at all levels of accelerated ageing and there was reduction in percentage of vigour index to 46 and 59% after 8 and 12 days of ageing respectively. Similar trend was observed by Ravichandran(1991) in groundnut and soybean and Viswanatha Reddy (1995) in brinjal.

The seedling drymatter production is the reflection of seed vigour. It also indicates the efficiency of seed to convert its weight to biomass production after germination. The dry matter production was reduced to 15 % and 30% after 8 and 12 days of ageing.

During storage, the protein become less and breakdown showing an increase in amino acid. Protein denaturation was yet another reason for deterioration of physiological vigour in the seed. In the present study, the decrease in protein content was from 19.8 to 15.9% and 13.8 % after 8 days and 12 days of ageing respectively.

Among the differentially aged seed, mildly (8 days), moderately (10 days) and severely aged (12 days) seeds of cowpea were raised in the field. Foliar spray of water, DAP 2%, Cytocyme1 % and Vit. E 1% were given two times once at 50% flowering and second spray after 15 days of first spray to uplift the performances of aged seed progenies.

As the seed aged, the field emergence, plant height, no of leaves, leaf area, total chlorophyll, dry weight of root, stem, leaf and crop growth rate reduced significantly in differentially aged seed progenies (Table 2).

Ageing is regarded as the series of events concerned with cellular disassembly in the leaf and the mobilization of materials released during the process.

More pod length is accommodating more no, of seeds and also better development of seeds with higher seed weight, which will lead to higher yield.

The pod length was reduced in mildly, moderately and severely aged seeds (Table 3).

Wu (1977) reported that reduced pod production, pod weight and delay in maturity in low vigour plants were obtained from seeds that were treated by the accelerated ageing procedure. Yield of any crop is determined by the isolate intercepted by the crop community, the efficiency with which intercepted isolation is converted to dry matter and the proportion of the biomass that is economic yield (Charles Edwards, 1982). Yield is the function of the characters of the individual plant, the plant community, the environment and the integration of the growth processes over time in a changing environment. Yield was significantly reduced due to severity in ageing. However, among the foliar spray, DAP 2 % enhanced the pod length, no of fertile seeds and yield. Gurusamy (1996) reported that DAP recorded more curd length and yield in cauliflower, increased yield in soybean (Sabir-Ahamed, 1989) and in pea (Paramasivam, 1990).

Table 1. Pattern of deterioration of seed and seedling characteristics under accelerated ageing in cowpea

Ageing	Germination (%) (Arcsine)	Root length (cm.)	Hypocotyl length (cm.)	Epicotyl length (cm.)	Vigour index	DMP (mg.)	Protein content (%)
A_0	98(83)	18.1	13.5	20.9	5158	52.5	19.8
A_1	88(69)	16.5	11.9	19.5	4227	48.4	19.6
A_2	81(64)	15.1	11.9	18.1	3696	46.5	17.9
A_3	74(59)	14.6	10.9	17.5	3226	46.1	17.6
A_4	72(58)	11.9	8.8	17.4	2768	44.6	15.9
A_5	66(54)	9.8	8.7	17.1	2379	41.8	15.7
A_6	64(53)	6.9	8.4	16.6	2085	36.3	13.8
CD(p=0.05)	4.41**	0.65**	0.69**	0.61**	176.9**	3.54**	0.46**

Table 2. Field performance of differentially aged seed progenies of cowpea (Recorded before foliar spray)

Ageing	Field emergence (%)	Plant height (cm.)	No.of leaves	Leaf area (cm.^{2}/plant)	Total chlorophyll (mg/g)	Root dry weight (g)	Stem dry weight (g)	Leaf dry weight (g)	Crop growth rate (mg/cm^2/day)
A_0	86	115.7	10.8	771.3	2.151	1.3	6.0	3.9	0.88
A_1	78	100.3	9.6	670.6	2.016	1.2	5.5	3.7	0.81
A_2	70	107.2	8.8	646.5	1.919	0.9	5.3	3.5	0.74
A_3	62	129.4	7.4	518.0	1.857	0.7	4.8	3.0	0.69
CD(p=0.05)	10.22*	NS	15.74**	37.75**	0.302**	0.02**	0.22**	0.22**	0.03**

Table 3. Influence of ageing and foliar application of nutrients on pod length (cm.) in cowpea

Ageing / Treatment	A_0	A_1	A_2	A_3	Mean
T_0 - Water	13.6	13.2	12.5	12.1	12.8
T_1 - DAP 2%	14.9	14.2	14.0	13.5	14.1
T_2 - Cytocyme 1%	14.5	14.2	13.9	12.9	13.8
T_3 - Vit.E 1%	14.6	14.2	13.9	13.5	14.0
Mean	14.4	13.9	13.6	13.0	
CD(p=0.05)	A	T	AxT		
	0.12**	0.11**	0.25**		

Thus reduced No. of fertile seeds/pod (Table 4) with increased No. of sterile seeds/pod (Table 5) resulted in lower seed yield (Table 6) and low seed recovery (Table 7).

Table 4. Influence of ageing and foliar application of nutrients on No. of fertile seeds/pod in cowpea

Ageing / Treatment	A_0	A_1	A_2	A_3	Mean
T_0 - Water	12.2	11.1	9.6	8.6	10.3
T_1 - DAP 2%	13.4	12.8	11.9	10.7	12.2
T_2 - Cytocyme 1%	12.6	11.8	11.4	10.7	11.6
T_3 - Vit.E 1%	12.6	11.8	10.5	9.8	11.1
Mean	9.5	11.8	10.8	9.9	
CD(p=0.05)	A	T	AxT		
	0.59**	0.53**	1.19**		

Table 5. Influence of ageing and foliar application of nutrients on No. of sterile seeds/pod in cowpea

Ageing / Treatment	A_0	A_1	A_2	A_3	Mean
T_0 - Water	1.8	3.2	3.6	4.6	3.3
T_1 - DAP 2%	1.0	2.0	2.3	2.9	2.0
T_2 - Cytocyme 1%	0.8	1.9	2.1	2.9	1.9
T_3 - Vit.E 1%	1.3	2.1	2.4	3.9	2.4
Mean	1.2	2.3	2.6	3.5	
CD(p=0.05)	A	T	AxT		
	0.32**	0.28**	0.64**		

Table 6. Influence of ageing and foliar application of nutrients on seed yield (kg/ ac.) in cowpea

Ageing / Treatment	A_0	A_1	A_2	A_3	Mean
T_0 - Water	205	155	125	64	137
T_1 - DAP 2%	329	248	215	126	229
T_2 - Cytocyme 1%	295	213	199	94	200
T_3 - Vit.E 1%	248	201	181	82	178
Mean	269	201	181	82	
CD(p=0.05)	A	T	AxT		
	5.9**	3.2**	11.4**		

Table 7. Influence of ageing and foliar application of nutrients on seed recovery (%) in cowpea

Ageing / Treatment	A_0	A_1	A_2	A_3	Mean
T_0 - Water	205	155	125	64	137
T_1 - DAP 2%	329	248	215	126	229
T_2 - Cytocyme 1%	295	213	199	94	200
T_3 - Vit.E 1%	248	201	181	82	178
Mean	269	201	181	82	
CD(p=0.05)	A	T	AxT		
	5.9**	3.2**	11.4**		

REFERENCES

Charles-Edwards, D.A. (1982). Physiological determinant of Crop growth. Academic Press Australia. North Ryde, NSW.

Chauhan, K.P.S., Negi, M.C.S. and Sen-Mandi, S. (1985). Accelerated ageing of seeds in hot water, germination characteristics of aged wheat seeds. *Seed Sci.& Technol.*, 13: 683-690.

Delouche, J.C. and Baskin, C.C. (1973). Accelerated ageing technique for predicting the relative storability of seed lot. *Seed Sci. & Technol.*, 1: 427-452.

Gurusamy, C. (1996). Studies on the pattern of seed development, maturation and deterioration in Cauliflower (*Brassica oleracea* L.var Botrytis). M.Sc.(Ag.) Thesis, TNAU, Coimbatore.

Heydecker, W. (1972). Vigour. In: Viability of seeds (ed.) E.H. Roberts, Champman and Hall, London. pp. 229-252.

Matthews, S. (1985). Physiology of seed ageing. *Outlook on agriculture.*, 14: 89-94.

Paramasivam, V. (1990). Studies on development, maturation, quality and storability of pea (*Pisum sativum* L.) seeds. M.Sc.(Ag.) Thesis, TNAU, Coimbatore.

Ravichandaran, G. (1991). Studies on the pattern of seed deterioration, the performance of progenies of differentially aged seeds in the field and the quality of their resultant seeds in maize, groundnut and soybean. M.Sc.(Ag.) Thesis, TNAU, Coimbatore.

Sabir-Ahamed, A. (1989). Studies on production of quality seed and storage in soybean (*Glycine max*.L.Merill) M.Sc. (Ag.)Thesis,TNAU, Coimbatore.

Thiagarajan, C.P. (1992). Assessment of quality of chilli seed (*Capsicum annum* L.) produced in different seasons by accelerated ageing test. *Madras Agric. J.* 79(11): 610 - 613.

Viswanatha Reddy. K. (1995). Studies on pre sowing seed treatment in differentially aged seeds of Brinjal (*Solamum melongena* L.) cv.PKM. M.Sc. (Ag.) Thesis, TNAU, Coimbatore.

Wu.T.Y. (1977). Effect of seed deterioration on yield components and physiological characters of soybean. *J.Agri.Res.China.*, 26 (4): 307-13.

□□□

Chapter 26

Studies on the Effective Role of Biofertilizers on the Germination, Growth and Yield of Greengram (*Vigna Radiata* (L.))

RAJASEKARAN,S., P.SUNDARAMOORTHY, K. SANKAR GANESH, K. SUMATHI AND L. BASKARAN
Environmental Biology Laboratory, Department of Botany
Annamalai University, Annamalainagar - 608 002, India

Introduction

Biofertilizers are cost effective and ecofriendly which improve soil quality and the yield of crop. They greatly provide benefit to farmers with only very small input cost (Kumudha and Gomathinayagam, 2007). The continuous application of inorganic fertilizers without organic manures has lead to poor soil fertility and affected the yield. Careful and scientific nutrient management can help in better crop yields (Jawahar and Suresh, 2007). The increase in these parameters by the application of nitrogen may be due to the fact that plant growth is stimulated by nitrogen supply. However, nitrogen application above a threshold amount may cause decrease in these parameters (Tutenkci *et al.* 2006). Biofertilizers have attracted greater attention particularly in developing countries like India as a substitute for costly chemical fertilizers. Nitrogen is an essential plant nutrient. That is most commonly deficient, contributing to reduced agriculture yields throughout the world. Pulses have the capacity to fix the atmospheric nitrogen by forming nodulation on roots, which protects the nitrogenase activity (Panwar and Vijayalaxmi 2005). Pulses occupy a significant place in India especially for vegetarian people as it is considered as a rich source of protein (Senthilkumar and Krishnaprabu, 2007). Hence, an attempt has been made to study the effective role of biofertilizers on the germination, growth and yield of greengram (*Vigna radiata* L.).

Materials and Methods

The seeds of greengram (*Vigna radiata* L.) were collected from the Pulses Research Station, Vamban (TNAU), Pudukottai district of Tamil Nadu. The biofertilizers were collected from authorized private Agrocentres, Mayiladuthurai. The pot culture experiment was conducted in the earthen pots filled with 5 kg of soil mixed with 100 gram of biofertilizers such as *Rhizobium leguminosorum, Bacillus megaterium and Bacillus mucilaginous* singly and in combination. The disease free seeds were sown in the pot and the water was irrigated two times per week. The treatment detail such as T_1-Control, T_2 - *Rhizobium leguminosorum*, T_3 - *Bacillus megaterium*, T_4 - *Bacillus mucilaginous*, $T_5 - T_2 + T_3$, $T_6 - T_2 + T_4$, $T_7 - T_3 + T_4$ and $T_8 - T_2 + T_3 + T_4$. The morphological parameters such as plant height, dry weight and yield were recorded. The micro and macronutrients such as total N, K and Na (Williams and Twine 1960), Ca and Mg (Yoshida *et al.* 1972), I, Zn and Mn (Devries and Tiller, 1980) of plants were also estimated.

Results and Discussion

The seedling parameters viz., germination percentage, root length, shoot length, fresh weight and dry weight of greengram grown with biofertilizers are presented in Table 1. The higher values were recorded in the combined application of biofertilizers than single application. The plant height, dry weight and yield parameters were higher in combined application, when compared with single application (Table 1). The increase in growth attributes might be due to the microorganisms that are used as biofertilizers which stimulate the plant growth by providing necessary nutrients by their colonization at the rhizophere or by their symbiotic association (Kumudha and Gomathinayagam 2007). Similar results were recorded in the earlier studies of Kumutha (2005) and Jain *et al.* (2007). Similarly, the micro and macronutrients are higher in the application of *Rhizobium* sp., *Bacillus megaterium* and *B. megaterium.* The lower nutrient content was observed in control plants (Table 2). The enhancement in levels of micronutrients due to biofertilizer inoculation might be due to symbiotic effect of biofertilizers with host tree seedlings which effects an improvement in micronutrient uptake and mineral status (Buwalda *et al.*, 1983).

Table 1. Effect of biofertilizers on growth and yield of *Vigna radiata* L.

Treatments	Germination %	Root length (cm)	Shoot length (cm)	Fresh weight (g)	Dry weight (g/seedling)	Plant height (cm)	Plant dry weight (g)	No. of root nodules	Number of pods per plant	Number of seeds per pod	100 grain weight (g)
T_1 - Control	85 (±2.54)	8.210 (±0.246)	10.541 (±0.316)	1.541 (±0.046)	0.521 (±0.016)	26.540 (±0.796)	20.123 (±0.603)	11.255 (±0.337)	10.221 (±0.306)	5.231 (±0.159)	2.928 (±0.087)
T_2 – *Rhizobium leguminosorum*	88 (±2.55)	11.833 (±0.354)	11.528 (±0.348)	1.641 (±0.049)	0.536 (±0.016)	28.135 (±0.844)	21.351 (±0.640)	11.347 (±0.340)	11.241 (±0.337)	5.388 (±0.161)	2.954 (±0.088)
T_3 – *Bacillus megaterium*	87 (±2.61)	10.521 (±0.315)	11.352 (±0.340)	1.622 (±0.048)	0.528 (±0.158)	27.710 (±0.831)	21.341 (±0.638)	11.322 (±0.339)	11.200 (±0.336)	5.230 (±0.156)	2.930 (±0.087)
T_4 - *Bacillus mucilaginous*	86 (±2.58)	10.320 (±0.309)	11.282 (±0.338)	1.589 (±0.047)	0.511 (±0.015)	27.655 (±0.829)	21.281 (±0.638)	11.318 (±0.339)	11.200 (±0.336)	5.230 (±0.156)	2.930 (±0.087)
T_5 – T_2 + T_3	95 (±2.85)	12.000 (±0.360)	13.120 (±0.393)	1.920 (±0.057)	0.569 (±0.017)	29.840 (±0.895)	22.253 (±0.667)	11.655 (±0.349)	12.358 (±0.370)	6.741 (±0.202)	3.542 (±0.100)
T_6 – T_2 + T_4	93 (±2.79)	11.521 (±0.345)	13.100 (±0.393)	1.842 (±0.055)	0.550 (±0.165)	29.775 (±0.893)	22.110 (±0.663)	11.549 (±0.346)	12.100 (±0.363)	6.566 (±0.196)	3.489 (±0.104)
T_7 – T_3 + T_4	92 (±2.76)	11.348 (±0.340)	13.00 (±0.390)	1.811 (±0.054)	0.547 (±0.016)	29.571 (±0.887)	22.000 (±0.660)	11.533 (±0.345)	12.000 (±0.360)	6.328 (±0.189)	3.377 (±0.101)
T_8 – T_2 + T_3 + T_4	99 (±2.97)	13.442 (±0.403)	14.611 (±0.438)	2.133 (±0.063)	0.639 (±0.019)	31.321 (±0.939)	22.654 (±0.679)	11.974 (±0.359)	13.541 (±0.406)	7.211 (±0.216)	3.859 (±0.115)

± Standard deviation

Table 2. Effect of biofertilizers on micro and macro nutrient contents of *Vigna radiata* L. (Harvest stage)

Treatment	N	P	K	Na	Mg	l	S	Zn	Cu	l	B	Mo
T_1 - Control	154.217 (± 4.62)	20.163 (±0.604)	132.320 (±3.969)	60.421 (±1.812)	25.800 (±0.774)	180.120 (±5.403)	34.521 (±1.0356)	14.365 (±0.430)	12.810 (±0.384)	180.126 (±5.403)	12.120 (±0.363)	16.528 (±0.495)
T_2 – *Rhizobium leguminosorum*	155.112 (±4.623)	20.215 (±0.606)	133.100 (±3.993)	60.527 (±1.815)	26.437 (±0.784)	180.321 (±5.406)	34.545 (±1.036)	14.391 (±0.431)	12.831 (±0.384)	180.231 (±5.406)	12.131 (±0.363)	16.549 (±0.496)
T_3 – *Bacillus megaterium*	154.684 (±4.640)	20.130 (±0.603)	132.689 (±3.980)	60.434 (±1.813)	26.100 (±0.783)	180.21 (±5.406)	34.520 (±1.036)	14.389 (±0.431)	12.822 (±0.384)	180.221 (±5.406)	12.128 (±0.363)	16.530 (±0.495)
T_4 - *Bacillus mucilaginous*	154.563 (±4.636)	20.122 (±0.603)	132.575 (±3.977)	60.348 (±1.810)	26.000 (±0.780)	180.159 (±5.404)	34.510 (±1.035)	14.380 (±0.431)	12.810 (±0.384)	180.210 (±5.406)	12.120 (±0.363)	16.529 (±0.495)
$T_5 – T_2 + T_3$	156.785 (±4.703)	21.356 (±0.640)	133.630 (±4.000)	61.579 (±1.847)	26.435 (±0.793)	182.349 (±5.469)	35.210 (±1.056)	15.490 (±0.464)	13.110 (±0.393)	180.628 (±5.418)	12.359 (±0.370)	16.261 (±0.487)
$T_6 – T_2 + T_4$	156.694 (±4.700)	21.341 (±0.640)	133.592 (±4.007)	61.321 (±1.839)	26.371 (±0.791)	182.325 (±5.469)	35.114 (±1.0534)	15.382 (±0.461)	13.100 (±0.393)	180.610 (±5.418)	12.335 (±0.370)	16.254 (±0.487)
$T_7 – T_3 + T_4$	156.588 (±4.697)	21.118 (±0.633)	133.541 (±4.006)	61.280 (±1.838)	26.288 (±0.788)	182.300 (±5.469)	35.000 (±1.050)	15.361 (±0.460)	13.000 (±0.390)	180.600 (±5.418)	12.288 (±0.368)	16.248 (±0.487)
$T_8 – T_2 + T_3 + T_4$	160.341 (±4.810)	22.521 (±0.675)	134.344 (±4.030)	62.541 (±1.876)	26.641 (±0.799)	183.459 (±5.503)	36.100 (±1.083)	16.189 (±0.485)	13.574 (±0.407)	181.211 (±5.433)	13.125 (±0.393)	16.685 (±0.500)

± Standard deviation

Acknolwedgement

The authors are thankful to Rajiv Gandhi National Research Fellowship Scheme, University Grant's Commission, New Delhi for financial assistance for the research.

REFERENCES

Buwalda, T.D, Stribley, D.P. and Tinker, P.B. (1983). Increased uptake of anions by plants VAM. *Pl. Soil*, 71: 463-467.

De vries, M.P.C. and Tiller, K.G. (1980). Routine procedures for determining Cu, Zn, Mn and Fe in plant materials. Commonwealth Scientific and Industrial Research Organization, Australia, p. 12.

Jain, A.K., Kumar, S. and Panwar, J.D.S. (2007). Role of *Rhizobium*, phosphorous and micronutrients on growth and nodulation of greengram (*Vigna radiata* (L.) Wilczek.). *Ad. Plant Sci.*, 20(11): 337-339.

Jawahar, D. and Suresh, S. (2007). Effect of biofertilizers and enriched FYM on the yield of cotton under dry farming in vertisol. *Ad. Plant Sci.*, 20(11): 473-475.

Kumudha, P. (2005). Studies on the effect of biofertilizers on the germination of *Acacia nilotica* Linn. *Seeds Ad. Plant Sci.*, 18(11): 679-684.

Kumudha, P. and Gomathinayagam, M. (2007). Studies on the effect of biofertilizers on germination of *Albizia lebbek* (L.) Benth. Seeds. *Ad. Plant Sci.*, 20(11): 417-427.

Panwar, J.D.S. and Vijayalaxmi, (2005). Biological nitrogen fixation in pulses and cereals. Development in physiology, biochemistry and Molecular biology of plants, Editor. Bandana Bose and A. Hemantaranjan. New India Publishing Agency, New Delhi, pp. 125-158.

Senthilkumar, N. and Krishnaprabu, S. (2007). Studies on the role of plant growth regulators on yield and quality of greengram (*Vigna radiata* L. Wilczek) under graded levels of nitrogen. *Ad. Plant Sci.*, 20(11): 469-471.

Tutenkci, S., Erman, M. and Sonmez, F. (2006). Effect of nitrogen application and rhizobium inoculation on the yield and nutrient uptake in Lucerne (*Medicage sativa*). *Bull Pure Appl. Sci.*, 25, B (1) p. 45-49.

Williams, C.H. and Twine, V. (1960). In modern methods of plant analysis, K Peach M.V. Tracey (Eds.) Vol. V, Springer Verl., Berlin, p. 3-5.

Yoshida, S., Fordo, D., Cork, J. and Gomez, K. (1972). Laboratory manual for physiological studies of rice 3rd edn: International Rice Research Institute, Phillippines, p. 11-23.

□□□

Chapter 27

Physiological Aspects of Grain Yield in Soybean

VIJAYARAGHAVAN, H.,* S.VINCENT* AND K.BALAKRISHNAN **
*Department of Crop Physiology, Tamil Nadu Agricultural University
Coimbatore - 641 003 **Department of Plant Breeding and Genetics, AC&RI, Madurai

Introduction

Soybean as the 'golden bean' or the 'miracle bean', the western world provided a massive push to its growth during the early part of the century. The use of soybean dates back to the beginning of China's agricultural age. The utilization of this crop has been mentioned in Chinese medical compilations dating back 6000 years. For centuries, soybean has meant meat, milk, cheese, bread and oil to the people of China, Japan, Korea, Manchuria, the Philippines and Indonesia. This explains why this crop has often been mentioned in these countries as the 'Cow of the field', or 'Gold from soil'. The versatility of soybean has of late been recognized in the West, that the West has been a late entry into the area of soybean production could also be illustrated by the fact that even in 1921, China produced about 80 percent of the world's soybean output.

Many forces were operative in motivating India to be an active participant in the soybean development race since the beginning of the 1970s. Even though

India's share in the world's total area under soybean is about one percent at present, development as well as the expansion of this crop in the country during the last 15 years is rated as one of the striking occurrences in the agricultural development process. Area under soybean in India had increased from about 32 thousand hectares in 1971-72 to about 814 thousand hectares in 1983-84, marking a twenty fivefold increase. In addition to the rapidity of this expansion, soybean development in India involves another facet: selectivity in its spread. Out of 814 thousand hectares under soybean in the country during 1983-84, about 616 thousand hectares were in the state of Madhya Pradesh in Central India.

The absolute performance of India both in soybean area and production expansion has been considerable. During this period, there was about an eighteen-fold increase in area and a seventeen-fold increase in production. Only Argentina recorded a level of absolute performance, which exceeded that of India.

Table 1. World area, production and yield of soybean

Year	Area (1000 ha)	Production (000 MT)	Yield Kg/ha
1994	62145	136170	2191
1995	62470	126868	2031
1996	61607	130516	2119
1997	67117	143404	2137
1998	70690	158327	2240

Table 2. Comparative area, production and productivity of some soybean growing countries

Country	Area (1000 ha)	Production (1000 t)	Productivity (Kg/ha)
Argentina	6894	18718	2715
Brazil	13251	31357	2366
Canada	980	2737	2793
China	8205	13757	1697
India	6350	6100	961
Indonesia	1091	1306	1198
USA	28657	75028	2618
Asia	16824	22768	1353
World	70690	158327	2240

On the world basis, India's soybean area and production is 8.3 and 3.7 per cent respectively. In terms of contribution of total oilseed crops in India, it is third. It contributes 20.8 per cent to the total oil seeds production of the country. From 1981-82 to 1990-91, India has experienced a tremendous growth in area and production of soybean which showed increases of 312 and 458 per cent respectively. Not withstanding the spectacular increase in area and production, the productivity of the crop has remained low. The current level of soybean yield, being around 1 t/ha, is about half of the world average productivity.

Table 3. Comparative area, production and productivity of oil seed crops in India

Country	Area (1000 ha)	Production (1000 t)	Productivity (Kg/ha)
Castor	747.5	890.7	1192
Groundnut	7814.4	9023.6	1155
Linseed	845.3	324.8	384
Niger	558.8	157.3	281
Rapeseed and Mustard	6855.5	6942.3	1013
Safflower	707.0	432.3	611
Sesame	2060.2	672.9	327
Soybean	5227.2	5201.5	995
Sunflower	1995.1	1315.0	659
All India	26810.7	24960.4	931

India harvests two oilseed crops annually but produces only one soybean crop, planted during the monsoon season in June and harvested from late September. The country produced 7.14 million tonnes in 2006. India is expected to export a record 4 million tonnes of soymeal in the season beginning October 2007, buoyed by higher soybean output. High freight rates from South America have also made Indian feedstock competitive in markets in southeast Asia. India sells soymeal mainly to China, Vietnam, South Korea and Indonesia.

Table 4. Area, production and productivity of soybean in Tamil Nadu

Year	Area (l.ha)	Production (l. MT)	Productivity (Kg/ha)
1989-1990	0.02	0.01	307
1990-1991	0.01	0.01	479
1991-1992	0.25	0.12	494
1992-1993	NA	NA	NA

Contd...

1993-1994	0.31	0.08	232
1994-1995	0.38	0.32	842
1995-1996	0.23	0.25	1082
1996-1997	0.15	0.13	860
1997-1998	0.10	0.09	860
1998-1999	0.04	0.04	1049
1999-2000	0.02	0.02	869
2000-2001	0.03	0.02	944
2001-2002	0.10	0.01	872
2002-2003	0.10	0.03	824
2003-2004	0.20	0.03	874
2004 -2005	0.15	0.03	862
2005-2006	0.20	0.03	970

Soybean growing districts of Tamilnadu (2004 -2005)

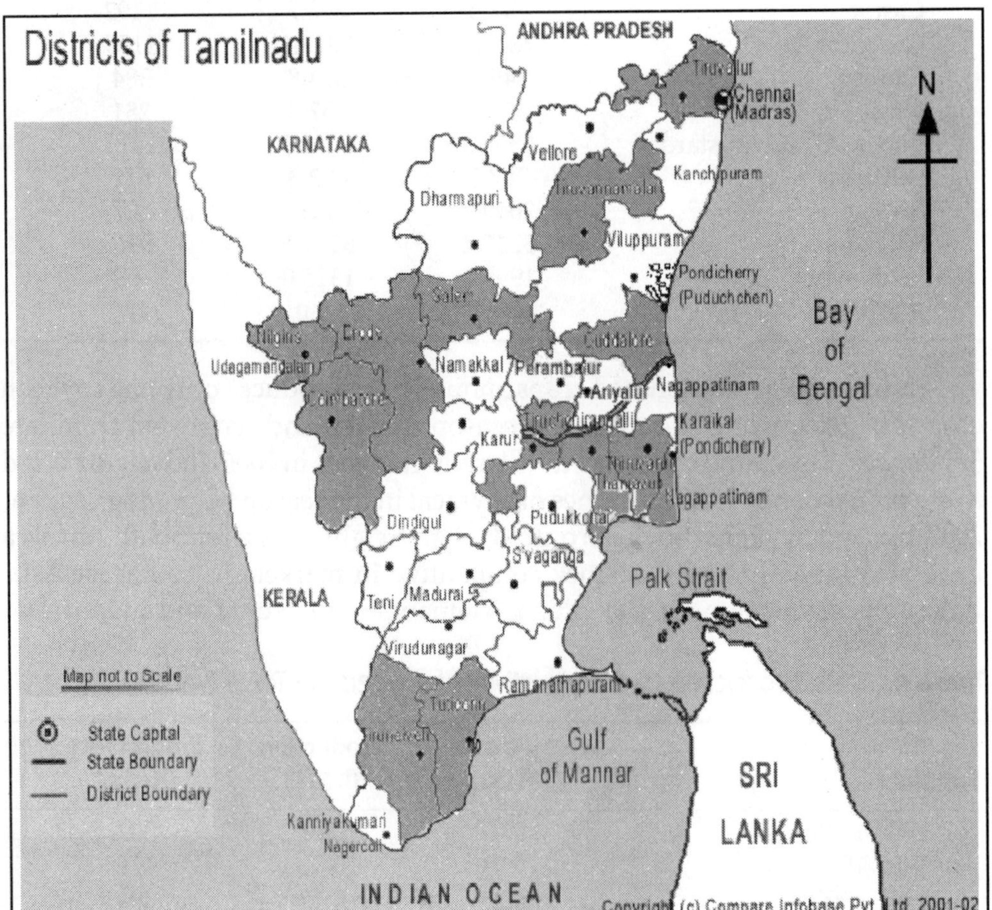

The objectives of this chapter are:

(1) To discuss the physiological traits that have been implicated in controlling yield, with emphasis on current research and how these traits are affected by environemental perturbations,

(2) To integrate many aspects of soybean growth and development and discuss how their interactions affect yield, and

(3) To postulate where future gains may be made in the understanding of the physiological processes controlling soybean productivity.

Vegetative Growth

The rapid establishment and maintenance of an optimum leaf area index (LAI) are important in maximizing photosynthetically active radiation (PAR) interception and canopy photosynthesis. The LAI of a soybean cultivar is determined by the size and number of the individual leaves and the rate of leaf senescence. Leaf number in return, is a function of the rate and duration of leaf appearance on the main stem and branches.

Leaf primordial are initiated on the shoulder of a shoot apical dome. The shoot apex of a seed embryo has one trifoliolate primordium. However, two more primordials develop by the time the seedling emerges. The plastochron interval (time between leaf appearance) remains constant during vegetative development but becomes shorter with the onset of flower evocation.

Global increases in air CO_2 concentration should enhance soybean development in areas with adequate soil nutrients and water to support the increased growth.

Studies have been conducted to examine the effects of such genes on leaf expansion and duration. In general, cultivated genotypes of soybean have a greater area per leaf than the wild - type strains. There probably is an optimum shape and / or area per leaf for maximum seed yields, as the large 5 - and 7 - leaf types are not found in currently produced cultivars.

Few of the above mentioned studies have compared cultivars for differences in rate or duration of vegetative growth and whether such differences affect seed yield. Frederick et al. (1991) reported that seed yield differences among soybean cultivars could not be attributed to differences in rate of leaf appearance. It has been reported that a critical leaf area index must be established for maximum yields under nonstressed conditions. However, it has been found that soybean can withstand substantial reductions in leaf area or vegetative dry matter without significant yield losses if growing conditions are favourable during reproductive development. Vegetative development will also be discussed below with respect to its effects on leaf photosynthesis, nitrogen metabolism, seed yield, and seed yield components.

Flowering and Maturity

Flowering date affects the duration of both vegetative and reproductive development. We discuss flowering and maturity before vegetative growth since many vegetative characteristics of soybean depend upon reproductive behavior. This unique relationship between reproductive and vegetative development has resulted in the selection of soybean as a model crop for many photoperiod and air temperature studies.

Soybean generally can be described as a short - day plant, with reproductive responses to short - day photoperiods being quantitative rather than qualitative or absolute. Soybean can perceive the photoperiod as early as the first trifoliate leaf stage. Flower development first occurs in a leaf axil(s), the midrib of the main stem and continues towards the top and bottom of the main stem and along the branches. The various phases of reproductive development are not equally sensitive to changes in photoperiod.

Studies comparing cultivars of similar maturity group classification have shown little association between seed yield and flowering or maturity date. Frederick *et al.* (1991)reported that the older cultivar Manchu flowered earlier than the newer cultivars Clark 63 and Willimas 82, whereas the older cultivar Dunfield flowered later. Physiological maturity dates of the older cultivars generally were 5-7 days earlier than those of the newer cultivars, although others have found no association between seed yield potential and harvest maturity date.

Delayed flowering has been reported to be associated with higher soybean yields, suggesting that future yield improvements may be made through the genetic manipulation of flowering date.

Soybean yield could potentially be increased by broadening the adaptability of current cultivars, allowing them to be planted earlier or over greater geographic areas.

Photosynthesis and Associated Traits

A. Genotypic Differences

Because pods and stems contribute little to the net CO_2 uptake by soybean, most of the drymatter in the seed must originate from leaves. This apparent relationship between leaf photosynthesis and yield has stimulated much interest concerning the genetic and environmental factors that affect photosynthesis and whether higher yields can be obtained by selecting for higher rates of photsyntehsis. Differences among species and cultivars in leaf photosynthetic CO_2 exchange rate (CER) have been documented, and CER, in some cases, has correlated well with plant productivity.

Carbon production by plants is not only related to the CER of the individual leaves, but also to the total leaf area of the plants and the duration of the leaf area. Light interception by a soybean canopy should increase up to a certain LAI at which PAR interception is maximum. At the critical LAI, maximum canopy photosynthesis and dry weight accumulation should occur.

B. Leaf Nutrition

Nitrogen is one of the most important elements needed by soybean for optimum vegetative and reproductive growth. Therefore, the amount of nitrogen per unit leaf area may be a significant determinant of potential CER, especially under nonstressed conditions. The relationship between leaf CER and nitrogen may be useful in developing models predicting soybean yield and growth.

Pettigrew *et al.* (1990) found that withholding phosphorus at flowering from hydroponically grown soybean reduced peak CER by 13.2%.

C. Environmental Stress Effects

Among the many biotic and abiotic stresses that crop plants are exposed to during the growing season, soil water deficits results in by far the greatest yield losses. Most crop plants experience plant water stress at some point during the growing season and show diurnal variation in plant water status even under irrigated conditions. Reducations in CER have long been recognized as contributing to yield decreases resulting from drought stress limits CO_2 flux to the sites of assimilation, thereby inhibiting leaf CER. However, recent studies conducted under controlled conditions have indicated that the inhibition of chloroplast metabolism may play a greater role than stomatal closure in CER reductions of water - stressed leaves. These nonstomatal factors include decreases in photosystem I and II activities, the inhibition of the CO_2 fixation pathways. Grann and Boyer(1990) reported that both stomatal and nonstomatal factors limit CER of plants grown under soil water deficit.

Differences in leaf water potential between old and new cultivars were observed by Frederick *et al.*(1990) under soil water deficit, but not consistently under irrigation. Air temperature extremes also significantly affect CER and soybean productivity. However, little information is available concerning whether cultivars differ in CER sensitivity to low or high temperature stress.

Nitrogen Metabolism and Senescence

Nitrogen in soybean originates from the biochemical processes of N_2 fixation in root nodules and from the uptake of soil nitrate (and soil NH_4 if available). Seed nitrogen comes from ongoing nitrogen assimilation and from the remobilization of nitrogenous compounds out of vegetative tissue.

Reductions in N_2 fixation as a result of drought can be a major limiting factor of soybean yield. Although drought stress reduces leaf expansion and appearance rates under field conditions, leaf nitrogen content and chlorophyll concentration are not significantly affected by moderate or slow - developing soil water deficits. Senescence of lower leaves may not change or be delayed by moderate or slow - developing stress. Cellular expansion and division and N_2 fixation have been reported to be more sensitive to drought stress than CER. Decreases in leaf growth with no change in nitrogen content suggest that reductions in leaf nitrogen accumulation and growth occur simultaneously under slow - developing drought stress, although no conclusive evidence exists showing a cause and effect relationship. Decreases in plant xylem water potential have been reported to collapse the water potential gradient needed for cellular expansion and inhibit N_2 fixation activity, which may explain the similar reductions in growth and nitrogen assimilation with soil water deficit. Maintainng leaf nitrogen as intact photosynthetic apparatus under drought stress would allow CER to return to normal when soil water contents improve.

Seed Yield, Components of Yield, and Harvest Index

A. Seed Yield

As for CER, seed yield is controlled by many genes and is a quantitatively inherited trait. The heritability for seed yield is usually less than 50%. Soybean yield is determined by the number of seeds produced per unit area and the weight of the individual seeds. Each reproductive structure on the soybean plant progresses through defined stages of development: flower initiation and growth, pollination, pod elongation, and seed growth within the pod. However, the pods on a plant develop sequentially rather than simultaneously, with the timing of pod development varying among nodal positions on the plant and location in a node. To standardize the description of soybean growth, systems have been established to characterize stages of reproductive development on a whole plant basis. Although most soybean cultivars are characterized as determinate, indeterminate, or semi determinate in growth habit, their development is usually indeterminate by definition since vegetative and reproductive growth usually occur simultaneously during certain portions of the growing season. Even determinate cultivars continue vegetative branch growth after the terminal flower inflorescence has formed on the main stem. The indeterminate growth habit and nonsynchronous pod development of soybean suggest that the physiological factors that control soybean yield are complex.

Yield Components

Seed yield is determined by the number of seeds per unit area and the weight of the individual seeds. The number of seeds per unit area, in return, is a function of the number of main stems per unit area, number of braches per main stem, number of pods on the branches and main stem, and the number of seeds per pod. These yield components are established in an ordered sequence. Although significant overlap may occur in their development, seed number and weight are often inversely correlated. Soybean has also the capacity to compensate for stress - induced reductions in one yield component by enhancing the development of subsequently formed yield components once the stress is relieved (i.e., yield component compensation). Under field conditions, where environmental stresses frequently occur, yield differences among cultivars may not only depend upon differences in individual yield component responses, but also on differences in yield component compensation.

The number of pods that develop on a soybean plant is influenced by the number of flowers that are formed and the percentage of flowers and small pods that are aborted. For modern cultivars, pod number is frequently the yield component most associated with yield.

It is well recognized that soybean aborts a large percentage of their flowers and small pods, with values generally over 50%. Flowers formed later in the growing season show higher rates of abscission than early forming flowers. The physiological factors that control abscission are not well defined. One hypothesis is that assimilate supply does not meet assimilate demand by developing pods, resulting in the inhibition of further pod development.

In summary, each of the yield components in soybeans makes a significant contribution to final seed yield. These components differ among cultivars in number and positioning the plant. The frequently reported inverse correlation between seed number and size suggests that the physiological mechanisms controlling seed number and seed size are tightly balanced in soybean. However, the exact mechanisms involved in this regulation are still not understood.

Close monitoring of environmental conditions at the various stages of soybean development, although more time consuming and expensive, may provide better insight into how cultivars differ in yield determination. Part of the genotype x environment interaction that frequently occurs for seed yield and yield components may be eliminated by testing cultivars only in the environments for which they are adapted. The physiological evaluation of cultivars in environments for which they are adopted would eliminate much of the genotype x environment interaction that typically occurs for such traits, thereby enhancing the understanding of the physiological factors that control soybean yield.

Harvest Index

Harvest index is usually measured at maturity after most of the leaves have been shed, and has been termed the "Apparent harvest index," Even though apparent harvest index values are higher than those of the actual harvest index, the two are highly correlated.

Harvest indices measured during reproductive development can be used to estimate the length of seed fill in soybean. If the increase in harvest index is linear during seed fill, the effective filling period for whole plants can be estimated from the final harvest index and the slope of the harvest index measured during pod fill. This estimation has been referred to as the duration of reproductive period. As for the other methods of estimating the length of seed fill, significant genotype x environment interaction can occur for reproductive period duration. Measurement of reproductive period duration may be valuable for research purposes, but probably not as a selection tool in breeding programs since a large number of immature seeds must be separated from their pods for the calculation of harvest index.

Physiological Modeling

During the past 25 years, numerous models have been developed to characterize soybean growth and yield as a function of environmental conditions and to quantify the interactions that occur between soybean and biotic stresses, such as diseases, insects, and weeds. The concept of crop modeling first arose when it was recognized that the interactions between reproductive development and environmental conditions needed to be quantified.

Soybean modeling provides the opportunity for removing genotype x environment interaction effects from field data and can be used as a standard analysis of research results. Current emphasis is on pest effects which must be taken into consideration when conducting field research. For modeling procedures to be rapidly adopted by crop scientists, the logic and computer analysis involved must become a part of our educational system. The success of an approach such as modeling should be measured in terms of the number of researchers who participate and use the available technology, as well as its contribution to scientific knowledge and number of products developed.

Chromosome Mapping

Many genes control soybean growth and seed yield. These genes are expressed at the physiological, morphological and phenological levels of organization. Cryptogenic studies utilizing DNA marker sequences have resulted in at least the partial mapping of genes controlling several agronomic traits. Gene linkage analysis utilizing restriction - fragment length polymorphism (RFLP) genetic markers has the potential to make significant advances in chromosome mapping. Many studies are currently being conducted to map soybean chromosomes and to gather information concerning the location and expression of various genes. Crop geneticists are utilizing these new techniques to examine genes controlling traits such as oil and protein synthesis. Physiologists may enhance this research by identifying genetic differences among soybean cultivars for individual traits. Physiological modeling will also play an important role by integrating information from the various levels of plant organization, from cellular to whole plant.

CONCLUSION

In summary, data in the literature support our hypothesis that the intensity of nitrogen demands (seed nitrogen demand per plant / nitrogen assimilated per plant / unit time) is a major factor limiting soybean seed yield. Increases in nitrogen demands per plant through increases in seed number per plant, seed growth rate, and / or seed nitrogen concentration increase the intensity of nitrogen demand. In contrast, factors that increase nitrogen supply to the plant, such as increases in root density or nitrogen assimilation, reduce the intensity of nitrogen demand, environmental conditions during seed fill that favor high nitrogen assimilation (such as adequate soil water conditions) or decreased seed growth rates (such as long - day photoperiods) reduce the intensity of nitrogen demand per plant and slow the rate of leaf senescence. This model also demonstrates that the physiological factors that determine seed yield are complex and involve vegetative, reproductive, root, and nodule structures. Crop modeling will enhance the understanding, of an interactions, that occur between these plant structures.

The interactions that occur between nitrogen demand by developing seeds, nitrogen assimilation during seed fill, and the initiation of leaf senescence may be mediated through chemical signals. However, increases in the intensity of nitrogen demand may result in relatively more nitrogen partitioned to developing seeds and less to leaves. The resulting changes in the ratio of nitrogen influx to efflux for leaves may hasten the senescence processes. The data discussed in this chapter suggest that the interactions that occur among seeds, leaves, and roots are established during the early stages of seed fill. Research is needed to identify these mechanisms of regulation of soybean nitrogen metabolism. Research is also needed to test this model as a function of genotype and environmental conditions and also in relation to the nitrogen metabolism of other crop species.

In most cases, yield improvement has resulted from increases in seed number per plant. The higher intensity of nitrogen demand of such modern cultivars is met by greater vegetative dry matters and root densities. For irrigated culture or areas with soils of high water -retention capacity, future yield improvements may be obtained through continued increases in vegetative dry matter and root densities, thereby increasing nitrogen assimilation capacity during the growing season and nitrogen for redistribution during seed fill. However, this approach may have its limitation. In areas where short - term stresses frequently occur that limit nitrogen assimilation, genotypes with few seeds per plant and /or slow seed growth rates should be developed due to their low intensity of nitrogen demand. This approach of developing specific ideotypes to specific production environments has the potential to result in significant soybean seed yield increases.

REFERENCES

Pettigrew, W.T., Hesketh, J.D. and Peters, D.B. (1990). CO_2-saturated photosynthesis as affected by phosphate stress. *Photosynthetica*. 24 : 209

Frederick, J.R., Woolley, J.T., Hesketh, J.D. and Peters, D.B. (1990). Water deficit development in old and new soybean cultivars. *Agron. J.*, 82 : 76.

Frederick, J.R., Woolley, J.T., Hesketh, J.D. and Peters, D.B. (1991). Seed yield and agronomic traits of old and modern soybean cultivars under irrigation and soil water - deficit. *Field Crops Res.*, 27 : 71.

Grann, T. and Boyer, J.S. (1990). Very high CO_2 partially restores photosynthesis in sunflower at low water potentials. *Planta*. 181 : 378.

□□□

Chapter 28

Physiological Constraints in Productivity of Grain Legumes

BALAKRISHNAN K.[1], R.BALASUBRAMANIAN[2] AND H.VIJAYARAGHAVAN[3]
[1]Professor of Crop Physiology, [2]Professor of Agronomy
Agricultural College & Research Institute, Madurai-625104
and [3]Professor & Head, Dept. of Crop Physiology, TNAU, Coimbatore

Introduction

Beside agronomic limitations, low productivity of pulses is also due to inherent physiological limitations observed in these crops. As pulses are cultivated through out the year in one or other part of the country mostly under dry land / rainfed conditions (*kharif* pulses) and also on residual moisture situations (*rabi* pulses) without suitable agronomic practices, the inherent physiological limitations are compounded and magnified by the poor agro-climatic conditions. The analysis of various physiological processes and morphological features will result in understanding of the various yield limiting factors in pulses. There are a good number of well researched and documented physiological limitations restricting the productivity of pulses.

Poor Germination and Seedling Vigour

The pulse crops, in general, exhibit a very slow growth rate at germination and seedling stage. This deprives the plants the ability to make the best use of ideal conditions at germination and seedling stage. Dry matter accumulation is very slow in most of the long duration pulses in the initial stages of growth. Poor seedling vigour is primarily due to slow development of leaf area. Slow development of canopy leads to poor interception of radiation leading to slow growth rate. The lack of seedling vigour in pigeonpea and chickpea delays crop canopy (Khanna-Chopra and Sinha, 1990) resulting in low light interception and low yields. So there is a need to enhance seedling vigour. Soaking seeds before sowing increases the germination percentage and reduces the time required for germination.

Most of the *rabi* pulses like lentil, chickpea and lathyrus are grown on residual soil moisture. At times, the moisture is limiting factor in seed zone for proper germination due to early withdrawal of monsoon. Under such conditions, seed priming (soaking seeds for 6-8 hr in water) has proved quite beneficial in improving plant stand and early vigour which results in improving production. Pulse seeds are pre hardened with 100 ppm zinc sulphate (Anon. 2005).

Depth of Sowing

Depth of sowing is an important factor influencing the emergence and establishment of crops especially grown under dryland conditions. Various factors influencing seedling depth include soil texture, date of sowing, climatic conditions and seed size. Under hot dry conditions, sowing of the large seeded pulse crops at depth of up to 7.5 cm is advisable especially on coarse textured soils. On fine textured soils and in moist conditions, sowing depths at 5 cm are more satisfactory (Smart, 1976). For smaller seeded pulses, sowing depths of 2.5 - 4.0 cm are probably adequate for moist conditions. Deep sowing under hot seasons can give the protection from prevailing high temperatures. The seed should be placed at such a depth where it is likely to be induced to germinate by light showers. It is the most advantageous if the seed can be so placed that it does not germinate until a substantial amount of rainfall has been received. Sowing of seeds too deep in wet conditions, give rise to poor germination as in such conditions temperature may be relatively low, aeration may be inadequate and perhaps also diffusible materials may be lost from the seed to the surrounding soil solution with adverse results (Singh *et al.*, 2005).

Experiments conducted at Ludhiana also showed beneficial effects of deep sowing (10 - 12 cm) on the growth and yield of chickpea (PAU, 1981). Furthermore, deep sowing not only improves root and shoot growth but also reduces wilt attack under rainfed conditions.

In pigeonpea, the optimum depth of seeding seemed to be 5 - 6 cm. Tayo (1983) found that emergence declined with the increase in sowing depth to

8 cm. The grain yield reduced significantly by sowing pigeonpea 8 cm deep. In the case of too deep sowing, seedlings are unable to emerge, possibly due to the reason that the reserves in their cotyledons are exhausted whereas in shallower depths seedlings are exposed to water deficits if dry weather and hot temperature prevails soon after sowing.

Time of Sowing

Pulses are extremely sensitive to the environment. Therefore, flowering, plant height and yields of biomass and grains vary markedly between locations and seasons. Thus, understanding the prevailing environment and crop growth is important first, because the time from sowing to flowering and maturity determines the duration of biomass accumulation and, second the duration of different developmental phases, principally the vegetative and reproductive phases, affects the partitioning of biomass and consequently the ratio of grain and straw yield. The too extended and reduced vegetative periods may be detrimental to high yields. The optimum sowing time, a non-monetary input, plays a vital role in realising the yield potential of an improved variety as production efficiency of a variety greatly differs under different sowing dates depending upon the foliage characteristics and canopy structure. The variety which quickly attains optimum leaf area index and retains it for a longer period is considered to be more efficient as it can take better advantage of solar energy available during the growing season. Therefore, sowing time varies with variety and agro-climatic conditions. The effect of date of sowing under field conditions is indeed a reflection of the response to photo-thermal period (Balakrishnan, 1986).

Pigeonpea is basically a short day plant suitable for growing in summer and rainy seasons coinciding maturity with short-day length. April for summer sowing, June for kharif or rainy season and September for rabi or winter season has been found optimum time of sowing (Balakrishnan and Natarajaratnam, 1992). Padhi (1995) reported that 1 June sowing recorded the highest grain yield with both narrow and wider spacing in different varieties and sowing beyond this date reduced the yield.

Greengram is grown in all the three seasons, *viz.*, rainy (*kharif*), winter (*rabi*) and summer. However, it is seldom grown in the same agro-climatic zone in all the three seasons. Thus, it is grown in different regions in different seasons. The kharif greengram is sown with the onset of monsoon during June-July. The summer crop is grown under irrigated conditions for which sowing is done during March. Sowing in the last week of March gives the highest grain yield as compared to early or delayed sowing (Chandra, 1984). This is popular in many parts of the country where there is assured irrigation facility and intensive agriculture is practiced.

Blackgram is basically a hot season crop exhibiting tolerance to higher temperatures and susceptibility to cold and frost. In northern parts of the country, where the temperatures during winter are quite low, it is cultivated only during the rainy season (Chaudhury *et al.*, 1985). However, in the eastern states it is also grown during winter. In the central and southern states where there is not much variation in climate, it is cultivated both during winter and rainy season (Jeswani and Baldev, 1990).

Micronutrient Deficiency

Pulse crops respond well to application of micronutrients like zinc, boron, molybdenum, iron, copper and manganese under deficient conditions. Pulses have specific need for some of the micronutrients for carrying out specific activities. Genetic variations with respect to susceptibility of micronutrients deficiency have also been observed in several pulses (Ahlawat and Ali, 1993). Aulakh and Patel (1991) reviewed the effect of zinc application on the grain yield of different pulse crops at different locations. The yield gain of 400 kg/ha in pigeonpea was recorded with the foliar spray of 0.5% zinc sulphate at flower initiation stage over control (Paste and Jana, 1988). Multi location studies on micronutrients revealed considerable yield increase (450 kg/ha) in chickpea due to foliar spray of 0.5 kg ferrous sulphate per ha over control (Takkar and Nayyar, 1986). Molybdenum application improved nodulation and productivity of mungbean (Ali and Mishra, 2000). Sakal *et al.* (1998) reported that application of 2 kg boron per ha increased the grain yield of chickpea from 1250 to 2000 kg/ha and lentil from 1250 to 1600 kg/ha. Seliga (1998) reported highest copper use efficiency for pod and grain production in faba bean. Three foliar sprays of 0.5% manganese sulphate at weekly intervals after irrigation were found to significantly increase the grain yield of mung bean in a manganese deficient soil (Khurana *et al.*, 2002).

Lower Photosynthetic Efficiency

The total photosynthate availability in a plant depends upon the photosynthetic surface and the rate per unit surface of area, minus respiration and photorespiration. The main photosynthetic characteristics of pulses are described (Sinha *et al.*, 1988).

Photosynthetic characteristics of grain legumes

Photosynthetic rates	Normally 20 – 30 mg CO_2 m^2 / hour
Temperature range	Optimum between 25 and 30°C
Light saturation	5,000 f.c. or 50,000 lux
CO_2 compensation point	About 50 ppm
Photorespiration	Present
Main carboxylating enzymes	RuBP carboxylase
Kranz anatomy	Absent

The important points are, the pulses have a temperature optimum of 25 - 30°C and their light saturation occurs around 50,000 lux. In the cultivation of these crops, temperature plays a very significant role which has not so much been appreciated. An increase in temperature from 25 to 35°C enhances photorespiration by 100 per cent. The net photosynthesis is dependent upon gross photosynthesis rate minus the rate of respiration and photorespiration. Increase in temperature leads to greater increase in photorespiration, but not of photosynthesis, therefore, the photosynthesis would be adversely affected. This, therefore, calls for greater attention to studies on the effect of temperature on photosynthesis in these crops.

The rate of photosynthesis in mungbean, cowpea and chickpea was examined at different stages of growth. In both mungbean and cowpea, the rate of photosynthesis was maximum at the time of flowering and there was considerable variation among the varieties. The rate of photosynthesis was evaluated in mungbean, cowpea, pigeonpea and peas (Sinha, 1977).

Flower and Fruit Shedding

Most pulses produce a large number of flower buds and flowers, but have relatively poor fruit setting. There is a large scale abscission or shedding of flowers and fruits due to various physiological problems. Sinha *et al.* (1988) compared the sink potential of wheat and cowpea at the individual plant and population. It would be clear from these tables that the sink potential of cowpea is far greater than that of wheat, but its realization is poor. Therefore, poor fruit set is an important consideration in the productivity of pulses in general.

Sink potential in wheat and cowpea (per plant basis)

Wheat	Cowpea
2.5 tillers	10 inflorescences
125 flowers	10 flowers
125 ovules	750 ovules
5 g	75 g
40 g / 1000 grains	100 g / 1000 grains

Sink potential and its realization in wheat and cowpea-unit area basis

	Wheat	Cowpea
Population	500 tillers / m^2	16 plants / m^2
Sink	50 florets / tiller	75 flowers / plant
	50 ovules / tiller	750 ovules / tiller
	25,000 ovules / m^2	12,000 ovules / m^2
Sink	1 kg / m^3 grains	1.2 kg / m^3 grains
	40 g / 1000 grains	100 g / 1000 grains
	10 tonnes / ha	12 tonnes / ha
Realization	7.5 tonnes / ha	2 tonnes / ha
Per cent realization of sink potential	75	16.6

A large number of researchers have reported lower fruit set in pulses; 4 - 11.6% in pigeonpea (Sinha, 1977), 14 - 45% in chickpea (Argikar, 1970; Savithri and Ganapathy, 1990), 11 - 31.5% in mung bean (Savithri *et al.*, 1978), 9 - 11% in broad bean (Tamaki *et al.*, 1973) and 16 - 18% in cowpea (Sinha, 1977).

The most poossible reason attributed for this phenomenon is that the plant is unable to meet the carbon and nitrogen demands of emerging flowers and fruits. The greater requirement of these needs is their continuous translocation from vegetative parts. This eventually results in senescence of vegetative tissues and reduction in duration of seed filling period and subsequently lower seed yield (Luthra *et al.*, 1983). Gates *et al.* (1983) attributed higher flower and pod shedding to intra-plant competition for assimilates between reproductive and vegetative organs.

It has been observed in many grain legumes that spray of NAA @ 40 ppm twice at first appearance of flowers and after a fortnight significantly reduced the shedding of flowers and increased the fruit set (Anon. 2005).

Source Sink Relationship

Most of the pulses produce a large number of flowers but only a few of them produce pods and seeds. Rao and Ghildhayal (1985) reported that in mung bean source might be the limitation for higher yield. They concluded that the sink itself was instrumental in hastening the decrease in photosynthesis and leaf senescence by affecting directly mobilization and utilization of leaf nitrogen and thus indirectly nodulation.

The decline in yield parameters like pods, number of seeds and seed weight in the late-formed pods indicate that pod filling was limited by supply of assimilates or other nutrients. Higher the seed weight, fewer will be the number of seeds produced. Towards the apex, both fruit and seed weight decline and some of the fruits do not contain any seeds. Indeterminate plants produce flowers continuously but such flowers later on do not develop into fruits or if they do, do not produce any seeds due to the limitations of leaves at the upper flowering nodes. Further, the first formed fruits and seeds may exert some regulatory effect on fruit formation and seed development by regulating leaf area or flower biology.

The potential constraint for higher production in short duration determinate type of pulses is the production of low biological yield. Low grain yield was associated with low total biological yield. Therefore, higher biomass production and reduction in the number of late formed branches may help in increasing grain yield in indeterminate type of faba bean (Silim and Saxena, 1990). Productivity of determinate type can be increased by higher plant density (Gates *et al.*, 1983). The potential advantage of determinate growth is the reduction in the competition from vegetative sink to the reproductive sink for assimilates, resulting in improved productivity..

In rainfed areas, pulses generally suffer with inadequacy of N at the reproductive phase due to moisture stress and degeneration of root nodules. Foliar spray of 2 per cent urea helps in improving N content of leaves making them photosynthetically more active. A large number of frontline demonstrations carried out on chickpea, lentil, mungbean and urdbean in different regions have shown spectacular increase of 13 - 20 per cent in productivity of these pulses. This could prove to be one of the simplest and low cost technologies for improving production of pulses in the rainfed areas (Ali and Kumar, 2007).

Poor Partitioning / Harvest Index

The harvest index (HI) of pulse is varying between 10 and 40 per cent. Harvest index is influenced by many factors including the genotype, environment and agronomic management. In pigeonpea, harvest index may range from as low as 10 per cent to as high as 52 per cent (Lawn and Troedson, 1990). Harvest index is influenced by the duration of the crop. In blackgram, the HI ranges from 28 to 40 per cent among genotypes and this characteristic is important for grain yield (Chandra Babu *et al.*, 1992).

The translocation of photosynthates from source to sink depends on duration of vegetative and reproductive phases. Further, it depends on the relative partitioning of current assimilation and extent of remobilization of stored assimilates to reproductive parts. Usually the partitioning will be more if the total proportion of reproductive growth is longer and remobilization of assimilates is significant from vegetative parts to reproductive organs. The partitioning of current assimilates into vegetative and reproductive parts also depend on the nature of the crop. The current assimilates are shared between both vegetative and reproductive growth in indeterminate type while the share of current assimilates in vegetative growth is less in determinate type. The partitioning is further influenced by photoperiod. Under adverse photoperiod the partitioning of assimilates to reproductive organs is reduced and initiates vegetative growth in the form of prolonged growth or vine formation takes place. Likewise, induction of optimum photoperiod results in termination of vegetative growth and initiation of reproductive growth, causing diversion of higher portion of current assimilates to reproductive organs. It also increases the duration of reproductive phase. It is highest when the crop is grown in day length at or below the critical photoperiod (Lawn and Wallis, 1987).

Since most of the pulse crops are indeterminate they do not have a distinct vegetative phase as in cereals. Thus, in pulses vegetative growth continues even during pod development. This results in competition between vegetative growth and pod and seed development for partitioning of assimilates. This can be an important plant characteristic limiting yield (Khanna-Chopra and Sinha, 1990). The greatest potential for physiological improvement in productivity of the tropical pulses lies with increasing the proportion of biomass partitioning into seed.

Nitrogen Partitioning

Pulses require a large quantity of nitrogen for their growth and grain yield. They meet part of their nitrogen requirement through biological nitrogen fixation (BNF). The reduction in nodulation and nitrogen fixation during pod filling stage may be due to preferential mobilization of assimilates for seed development which results in lower availability of photosynthesis to nodule growth, maintenance and development (Sinha, 1977). As the current nitrogen fixation declines, the nitrogen requirement of the developing seeds is met by translocation of nitrogen from the subtending leaves. The enzyme protein mobilized from the leaves is Rubisco, the loss of which in leaves tends to reduce photosynthetic efficiency of subtending leaves (Grover *et al.*, 1985). As the pod develops, the leaf undergoes senescence and is finally shed. This phenomenon of monocarpic senescence is observed in chickpea, pigeonpea and cowpea but not in mung bean (Paharia, 1986). Further, it is not only inadequate BNF which is responsible for lower production in pulses, but also poor mobilization of nitrogen from vegetative parts to reproductive parts which results in a large quantity of carbon and nitrogen left unused in vegetative parts.

As seeds of pulses are rich in proteins, their nitrogen requirement is much higher as compared to cereals. Nitrogen content in seed is several times higher than in other plant parts viz., leaves, stems, roots etc. The proportion of total plant nitrogen recovered in seed exceeds grain HI (*i.e.*, nitrogen harvest index or NHI), eg. Level of nitrogen in the seeds ranges from 3 - 7 per cent, while in plant residues it is 0.75- 1.5 per cent. Thus for most pulses a higher HI necessitates an even higher NHI. Hence the comparison of pulses with cereals is grossly unjustified. The greater nitrogen demand for seed protein inspite of the ability of the pulses to draw adequate nitrogen both from soil and biological nitrogen fixation is said to be one of the significant constraints for higher productivity (Sinclair and de Wit, 1975).

As reported above, a great proportion of the nitrogen requirement for seed is met by remobilization from leaf and stem as reported in pigeonpea (Lawn and Troedson, 1990) and chickpea (Davies *et al.*, 2000), which results in self destruction of leaves. Ganeshaiah and Uma Shankar (1981) pointed out that lack of carbohydrate during seed filling period is a major limitation in the ability of pulses to take up N and suggested that rates of photosynthesis during seed-fill period be increased, thereby enhancing N uptake utilization for seed development.

Water Use Efficiency

Water relation of legumes is of special significance since they are mostly grown under rainfed conditions. In India, barely 8 per cent of the pulse area is irrigated. Legumes have deep root systems but escape drought by early completion of life cycle in 60 to 70 days. Water stress affects leaf area, leaf area index and other physiological traits, and including legume-rhizobium symbiosis. Water use efficiency is poor in legumes essentially due to lower rate of dry matter production. In general, WUE of legumes is 500 kg water per kg dry matter while for cereals, it is 300 to 350 kg water per kg dry matter. Legumes at wilting point do not develop substantial negative leaf water potential (Y_1 = -1.5 to -2.5 MPa) as against cereals (Y_1 = -2.5 to 8.5 MPa). Legumes also do not tolerate water logged conditions. In fact excess water has deleterious effect on growth, yield and N fixation (Chaturvedi *et al.*, 1980).

Chaudhary *et al.* (1985) obtained a 66 per cent increase in mung bean grain yield with rice straw mulch, 75 per cent with black polythene mulch and 26 per cent increase with black polythene + straw. Chakravarthy *et al.* (1986) reported that 6 tonnes ha^{-1} wheat straw mulch applied to mungbean at the first irrigation (20 days after sowing) eliminated the necessity for the second irrigation on the 40th day as generally recommended in a semi-arid region during the summer season. Mulching has been found to increase the dry matter production of mungbean and cowpea in addition to water use efficiency. The use of organic matter as surface mulch and incorporation into the soil in subsequent years has been found useful in reducing run off and soil loss and increasing crop yields (Singh *et al.*, 2005).

Plant Type Concepts

Efforts are needed to redesign pulse varieties with appropriate growth habit and efficient source-sink kinetics. This has amply been demonstrated in cereals, particularly wheat and rice, that physiologically more efficient plant types can bring miracles as far as grain yield is concerned. Even among pulses, development of semi-leafless plant type in pea has resulted in enhanced photosynthetic efficiency and yield. Pulses still have wild traits like indeterminate growth habit, which causes intra-plant competition for photosynthate resulting into poor portioning. It is presumed that the determinate types under good management would partition photosynthate to yield components with greater efficiency through early seedling vigour, fast leaf development, synchronous flowering, slow senescence and increased rate of dry matter production (Lawn, 1990).

Tall plant types have been mentioned as an ideal plant for chickpea. Further, it was suggested that a relatively erect plant type with shorter internodes might be a suitable ideotype in chickpea (Saxena and Johansen, 1990). Determinate plant types of less than 100 cm height having open canopy of few branches, small leaves, small pods, few seeds per pod, seed size not exceeding 10 g/100 seeds, open inflorescence and deep rooting are some of the characters suggested for desirable plant type for redgram by Ariyanayagam *et al.* (1997). The ideal plant type for pure cowpea crop was conceived by Singh and Sharma (1997) as follows, erect / semi-erect plant type with medium sized leaves and basal branches as anchors to avoid lodging, 60 to 70 days of crop duration with near synchronous maturity, long peduncles with pods above the plant canopy for easy harvesting by manual or mechanical means, medium seed size with rough to smooth seed coat to avoid seed quality problem during maturity, harvesting and threshing. Plant type with determinate growth habit of 50 to 60 cm height and greater early vigour is desirable for spring / summer and *rabi* seasons while semi-determinate to indeterminate plant type of 70 to 80 cm height are considered as ideal plant type in blackgram for *kharif* season. Plants with pods of 6 to 8 seeds and 4 g / 100 seed weight may also be considered in blackgram (Satyanarayana *et al.*, 1991). The plant type concept has been suggested by Kuo *et al.* (1978).

CONCLUSION

The present - day agriculture is changing very fast. Thus, crop cultivation concepts with respect to cultivation techniques need more attention. A greater understanding regarding soil and crop management and environment is essential to sustain crop productivity. Pulses due to low yields require more appropriate technology under changed situation Systematic efforts are needed for identifying the various types of yield barriers *viz.*, poor seed germination, low photosynthetic efficiency, low pod setting, poor source-sink relationship, low HI, low water use efficiency and lack of suitable plant types. All these issues could be addressed by joint and co-ordinated efforts of Plant Breeders, Crop Physiologists and Agronomists. These approaches will help in imparting the much desired respite to pulse crops against unfavourable environment and increase inherent capacity for higher productivity.

REFERENCES

Ahlawat, I.P.S. and Ali, M. (1993). Fertilizer management in pulses. In: Fertilizer Management in Food Crops (ed.) H.L.S.Tandon, pp. 114-138, New Delhi, Fertilizer Development and Consultation Organisation.

Ali, M. and Kumar, S. (2007). Promising technologies for improving pulse production. *Indian Farming* (July): 15 - 18.

Ali, M. and Mishra, J.P. (2000). Nutrient management in pulses and pulse based cropping systems. *Fertilizer News*, 45(4): 57-69.

Anonymous. (2000). Expert committee report on pulses. Technology mission on oilseeds and pulses, New Delhi.

Anonymous. (2005). Crop production guide. TNAU, Coimbatore.

Argikar, G.P. (1970). Gram. In: Pulse Crops of India (Eds P.Kachroo and M.Arif), pp.54-135. New Delhi: ICAR.

Ariyanayagam, R.P., Singh, N.B., Jain, K.C. and Asthana, A.N. (1997). Plant types for higher productivity of pigeonpea. In: Recent advances in pulses research. A.N.Asthana and Masood Ali (eds.). Indian Institute of Pulses Research, Kanpur. P.89-108.

Aulakh, M.S. and Patel, M.S. (1991). Soil related constraints and management for oilseed and pulse production in India. In: Soil related constraints in crop production, pp. 129-144, New Delhi. Indian Society of Soil Science.

Balakrishnan, K. (1986). Seasonability-its physiological impact on redgram. Ph.D (Thesis), TNAU, Coimbatore.

Balakrishnan, K. and Natarajaratnam, N. (1992). Effect of sowing dates on yield components in pigeonpea. *Madras agric. J.*, 79: 311-375.

Chakravarthy, N.V.K., Sastry, P.S.N. and Tomar, O.P.S. (1986). A study of biomass production of summer mung (*Vigna radiata* (L.) Wilczek) as influenced by modification of micro environment. *Indian Journal of Plant Physiology,* 29: 211-218.

Chandra, S. (1984). Increasing pulse production in India. In: Proc. Symposium on increasing pulses production in India- constraints and opportunities, New Delhi. Pp.23-39.

Chandrababu, R., Sadasivam, R. and Natarajaratnam, N. (1992). Dry matter accumulation and HI in blackgram. *Madras Agric. J.*, 79: 43-44.

Chaturvedi, G.S., Aggarwal, P.K. and Sinha, S.K. (1980). Growth and yield of determinate and indeterminate cowpea in dryland agriculture. *Journal of Agricultural Science,* Cambridge, 94: 137-144.

Chaudhury, T.N., Chopra, U.K. and Sinha, A.K. (1985). Root growth, leaf water potential and yield of irrigated summer mung bean in relation to soil water status and soil temperature under various mulches. *Field Crops Res.*, 11: 325-333.

Davies, S.L., Turmer, N.C., Palta, J.A., Siddique, K.H.M. and Plummer, J.A. (2000). Remobilisation of carbon and nitrogen supports seed filling in chickpea subjected to water deficit. Australian Journal of Agricultural Research 51: 855-866.

Ganeshaiah, K.N. and Uma Shankar, R. (1981). Auto-parasitization of leaves in pulses – N limiting or C limiting? *Current Sci.*, 50: 942-944.

Gates, P., Smith, M.L. and Boulter, D. (1983). Reproductive physiology of *Vicia faba* L. In : The Faba bean, *Vicia faba* L. (Ed. P.D.Hebblewaite), London, Buntherworks. pp. 133-142.

Grover, A., Koundal, K.R. and Sinha, S.K. (1985). Senescence of attached leaves regulated by developing pods. *Physiologia Plantarum*, 63: 87-92.

Jeswani, L.M. and Baldev, B. (1990). Advances in pulse production technology. New Delhi, ICAR. Pp.167-70.

Khanna-Chopra, R. and Sinha, S.K. (1990). What limits the yield of pulses? Plant processes or plant type. In: Proceedings of the International Congress of Plant Physiology, Vol. 1 (Eds. S.K.Sinha, P.V.Sane, S.C.Bhargava and P.K.Agrawal), pp. 268-278, New Delhi: Society for Plant Physiology and Biochemistry.

Khurana, M.P.S., Bansal, R.L. and Nayyar, V.K. (2002). Relative efficiency of different carriers of manganese in moong. In: Proceedings of 89th Indian Science Congress, Section of Agricultural Sciences, p.15, Kolkata. The Indian Science Congress Association.

Kuo, C.G., Wang, L.J., Chenge, A.C. and Chou, M.H. (1978). Physiological basis for mungbean yield improvement. In: The first International mungbean symposium. AVRDC, Taiwan.

Lawn, R.J. (1990). Physiological constraints to the productivity of tropical grain legumes and prospects for improvement. In: Proceddings of the International Congress of Plant Physiology, p. 246-260.

Lawn, R.J. and Troedson, R.J. (1990). Pigeonpea: Physiology of yield formation. In: The Pigeonpea (Eds. Y.L.Nene, S.D.Hall and V.K.Shella), pp. 179-208, Wallingford:CAB International.

Lawn, R.J. and Wallis, J.H. (1987). Limits proposed by climatological factors. In Food Legumes Improvement for Asian Farming Systems, Proceedings No. 18 (Eds. E.S.Wales and D.E.Byth), pp. 83-94, Canbarra:AICAR.

Luthra, Y.P., Sheoran, I.S. and Singh, R. (1983). Relative photosynthetic efficiency of subtending leaves, pod walls and seeds of pigeonpea. *Photosynthetica,* 17: 210-215.

Padhi, A.K. (1995). Effect of sowing date and planting geometry on yield of redgram. *Indian J. Agron.,* 40: 87-100.

Paharia, C.S. (1986). Response of *Vigna radiata* (L.) Wilczek to water stress and sink size. Ph.D Thesis, Indian Agricultural Research Institute, New Delhi.

Paste, A.M. and Jana, P.K. (1988). Effect of phosphorus and zinc on pigeonpea varieties grown during winter. *Indian Journal of Agronomy,* 33: 399-404.

PAU (Punjab Agricultural University) (1981). Present Status and Future Strategies of Pulses Research in Punjab, Ludhiana: Pulses section, Punjab Agricultural University.

Prasad, R. (2002). Text book of field crops production. ICAR, New Delhi. Pp 252-396.

Rao, T.R.K. and Ghildayal. M.C. (1985). Analysis of photosynthetic source and sink relationship in mungbean (*Vigna radiata* L. (Wilczek)). *Indian Journal of Plant Physiology,* 28: 135-144.

Sakal, R., Sinha, R.B., Singh, A.P. and Bhogal, N.S. (1998). Responses of some rabi pulses to B, Zn and S application in farmers' fields. *Fertilizer News,* 43(2): 39-43.

Satyanarayana, A., Seenaiah, P., Murthy, S.S. and Naidu, N.O. (1991). Improvement of blackgram through new plant types. In: Proceedings of Golden Jubilee Symposium by the Indian Society of Genetics and Plant Breeding, NewDelhi.p.12-15.

Savithri, K.S. and Ganapathy, P.S. (1990). Realisation of yield potential in chickpea (*Cicer arietinum* L.) In: Proceedings of the International Congress of Plant Physiology, vol.I (Eds. S.K.Sinha, P.V.Sane, S.C.Bhargava and P.K.Agrawal), pp.313-319, New Delhi.

Savithri, K.S., Ganapathy, P.S. and Sinha, S.K. (1978). Fruit and seed development in mungbean (*Phaseolus aureus* Roxb). *Journal of Agricultural Science*, Cambridge, 90: 551-556.

Saxena, N.P. and Johansen, C. (1990). Chickpea ideotype for genetic enhancement of yield and yield ability in south Asia. In: Proceedings of the Second International workshop in chickpea improvement, 4-8, December, 1989, ICRISAT, Patancheru, India. P.81-85.

Seliga, M. (1998). Nitrogen fixation in several grain legume species with contrasting sensitivities to copper nutrition. *Acta Physiology Plantarum*, 20: 263-267.

Silim, S.N. and Saxena, M.C. (1990). Composition of dry matter accumulation, partitioning and yield of determinate and indeterminate faba bean (*Vicia faba* L.) In: Proceedings of the International Congress of Plant Physiology, Vol. 1. (Eds. S.K.Sinha, P.V.Sane, S.C.Bhargava and P.K.Agrawal), pp. 289-298, New Delhi: Society for Plant Physiology and Biochemistry.

Sinclair, T.R. and de Wit, C.T. (1975). Photosynthates and nitrogen requirement for seed production in various crops. *Science*, 189: 565-567.

Singh, B.B. and Sharma, B. (1997). Restructuring cowpea for higher yield. In: Recent advances in pulses research. A.N.Asthana and Masood Ali (eds.). Indian Institute of Pulses Research, Kanpur. P.139-156.

Singh, G., Sekhan, H.S. and Kolar, J.S. (2005). Pulses. Agrotech Publishing Company, Udaipur. p.235-240.

Sinha, S.K. (1977). Food Legumes: Distribution, Adaptability and Biology of Yield. FAO Plant Production and Protection Paper, 3. Rome. FAO.

Sinha, S.K., Bhargava, S.C. and Baldev. (1988). Physiological aspects of pulse crops. In: Pulse Crops (eds.). Baldev, B., Ramanujam,S. and Jain,H.K. Oxford and IBH, NewDelhi.

Smart, J. (1976). Tropical pulses, London, Longman.

Takkar, P.N. and Nayyar, V.K. (1986). Integrated approach to combat micronutrient deficiency. In: Proceedings of FAI seminar on growth and modernization of fertilizer industry, PS 111/2 pp.1-16, New Delhi, Fertilizer Association of India.

Tamaki, K., Asanuma, K. and Naka, J. (1973). Physiological status of growing process of broad bean plants VII. Effect of plant density on growth and seed production. Technical Bulletin of the Faculty of Agriculture, Kagwa University, 25: 1-11.

Tayo, T.O. (1983). Growth, development and yield of pigeonpea (*Cajanus cajan* L.) Millsp.) in the lowland tropics. 4. Effect of sowing depth. *Journal of Agricultural Science*, Cambridge 101: 435-440.

□□□

Chapter 29

Studies on the Effect of 28 - Homobrassinolide on Biochemical Parameters and Yield of Groundnut Cv. Vri 2.

PRAKASH, M., S. SUGANTHI, J. GOKULAKRISHNAN, B. SUNILKUMAR AND T. SABESAN
Department of Agricultural Botany, Faculty of Agriculture, Annamalai University
Annamalainagar - 608 002

Introduction

Among the various brassinosteroids tested, brassinolide (BL) and its epimer 24- epibrassinolids (EPR) and 28 – homobrassinolide (HBR) were found to be more effective than the other compounds in short term bioassays generally involving excised plant parts (Takematsu *et al.*, 1983 a, b). Brassinolides induced plant growth was reported to be associated with increased metabolic processes like photosynthesis (Sairam, 1994) and protein synthesis (Kalinch *et al.*, 1985). Brassinosteroids are reported to occur at low concentrations in all plants tested so far and are considered essential for normal growth and development (Sasse, 1997). Brassinosteroids were found to enhance the growth and yield of groundnut (Vardhini and Rao, 1998). In the present study the effect of HBR on growth, physiological, biochemical and yield attributes of groundnut was investigated.

Materials and Methods

The present experiment was carried out to study the influence of 28-homobrassinolides on growth, physiological, biochemical and yield parameters of groundnut cv VRI 2. Field experiments were conducted at TANCOF farm at Neyveli, Tamilnadu. 28-homobrassinolide was purchased from Godrej Agrovet Ltd., Mumbai (in the form of Godrej Double). Seeds were purchased from Regional Research Station, TNAU, Vridhachalam. Groundnut plants were raised in the plots of 4 x 3 m in the field and sprayed with 28-homobrassinolide on 25, 30, 35 and 50 DAS as per the following schedule of treatments.

T_1 - Control (Water spray)
T_2 - BR 0.25 $mg.l^{-1}$ on 25 and 35 DAS
T_3 - BR 0.50 $mg.l^{-1}$ on 25 and 35 DAS
T_4 - BR 1.00 $mg.l^{-1}$ on 25 and 35 DAS
T_5 - BR 0.25 $mg.l^{-1}$ on 30 and 50 DAS
T_6 - BR 0.50 $mg.l^{-1}$ on 50 DAS

The experiment was conducted in Randomized block design with three replications. Observations were recorded by selecting ten plants at random from all treatments of each replication. Sampling was done on 35 and 50 DAS for growth, physiological and biochemical parameters. The yield and yield attributing characters like number of pods per plant, pod weight, 100 seed weight, plot yield, shelling percentage and dry matter production were recorded. The leaf area index was calculated as per the formula of Williams (1946). Chlorophyll content was estimated by the procedure of Yoshida *et al.*, (1971). Soluble protein was estimated by the method of Lowry *et al.*, (1951). Sugar content was estimated by the procedure suggested by Somogyi (1952). The data was subjected to statistical analysis and results are discussed.

Results and Discussion

Generally the 28-homobrassinolide treated plants recorded higher values for physiological, biochemical and growth parameters when compared to the control. Regarding the growth and physiological parameters, the plants sprayed with 28-homobrassinolide @ 1.00$mg.l^{-1}$ on 25 and 35 DAS recorded higher values. The chlorophyll content and soluble protein was maximum in the plants sprayed with 1.00 $mg.l^{-1}$ 28-homobrassinolide twice on 25 and 35 DAS. This result was in consonance with the findings of Vardhini and Rao (1998). While considering the biochemical parameters like total sugar content and nitrogenase activity the plants undergone the 1.00 $mg.l^{-1}$ 28 homobrassinolide sprayed twice recorded maximum values when compared to the control. Plant growth regulator promoted nodulation, nitrogen fixation and nitrogenase activity has been already established by many research workers. In this present study also, BR treated plants recorded higher values for nitrogenase activity than the control plants and were found higher in plants treated with BR@ 1.00 mg l^{-1} on 25 and 35 DAS. Foliar application of brassinosteroids also increased nitrogenase activity (Vardhini and Rao, 1999). This confirms the result of the present study. This may be due to the fact that BR may interact with other plant growth regulators and thereby they would have initiated several physiological and biochemical processes leading to enhanced levels of these parameters.

Table 1. Effect of 28-homobrassinolide on certain growth and physiological parameters of groundnut

Treatment	Number of leaves		Leaf area index		Dry matter production ($g.pl^{-1}$)		Total chlorophyll ($mg.g^{-1}$ fr.wt)	
	35 DAS	50 DAS	35 DAS	50 DAS	35 DAS	50 DAS	35 DAS	50 DAS
T_1	31.00	50.40	0.841	2.019	6.60	11.44	1.645	1.544
T_2	31.93	52.27	1.004	2.185	8.43	12.68	1.764	1.627
T_3	31.80	53.62	1.106	2.233	8.85	12.94	1.847	1.752
T_4	34.87	54.11	1.301	2.296	9.47	13.92	2.104	2.002
CD(0.05)	1.476	2.331	0.230	0.229	1.435	0.703	0.060	0.043

Table. 2. Effect of 28- homobrassinolide on certain biochemical parameters of groundnut

Treatment	Total sugars ($mg.g^{-1}$ fr.wt)		Soluble protein ($mg.g^{-1}$ fr.wt)		Relative water content (%)		Nitrogenase activity (µmol per g root weight per hour)	
	35 DAS	50 DAS	35 DAS	50 DAS	35 DAS	50 DAS	35 DAS	50 DAS
T_1	83.76	93.62	5.89	6.069	45.13	43.75	5.98	5.61
T_2	89.46	96.81	6.06	6.31	47.31	46.44	6.22	6.22
T_3	88.64	99.98	5.80	6.14	47.33	47.10	6.06	5.98
T_4	91.93	102.56	6.47	6.59	49.72	49.34	6.31	6.81
CD(0.05)	3.362	2.814	0.561	1.106	1.262	1.954	0.614	0.461

Table. 3. Effect of 28-homobrassinolide on yield and yield attributes of groundnut

Treatment	Number of pods per plant	Pod weight ($g.pl^{-1}$)	100 seed weight (g)	Plot yield (kg)	Shelling percentage	Dry matter production
T_1	11.93	9.61	41.04	7.43	72.54	13.75
T_2	13.27	11.23	43.23	8.23	73.89	14.47
T_3	14.20	12.55	44.47	8.03	74.22	15.66
T_4	17.60	14.32	45.97	9.43	75.49	18.70
T_5	16.20	13.47	44.14	9.03	74.42	18.03
T_6	13.07	12.37	44.24	8.13	74.00	17.35
CD (0.05)	0.848	1.225	0.545	0.268	0.359	0.823

The yields parameters like number of mature pods per plant, pod weight and 100 seed weight also excelled in its performance with the application of 1.0 $mg.l^{-1}$ brassinolide twice on 25 and 35 DAS. This in turn increased the dry matter production and plot yield. A similar report of increased crop yields with HBR application was reported by Ramraj et al. (1997) in cotton, wheat, rice mustard and groundnut. Sairam (1994) also reported that HBR application to seeds and yound seedlings of wheat stimulated metabolism and growth leading to increase in number of ears per plant, grains per ear and grain size. Since the effects of brassinolide in enhancing cell division and cell elongation are well established particularly in young meristematic tissues (Mandava, 1988). Application of BR might have influenced growth and developmental processes of crop plants leading to increased crop yield.

In the present study also, application of BR have increased number of pods per plant. It could be due to increased translocation of assimilates to developing pegs in groundnut, leading to increased pod yield through increased pod and seed size. Total dry matter production, the final product of the crop growth and development ultimately decides the yield of crop plants. By increasing the dry matter production, the possibilities of increasing the economic yield through enhanced translocation of assimilates can be attempted. In the present study also, it was found that application of BR increased total dry matter production. The present findings contribute for the effect of BR on growth, physiological, biochemical and yield parameters of groundnut.

REFERENCES

Kalinch, F. N., Mandava, N. B and Todhunter, J. A. (1985). Relationship of nucleic acid metabolism to brassinolides induced responses in bean. *J. Plant Physiol.*, 120: 207 - 214.

Lowry, O. H., Rosen Rough, N. J., Farr L. A and Randall, R. J. (1951). Protein measurements with Folin Phenol reagent. *J. Biol. Chem.*, 193: 265 - 275.

Mandava, N. B. (1988). Plant growth promoting brassinosteroids. *Ann. Rev. Plant Physiol. and Plant Mol. Biol.* 39: 23 - 52.

Pearce, R. B., Brown, R. H and Balster, R. E. (1968). Photosynthesis of alfalfa leaves as influenced by age and environment. *Crop Sci.*, 6: 677-680.

Ramraj, V. M., Vyas, B. N., Godrej, N. B., Mistry, K. B., Swami, B. N and Singh, N. (1997). Effects of 28-homobrassinolide on yields of wheat, rice, groundnut, mustard, potato and cotton. *J. Agrl Sci., Cambridge,* 128: 405 - 413.

Sadasivam, S and Manickam, A. (1996). Biochemical methods. New Age International Publishers, New Delhi.

Sairam, R. (1994). Effect of homobrassinolide application on plant metabolism and grain yield under irrigated and moisture stress condition of two wheat varieties. *Plant Growth Regul.*, 14: 173 - 181.

Sasse, J. M. (1997). Recent progress in brassinosteroid research. *Physiol. Plant.* 100: 696 - 701.

Somogyi., M. (1952). Notes on sugar determination. *J. Biol. Chem.*, 200: 145 - 154.

Takematsu, T., Y. Takeuchi and M. Koguchi (1983 a). New plant growth regulators. Brassinolide analogues, their biological effects and application to agriculture and biomass production. Chemical Regulation of Plants. 18: 2-15.

Takematsu, T., Y. Takeuchi and M. Koguchi (1983 b). Bioassay methods and physiology of brassinolides. Chemical Regulation of Plants. 18: 38-54.

Vardhini, B. V and Rao, S. S. R. (1998). Effect of Brassinosteroids on growth, metabolite content and yield of *Arachis hypogaea. Photochemistry.* 48: 927 - 930.

Vardhini, B. V and Rao, S. S. R. (1999). Effect of brassinosteroids on nodulation and nitrogenase activity in groundnut (*Arachis hypogaea* L.) *Plant Growth Regul.* 28: 165 - 167.

Watson, D. J. (1958). The dependence of net assimilation rate on leaf area index. *Ann. Bot.*, 22: 37-54.

Williams, R. E. (1946). The physiology of plant growth with special reference to the concept of NAR. *Ann. Bot.*, 10:41-71.

Yoshida, S., Forna, D. A and Cock, J. H. (1971). Laboratory Manual for physiological studies on Rice. *IRRI Publication, Philippines.*

□□□

Chapter 30

Groundnut Research on Drought Tolerance or Wue at NRCG: From Breeding to Farmer's Field

NAUTIYAL, P.C., M.S. BASU*, R. K. MATHUR, RUKAM SINGH AND T. RADHAKRISHNAN**

National Research Centre for Groundnut, Post Bag No. 5, Ivnagar Road, Junagadh, 362 001, Gujarat. * Ex-Director, NRCG, presently working at ICRISAT, Patancherue, Andhra Pradesh ** Present Address: National Research Centre for Oil Palm, (ICAR), Pedavegi-534 450, West Godavari District, Eluru (A.P.)

Summary

To increase crop productivity in rain-dependent system, both drought tolerant cultivars and better understanding of crop water use efficiency (WUE) is required. Therefore, studies were conducted to improve the genetic makeup of groundnut crop for efficient utilization of water in transpiration (T); exchange of transpired water for CO_2 fixation; more efficient in production of biomass and conversion of more of the biomass into kernels i.e. high harvest index (HI). Germplasm accessions were screened for WUE and relationship between specific leaf area (SLA) and carbon isotope discrimination ($^{13}C/^{12}C$, ?) was analysed. SLA was found related inversely with ? and used as surrogate of ? *vis-a-vis* WUE in segregating populations and subsequent experiments. Role of SLA in maintaining leaf water relations under normal and water-deficit conditions was evaluated. In addition, variation in drought tolerance index (DTI) was found significant, and low SLA types showed higher DTI under water-deficit and rain-dependent conditions.

Pot culture experiment was conducted to study the leaf water relations and photosynthetic rate as influenced by water-deficit in six genotypes varying in SLA (144-241 cm^{-2} g^{-1}). Low SLA types were found superior in maintaining photosynthetic rate, leaf relative water content (RWC), transpiration (T) and stomatal conductance (g_s) during progressive water-deficit stress than the high SLA. It is suggested that under water-deficit, there is a significant inverse relationship between SLA and RWC. The low SLA types were found to be drought tolerant in terms of total dry matter production in field studies and maintenance of higher RWC in pot-culture experiments. Thus the ability of the low SLA types to maintain higher RWC may form the basis for the difference in drought tolerance.

Field trial data were analysed based on the model i.e. Pod yield = T x TE x HI and results showed that high pod yield under water-deficit stress was associated with a combination of traits such as T, WUE, and HI, indicating that these traits are not physiologically linked. Parents were selected based on the high WUE and breeding was initiated to combine traits like high HI, and WUE. Finally, selection for both WUE and yield traits (HI) resulted in cultivars with improved performance in rain-dependent agriculture system. Breeding programmes initiated at four different locations in India and the material generated was evaluated for WUE, and promising cultures (about 182) were identified. Promising genotypes (about 6) were used in on farm demonstration of high WUE cultivars in farmer's participatory research and extension (FPRE). After two years of on farm trials in five villages in Junagadh, Jamnagar and Rajkot districts with 10 farmers in each village and farmers preferred genotypes were identified. In conclusion, WUE seems to be a desirable trait that can contribute to productivity when water resources are scarce, especially under rain-dependent system.

Introduction

Groundnut (*Arachis hypogaea* L.) is an important oilseed crop grown in an area of 8 million hectare of land mostly under rain-dependent conditions (about 80%). It has the inherent plasticity to cope up with harsh environmental conditions, especially the water scarce, but economic yield is reduced drastically, though there are genetic variations in plant adaptation and yields to environmental stresses (Joshi *et al.* 1993, Nautiyal and Joshi 2003, Nautiyal *et al.*, 2008). Crop productivity in past two-and-half decades increased very slowly and reached to the level of 980 kg ha^{-1} in 2005 from the level of 750 kg ha^{-1} in 1980s. Moreover, the genetic base of Indian groundnut cultivars is very narrow and there is urgent need to widen it by generating genetic material from more diversified parental stocks, and generating the intra-specific crosses between wild *Acharis* species and cultivars. Thus to increase productivity in rain-dependent cropping system, it is imperative to identify the morphological, physiological and molecular traits associated with desirable agronomical characters in available gene pool. Plant breeders usually select for high-yielding genotypes empirically, without particular attention to specific plant traits that might be associated with high WUE and drought tolerance. However, comparing low yielding lines with high yielding line shows clearly that many morphological, physiological and biochemical traits have been altered by the intense selection pressure for higher yields. These changes indicate that selection for high yielding genotypes has generated indirect selection pressure on the altered traits, if we exclude effects of gene that regulates the expression of two or more unrelated traits (pleiotropic gene). Similarly, identification of high and low WUE lines might reveal the specific traits and gene associated with it. Conventional breeding based on yield as a sole criterion is proving to be slow in achieving gains. Therefore, it has been postulated that it would be more efficient and successful to attempt improving growth and yield per resource use rather than selecting for yield alone (Gutschik, 1988).

Thus, any programme for drought tolerance or high WUE need to identify the source of drought tolerance or high WUE in the available gene pool. Recent studies of this type have shown a remarkable positive correlation between yield increase and stomatal conductance or T or TE and *vis-à-vis* WUE. Groundnut being a drought tolerant crop, still need basic and strategic research on its improvement for drought tolerance, because it may be one such trait that can contribute to productivity when water resources are limited (Wright *et al.* 1994). Cultivar variation for WUE and its association with carbon isotope discrimination ($^{13}C/^{12}C$, ?) has been demonstrated in groundnut grown in both the greenhouse (Hubick *et al.*, 1986) and in the field (Nagaswara Rao and Wright, 1993, Nautiyal *et al.*, 2002). Further, Nageswara Rao and Wright (1993) reported low cultivar by environment interaction for the relationship between SLA and WUE. The influence however, of environmental factors such as temperature and soil moisture-deficit on SLA has been reported (Vivekanandan and Gunesena, 1976). Association between SLA and carboxylation capacity has been studied, and it was suggested that the cultivars with low SLA have greater photosynthetic capacity per unit leaf area (Bowes *et al.* 1972). Nagaswara Rao *et al.* (1995) have recommended leaf thickness (low SLA) as a selection criterion for increasing WUE in groundnut.

The objective of the research was to improve groundnut WUE through trait based breeding programme and to screen for WUE by an easy, convenient and economic measurable parameter.

Materials and Methods

A. Field Experiments

Plant material and water-deficit stress treatments

The soil at the experimental site was a verticustochrept (pH of 8.5) with low available organic matter, nitrogen (N) and phosphorous (P). A set of 68 germplasm accessions including *Spanish, Valencia and Virginia* types was screened for WUE in two rainy (June-October) and one summer (January-June) seasons. Following water-deficit treatments were imposed.

T_1 = Adequate irrigation, replenished 100% of CPE at 4-day intervals using drip system.

T_2 = Simulated drought under rain-out shelters (ROS) during 40-75 days after sowing and replenished 25% of the cumulative evaporation at 5-day intervals during the treatment period only.

T_3 = Rain-dependent condition.

Above treatments were also used to study the effect of different soil moisture regimes using seven groundnut cultivars. The experiments were conducted in RBD with three replicates and the genotype represented three rows of 3 m length in each replicate and treatment. Seed to seed distance was 10 cm and row to row in case of *Valencia* and *Spanish* was 30 cm, whereas in case of *Virginia* types it was 90 cm. Stress treatments were managed at the same phenophase of the genotypes by staggered sowings. A healthy crop was maintained by following all the recommended agronomical practices including the fertilizers doses.

Growth Parameters

The plants from an area of 0.3m x 3 rows of each genotype were up-rooted for growth analysis at the sampling date i.e. 80 DAS for *Spanish* and *Valencia* types and 90 DAS for *Virginia* type. Leaf area was measured using an LI 3000 leaf area meter (LICOR, Lincon, USA) after drying the leaves to the constant weight. Specific leaf area (SLA) was calculated as SLA = leaf area (cm^2)/leaf dry weight (g). After recording the vegetative and pod dry weights, harvest index (HI) was calculated. Drought tolerance index (DTI), the ratio of the total dry matter (leaves + stem + pods) under water-deficit stress to that under irrigated conditions was calculated. The final harvest of *Spanish* and *Valencia* types was made at 110 DAS and for the *Virginia* type at 120 DAS. Pod weight recorded at final harvest at moisture content between 7 and 8% was adjusted for their energy content using a multiplication factor of

1.65. The harvest index (HI) was calculated as the ratio of total pod weight at the final harvest to the total biomass at the final harvest. Due to shedding of leaves prior to physiological maturity in groundnut, there is considerable loss in vegetative weight and hence the vegetative weight recorded at mid pod-fill (80-90 DAS) was used for determining total biomass production as suggested by Wright and Redden (1995).

Measurement of Carbon Discrimination

For estimation of delta i.e. $^{13}C/^{12}C$, 20 fully expended 3rd or 4th leaf from main axis in the plants selected randomly were collected and leaf area of 80 leaflets were measured using LI 3100 leaf area meter. The leaf samples were dried at 80°C for 3 days and weighed. One gram of fine powdered leaf samples was packed in glass vials and labeled. The samples were analysed for $^{13}C/^{12}C$ ratio by mass spectrometry in Canberra, Australia following Farquhar *et al.* (1982).

Measurement of Photosynthetic Rate and Transpiration in Field Trial

Photosynthetic measurements and leaf water relation were studied with 10 genotypes i.e. Girnar 1, GG 2, ICG 3845, ICG 3793, ICG 476, ICGS 76, ICGV 86635, ICGV 86031, ICGV 86707 and TMV_2NLM. Three leaflets of the second or the third leaf from the apex of the main stem were used for the measurement of single leaf photosynthetic rate, stomatal conductance (g_s) and transpiration by using an LI 6200 portable photosynthesis system (LI-COR, Lincoln, USA). All measurements were made between 0900 and 1000 hours, for three consecutive days.

Determining Transpiration Efficiency

Transpiration efficiency was calculated based on dry matter production per unit of water evaporated in the process of transpiration and ratio of carbon discrimination (Ä) and SLA by the following formulas:

TE = K/2.59 (2.59 is seasonal vapor pressure deficit (VPD) for Junagadh).

T = Adjusted biomass/TE

C delta = (0.037 x SLA) + 14.2

K = 14.4 - (0.53 x C Δ)

B. Pot Culture Experiments

Pot culture experiments were conducted in two contrasting seasons i.e. rainy and summer using seven genotypes. Earthen pots of 10 kg capacity were filled with a mixture of farm soil (black calcareous), sand and farm yard manure (2:1:1). Eight to 10 seeds were sown in each pot. About 25 DAS, extra seedlings were removed and only four seedlings were allowed to grow in each pot. The *Virginia* types were sown 15 days in advance of *Spanish and Valencia* types to impose water-deficit stress at similar phenophases. At 60 DAS, pots were arranged in a split-plot design, with soil moisture-deficit stress as the main treatment and cultivars as the sub-treatment. One set of pots was irrigated daily (C) while the other was subjected to progressive water-deficit stress (S) by withholding irrigation. Before initiation of the stress (at 60 DAS) five plants of each cultivar were harvested to determine total biomass (BIO) and SLA. On each day of observation, two pots of each cultivar were randomly selected and leaf samples from five to six plants were taken for the study of gas exchange parameters and leaf water relations. In rainy season, the observations were recorded for 7 days i.e. 0, 1, 3, 5, 7, 9, and 11 days after imposition of stress. Whereas, in the summer season due to high evaporative demand there was rapid decrease in the soil moisture content compared to that of rainy season. Hence, during the summer season experiment the observations were recorded for only five sampling days i.e. 0, 1, 3, 5, and 7 days after initiation of stress. Soil moisture between 5 and 15 cm depths was measured gravimetrically, in each pot for five sampling days after recording the photosynthetic rate.

Measurement of Photosynthetic Rate and Relative Water Content

Three leaflets of the second or third leaf from the top of the main axis were used for various measurements following Nautiyal *et al.* (1999). All measurements were made between 0900 and 1000 h. For measurement of leaf relative water content (RWC), 2 or 3 leaves from the top of each genotype replicated three times were collected in pre-weighted glass vials. After recording the fresh weight, the leaf samples were soaked for 4 h and turgid weight was recorded to determine RWC following Barrs and Weatherley formula (1962):

$$\text{Relative water content (RWC \%)} = \frac{\text{Fresh weight} - \text{dry weight}}{\text{Turgid weight} - \text{dry weight}} \times 100$$

Coinciding with the time of recording the photosynthetic rate, soil moisture was determined gravimetrically from each pot by collecting 100 g each from a depth between 5 and 15 cm. About 10 days after withholding water, when the RWC values in most of the cultivars was less than 50%, the experiment was terminated.

C. Breeding Programme

Selection protocols

The parents were selected on the basis of performance of TE, WUE and HI. Four crosses were made involving five selected parents and one locally adapted line. The best locally adapted line was crossed with the parent possessing the drought tolerant traits as mentioned earlier in this text. Thus following crosses were made to improve WUE in groundnut cultivars:

ICGS 76 x CSMG 84-1

CGS 44 x CSMG 84-1

ICGV 86031 x TAG 24

GG 2 x ICGV 86031 (Cross of locally adopted line with the parent selected)

The progeny's developed from these crosses were selected following trait based selection (indirect) to combine high WUE, HI and TE using selection index approach as well as empirical (direct) selection for pod yield approach. A simple crop analytical model has been used to analyse pod yield variations under water limited conditions into three functional components following the framework proposed by Passioura (1977), where:

Pod yield = Water (T) Transpired X Water-Use (WUE) Efficiency X Harvest (HI) Index

Through All India Coordinated Project on Groundnut (AICRPG) breeding programme was launched at four different locations i.e. ICRISAT (Andhra Pradesh), NRCG (Gujarat), Jalgaon (Maharastra) and Tirupati (Andhra Pradesh). Breeding populations emerged from these crosses of parents selected on the basis of high expression of three physiological traits i.e. TE, WUE and HI was initially set up. Both "empirical" and "trait" selection approaches were followed on these populations. Crosses were made to combine WUE and HI and the genetic material identified based on the physiological traits for high WUE. There were four crosses proposed at each location, including three common crosses and one cross including the best adapted local line. In order to maintain consistency between "empirical" and "trait based" selection protocols, it was proposed that the empirical procedure (direct selection based on pod yield) was followed in various generations i.e. in F_2, F_3 and F_4 generations. The selection progeny in F_2, was grown out in the target environment such as water-deficit stress and irrigated conditions. At the harvest of pod yield, total biomass and physiological components in the selected line were recorded. Thus, among the total lines developed promising cultures were identified (192 lines) and evaluated at multilocation in a plot size of 3 m x 4 rows (3.6m^2) with three replicates in a split plot design. The breeding material was tested at different locations for drought tolerance under the rain-dependent conditions for three consecutive years i.e. 1999-2002.

Selection of Promising Cultures

Among the advance breeding lines obtained after making crosses at four different locations in India, the promising advanced lines (numbers 192) were studied and multiplied at Junagadh between 2004 and 2006. Following lines based on high WUE, pod yield and HI were selected in Farmers Participatory Research and Extension (FPRE) approach.

ICR 3 (ICGS 76 x CSMG 84-1), cross made at ICRISAT

ICR 4 (ICGS 44 x CSMG 84-1), cross made at ICRISAT

JAL 42 (ICGV 86031 x TAG 24), cross made at Jalgaon

TIR 17 (ICGV 86031 x TAG 24), cross made at Tirupati

JUG 16 (ICGS 44 x CSMG 84-1), cross made at Junagadh

Testing of Selected Culture in on Farm Trials

Ten groundnut genotypes including five developed for high WUE and one local check (GG 20) were tested in on farm trials in four villages in Rajkot, Jamnagar and Junagadh districts during two *kharif* seasons. At each location initially ten farmers were selected. First year trial was conducted in mother-baby design and only two farmers were given full set of genotypes and remaining was considered for baby trial (three genotypes each). The plot size of mother trial was 25 m^2, while that of baby trial was 50 m^2. In the second year, all the participatory farmers were given a full set of genotypes and sowing was performed in about 200 m^2 each genotype. At harvest, yield data was recorded following the farmers' method and plants from an area of 0.33 m^2 was sampled and pod yield and harvest index (HI) were calculated following the oven drying method. A survey was conducted in the participatory villages and farmer's opinion was obtained regarding the suitability or otherwise of the genotypes for the cultivation in rain-dependent conditions and following are the traits preferred by the farmers: drought tolerant cultivar, retention of foliage up to final harvest and should be able to give quality fodder and "stay green" characteristics, fresh-seed dormancy, tolerance to stem rot, seed size medium to bold, foliar disease resistance, high oil content, synchronized flowering and higher number of pegs within the reach to enter into the soil, high peg strength, uniform pod size and maturity, high pod and fodder yield, and prolonged seed viability and high seedling vigour.

Results

Biomass Production and DTI

Field experiments showed significant impact of water-deficit on cultivars for yield parameters and biomass production; however effect of season was non-significant, therefore the mean values of individual parameter are presented. The reduction in total dry matter was highest in CSMG 84-1 and lowest in ICG 86031. On the other, reduction in total pod weight was higher in ICG 4747 and lower in ICGV 86031 and TAG 24. Mid-season drought simulated using the ROS (T_2) resulted in a reduction of total dry matter, though lower in TMV_2NLM and ICGV 86031 and higher in CSMG 84-1. Drought tolerance index based on total dry matter (DTI-BIO) and drought tolerance index based on pod yield (DTI-PY) showed significant cultivar variation in drought tolerance capacity (Table 1). Reduction in pod yield was higher in CSMG 84-1 and least in TMV_2NLM (18% in 1993, 20% in 1995 and 18% in 1996). The HI of the cultivars was reduced significantly under both T_1 and T_3. The HI under T_1 was highest in TAG 24 and least in ICGV 86031. TAG 24 also had highest HI under all treatments (0.39-0.42) (Table 2). The DTI-BIO was related inversely ($r= -0.71$, $P<0.01$) to SLA (Fig. 1a), but such a relationship was not evident between DTI-PY and SLA. Field study showed that SLA had a significant role in imparting drought tolerance in terms of TDM productivity. A positive correlation of SLA in T_1 with that in T_2 and T_3 ($r= 0.90$, $P<0.01$) suggested that the trait has low interaction with the environment (G x E). Relationship between total biomass and pod yield could not be established in any of the treatment, therefore to increase the productivity of groundnut, partitioning and HI needs to be exploited. SLA was significantly correlated with transpiration (T), Ä and LAI (data not shown). An inverse relationship was found between Ä and SLA under water-deficit condition (Fig. 2).

Gas Exchange Parameters

The average photosynthetic rate among 10 genotypes under normal irrigation (T_1) ranged between 10.69 ì mol m^{-2} s^{-1} in Girnar 1 and 19.76 ì mol m^{-2} s^{-1} in ICG 3793, whereas photosynthetic rate decreased under water-deficit (T_2) conditions and ranged between 4.41 ì mol m^{-2} s^{-1} in ICGV 86707 and 12.59 ì mol m^{-2} s^{-1} in TMV_2NLM (Table 3). Similarly, the transpiration rate ranged between 6.6 mol m^{-2} s^{-1} in GG 2 and 9.9 mol m^{-2} s^{-1} in TMV_2NLM under normal and 3.9 mol m^{-2} s^{-1} in ICGV 86707 and 8.4 in ICGS 76 followed by 7.7 mol m^{-2} s^{-1} in TMV_2NLM, under water-deficit. Stomatal conductance was highest in TMV_2NLM (0.60 mol m^{-2} s^{-1}) under normal and least in ICGV 86707 (0.10 mol m^{-2} s^{-1}) under water-deficit. Under stress condition, conductance was highest in ICGS 76 (0.27 mol m^{-2} s^{-1}) as compared to the lowest 0.10 mol m^{-2} s^{-1} in ICGV 86707. The leaf water content (RWC) under normal irrigation was between 92 and 94% and decreased to 76.8% in ICGV 86707. However, RWC was highest (89.7%) in ICGV 86031, under water-deficit. The difference between T_{leaf} and T_{air} also increased under water-deficit.

Physiological Interactions for Leaf Water Relations

In pot-culture experiment, RWC in the leaves of well-watered plants ranged between 93 and 95%. A decrease in the RWC was noticed even on the first day following withholding of irrigation water which was least in TMV_2NLM (1%) and highest in ICG 4747 (7.8%). With progressive soil moisture decline, RWC decreased gradually, and on 7th day was lowest in CSMG 84-1 (55%) and highest in TMV_2NLM (65%), followed by ICGV 86031 (64%). At peak stress, RWC in TAG 24 and ICGV 86031 was higher than in other genotypes. The mean RWC, measured over the treatment period in low SLA types was higher than in the high SLA types (Table 3 & 4). The ratio of RWC under stress to that of control was calculated and maximum difference in the cultivar ratio was found on 9th day, and was lower in Kadiri 3 and higher in TMV_2NLM (data not presented). Average stomatal conductance (recorded during the stress period) was higher in TMV_2NLM and TAG 24 (Table 3 & 4). The cultivars g_s ratio was distinct on 3rd day and ranged between 0.76 in ICGV 86031 and 0.35 in ICG 4747. Thus low SLA types under moisture-deficit conditions were able to maintain higher g_s than the high SLA type. Photosynthetic rates were found between 9.49 ì mol m^{-2} s^{-1}in ICGS 4747 ì mol m^{-2} s^{-1} and 13.93 in TMV_2NLM in control plants. On the first day of stress, the relative decrease in the photosynthetic rate was highest in ICG 4747 (23%). On the third day, reduction in the photosynthetic rate in general was around 25%, but in ICGV 86031 it was 15% only. At peak stress (11th day), most of the cultivar showed about 80% reduction in the photosynthetic rate, and average RWC was 40%. Average photosynthetic rate (measured over the stress period) was higher in the low SLA types, i.e. ICGV 86031 and TMV_2NLM than high SLA type, when the soil moisture was around 10%.

Relationship between RWC and SLA

In pot culture experiments, the RWC and SLA showed significant cultivar and seasonal variation in response to prevailing climatic condition and water-deficit. For example, SLA of the cultivar was high in the rainy season than the summer, though the cultivar ranking were maintained (Tables 3 & 4) (r=

0.90, P=0.01). In this experiment, higher VPD and temperature not only influenced the rate of development of stress but also the SLA values. SLA in general, was higher in the rainy season than in the summer season. Most of the variation, however, was in the relationship between SLA and RWC between experiments was nullified when the SLA values were multiplied with the mean VPD values for the respective seasons (Fig 3a & b). Cultivar difference in maintaining RWC under stress were quantified for individual cultivar by regressing the RWC against the soil moisture content measured on the daily bases. This analysis showed significant cultivar variation in RWC slope (Fig 3a & b). Various correlations between SLA and leaf water relations under normal stressed conditions was worked out (Table 5).

Hybridization Programme and Genetic Analysis

Crosses were made among the selected parents to generate the segregating population, which were further tested for WUE, TE and HI. Development of drought tolerant groundnut genotype with improved selection methods was performed at 4 different locations. The genetic material developed this way was made freely available to the various ACIRPG centers for testing and adoption of the cultivars by the farmers. Improved selection and technologies utilizing the trait-based approach in groundnut breeding programme was developed at each location. The genetic performance of selection methods *i.e.* empirical and trait based was done for various characteristics such as vegetative oven dry weight (veg. kg. ha), pod oven dry weight (pod. kg. ha), kernel oven dry weight (ker. kg. ha), SPAD readings, kernel harvest index (kernel HI), transpiration (T in mm) and transpiration efficiency (TE gm kg^{-1}) at Junagadh centre for the rainy and summer seasons is shown in tables 6 & 7. The best five genotypes for high WUE, transpiration efficiency (TE) and harvest index (HI) are shown in table 8.

Testing of High WUE Lines at Junagadh Location

Selection of the genotypes developed for high WUE was conducted at the Centre. One-hundred and seventy-six advanced breeding lines developed in "ACIAR-ICAR-ICRISAT" collaborative project on selection and breeding for WUE were revisited for WUE based on SLA, HI and total pod yield (Table 9). Finally, five lines i.e. ICR 3, ICR 4, TIR 17, JAL 42 and JUG 16 were selected from the material multiplied and evaluated at NRCG. Pod yield in genotypes varied with seasons and in general was higher in summer, but all the five genotypes selected gave much higher yield than the local check in both the seasons.

Table 1. Specific leaf area, adjusted pod yield (adj-pod), adjusted biomass (adj-bio), HI, DTI-PY and DTI-BIO of groundnut cultivars under control (T_1) and rain-dependent (T_3) conditions in two rainy seasons.

Variety	SLA (cm^2 g^{-1})		Adjusted pod weight (g m^{-2})		Adjusted biomass (g m^{-2})		HI		DTI-PY	DTI-BIO
	T_1	T_3	T_1	T_3	T_1	T_3	T_1	T_3		
ICGV 86031	155	169.5	314.85	264.75	877.6	782.85	0.24	0.22	0.84	0.9
TMV_2NLM	147.5	140.5	230.35	191.9	661.2	576.8	0.265	0.265	0.825	0.87
TAG 24	165.5	158.5	415.55	343.5	666.1	577.65	0.43	0.415	0.83	0.865
CSMG 84-1	197	183	300.45	244.95	791.15	626.3	0.28	0.28	0.815	0.785
ICG4747	177	177	301.5	206.45	949.25	793.2	0.195	0.16	0.685	0.83
Kadiri 3	183.5	167	344.1	274.4	968.2	795.95	0.225	0.235	0.85	0.825
ICG 476	227.5	201	293.95	235.15	602.6	502	0.355	0.32	0.795	0.825
S.E.										
Variety (v)	7.185		16.315		28.14		--			
Treatment (t)	3.835		8.715		15.04		--			

Source: Nautiyal *et. al.*, 2002, *Field Crops Research*, 74: 67-79.

Table 2. Specific leaf area, adjusted biomass, adjusted pod yield, HI, DTI-PY and DTI-BIO of groundnut cultivars under control (T_1), rain-out shelter (T_2), and rain-dependent (T_3) conditions in two rainy seasons.

Treatment	Variety ICGV 86031	TMV_2NLM	TAG 24	CSMG 841	ICG 476
SLA ($cm^2 g^{-1}$)					
T_1[a]	155	147.5	165	197	227.5
T_2[b]	142.5	137	157	189	201.5
T_3[c]	151.5	140.5	158.5	183	194
S.E.	Variety	5.71	Treatment	2.58	
Adjusted pod weight ($g m^{-2}$)					
T_1	314.8	229.85	265.5	300.45	293.95
T_2	229	185.65	284.45	186.65	217.2
T_3	264.75	191.25	343.6	244.95	235.15
S.E.	Variety	10.36	Treatment	8.02	
Adjusted biomass ($g m^{-2}$)					
T_1	792.8	661.2	666.1	791.15	602.6
T_2	676.65	572.9	534.55	579.25	465.1
T_3	716.15	576.8	577.65	626.3	502
S.E.	Variety	18.13	Treatment	14.04	
HI					
T_1	0.24	0.265	0.43	0.27	0.355
T_2	0.21	0.25	0.4	0.23	0.335
T_3	0.22	0.265	0.415	0.285	0.32
S.E.	Variety	0.09	Treatment	0.007	
DTI-PY					
T_2	0.73	0.805	0.685	0.72	0.715
T_3	0.84	0.825	0.83	0.815	0.795
DTI_BIO					
T_2	0.835	0.865	0.8	0.725	0.765
T_3	0.89	0.835	0.79	0.765	0.84

a: Control, irrigation given to replenish 100% pan evaporation at 5-day intervals

b: Drought simulated between 40 and 80 DAS, crop received irrigation equivalent to 25% of T_1.

c: Rain-dependent

Source: Nautiyal *et. al.*, 2002. *Field Crops Research*, 74: 67-79.

Table 3. Photosynthetic rate (m mol m^{-2} s^{-1}), transpiration (mol m^{-2} s^{-1}), stomatal conductance (mol m^{-2} s^{-1}), relative water content (%) and $T_{leaf} - T_{air}$ (°C) of different genotypes grown under field conditions in rainy season.

Genotypes	Photosynthetic rate (μ mol m^{2} s^{-1})		Transpiration (mol m^{2} s^{-1})		Stomatal conductance (mol m^{2} s^{-1})		Relative water content (%)		$T_{leaf} - T_{air}$ (°C)	
	T_1	T_2	T_1	T_2	T_1	T_2	T_1	T_2	T_1	T_2
Girnar 1	10.69	9.07	8.5	6.9	8.5	6.9	94.5	87.6	1.47	1.89
GG 2	11.84	5.68	6.6	4	6.6	4	92.7	81.5	1.88	2.44
ICG 3845	13.65	6.98	7.8	4.9	7.8	4.9	93.5	87.7	0.65	1.54
ICG 3793	19.76	8.35	9.2	6.8	9.2	6.8	93.4	89.2	0.86	1.34
ICG 476	11.66	11.16	7.1	6.1	7.1	6.1	92.1	87.4	1.86	1.89
ICGS 76	12.27	11.27	8.4	7.9	8.4	7.9	92.4	83.4	1.34	1.3
ICGV 86635	16.79	6.89	7.5	4.8	7.5	4.8	93.5	85.4	0.6	0.13
ICGV 86031	17.93	9.18	7.7	5	7.7	5	93.6	89.7	0.29	2.2
ICGV 86707	17.59	4.41	8.8	3.9	8.8	3.9	93.6	76.8	0.65	2.4
TMV2NLM	18.1	12.59	9.9	7.7	9.9	7.7	94	82.7	0.76	1.6
Mean	15.02	8.6	8.6	5.8	8.2	5.8	93.3	85.1	1.07	1.7
CD (P=0.05)	2.91		0.2		0.12		4.4		0.11	

Source: Nautiyal *et. al.*, 2002, *Field Crops Research*, 74: 67-79.

Table 4. Photosynthetic rate (ĩ mol m^{-2} s^{-1}), stomatal conductance[a] (g_s, mol m^{-2} s^{-1}) and leaf RWC (%) of groundnut cultivars with a range of SLA (cm^2 g-1)[b] under control (C) and progressive soil moisture-deficit stress (S) simulated in earthen pots in summer season.

Parameters	ICGV 86031	TMV_2NLM	TAG 24	CSMG 84-1	ICG 4747	Kadiri 3	ICG 476	SE
Rainy season								
SLA	140	144	163	185	193	205	210	1.15
Photosynthetic rate								
C[c]	12.8	13.9	11.6	10.8	8.5	13.4	11.1	0.28
S[d]	7	7.9	6.5	6.1	5.2	6.3	5.9	0.12
g_s								
C	0.36	0.36	0.38	0.33	0.21	0.31	0.39	0.003
S	0.17	0.24	0.19	0.15	0.12	0.15	0.19	0.04
RWC								
C	93.8	91.23	91.6	93.4	92.6	91.85	92.1	0.78
S	73.1	71.4	73.3	70	67.2	65.1	68.4	0.27
Summer season								
SLA	116	120	142	150	157	172	170	0.86
Photosynthetic rate								
C	12.7	12.2	14.9	14.8	10.8	12.42	11.6	0.41
S	6.0	6.1	6.4	5.0	5.8	5.5	5.4	0.15
g_s								
C	0.33	0.31	0.33	0.35	0.28	0.29	0.27	0.001
S	0.16	0.13	0.16	0.11	0.10	0.11	0.13	0.01
RWC								
C	94.7	93.3	93.5	94.8	93.2	93.0	92.7	1.11
S	68.3	64.9	63.3	62.2	56.0	56.4	60.3	1.52

a : Values are mean of the observations recorded for 7 days in the rainy seasons.
b: Values are mean of 5 plants at 60 DAS, ie before starting the stress treatment.
c: Control, plant received irrigation daily.
d: Soil moisture-deficit stress during observation period.
Source: Nautiyal *et. al.*, 2002, *Field Crops Research*, 74: 67-79.

Table 5. Correlation between values of SLA and each of RWC, photosynthetic rate and gs in groundnut cultivars grown under irrigated and water-deficit stress conditions.

Pair of parameters	Coefficient correlation, *r*	
	Stress	Irrigated
SLA and RWC		
Rainy season	-0.844*	0.084 NS
Summer season	-0.872*	0.559 NS
SLA and photosynthetic rate		
Rainy season	-0.780*	-0.454 NS
Summer season	-0.574 NS	-0.207 NS
SLA and g_s		
Rainy season	-0.525 NS	-0.316 NS
Summer season	-0.603 NS	-0.586 NS

Source: Nautiyal *et al.* (2002). *Field Crops Research*, 74: 67-79.

Farmer's Participatory Research and Extension (FPRE) Approach

In the year 2006, promising five advanced breeding lines i.e. ICR 3, ICR 4, JUG 16, JAL 42 and TIR 17 out of 176 genotypes grown and tested at Junagadh were selected for FPRE approach (Table 10). Among the genotypes, pod yield varied in different seasons and in general was higher in summer; however in all selected genotypes, pod yield was much higher than the local check. In FPRE approach, all the 5 genotypes along with local check (GG 20) were grown at four locations in two different districts of Gujarat during two rainy seasons. Data collected from all the four locations showed that pod yield in these genotypes was higher than that of the local check (GG 20) (Table 10). Genotype ICR 3 yielded 1610 kg ha^{-1}, which is approximately double of the yield of local check. The performance of ICR 3 was best at all the locations followed by JUG 16 and TIR 17. Farmer's preference was also for the genotype ICR 3, because in addition to the pod yield it possesses fresh-seed dormancy and also good amount of fodder. Performance of the genotypes at research station during two rainy seasons is shown in table 11.

Table 6. Genetic performance of selection methods in various traits at Junagadh location during summer season

Variable	Vegetative oven dried weight (kg hactare^{-1})	Pod oven dried weight (kg hactare^{-1})	Kernel oven dried weight (kg hactare^{-1})	Kernel HI	Spad	Transpiration (T) (mm)	Transpiration efficiency (TE) (gm kg^{-1})	Specific leaf area
σ_G^2 (emp)	2013367	409858	201914	0.002	3.55	10202	0.002	49.24
σ_E^2 (emp)	734882	219473	103597	0.0007	4.60	8145	0.008	71.39
h^2 (emp)	0.89	0.85	0.85	0.907	0.70	0.79	0.45	0.67
R (emp)	1340	589.7	415.2	0.048	1.58	89.76	0.03	5.76
σ_G^2 (trt)	1797698	514866	251083	0.003	4.11	12196	0.003	55.74
σ_E^2 (trt)	582565	135840	69850	0.0006	2.77	117399	0.011	87.93
h^2 (trt)	0.90	0.92	0.92	0.94	0.87	0.24	0.43	0.66
R (trt)	1274	687.9	479.3	0.05	1.83	53.83	0.035	6.04
RE = R (trt)/R (emp) (relative efficiency)	0.95	1.17	1.12	1.11	1.16	0.60	1.13	1.05
Kolmogorov-Smirnov D-statistic	0.14	0.13	0.15	0.13	0.17	0.49	0.10	0.10
Chi square calc. value	3.52	3	4.08	3	5.33	46.02	2.09	2.08
Chi square prob (prob. Of a value greater than chi sq. calculated)	0.17	0.22	0.13	0.22	0.07	<0.001	0.35	0.35

emp: Empirial selection method
trt: Trait based selection method
Source: Annual report ACIAR, project number CS1/97/114 "More efficient breeding of drought resistant peanuts in India and Australia"

Table 7. Genetic performance of selection methods in various traits at Junagadh location during rainy season

Variable	Vegetative oven dried weight (kg hactare^{-1})	Pod oven dried weight (kg hactare^{-1})	Kernel oven dried weight (kg hactare^{-1})	Kernel HI	Spad	Transpiration (T) (mm)	Transpiration efficiency (TE) (gm kg^{-1})	Specific leaf area
σ_G^2 (emp)	650002	91146	62339	0.002	2.29	2480	0.04	45.67
σ_E^2 (emp)	737307	309093	185035	0.002	3.20	1806	0.006	43.32
h^2 (emp)	0.73	0.47	0.50	0.71	0.68	0.80	0.96	0.76
R (emp)	686.8	206.8	177	0.04	1.25	44.67	0.20	5.89
σ_G^2 (trt)	1342133	112723	82339	0.003	1.77	4354	0.42	28.16
σ_E^2 (trt)	825936	256252	145361	0.001	3.06	2575	0.007	42.89
h^2 (trt)	0.83	0.57	0.63	0.87	0.63	0.84	0.95	0.66
R (trt)	1055	253.2	227.7	0.05	1.06	60.3	0.20	4.32
RE = R (trt)/R (emp) (relative efficiency)	1.54	1.224	1.29	1.4	0.85	1.35	0.10	0.73
Kolmogorov-Smirnov D-statistic	0.16	0.14	0.13	0.17	0.30	0.14	0.11	0.22
Chi square calc. value	4.69	3.52	3	5.33	17.52	3.52	2.52	9.19
Chi square prob (prob. Of a value greater than chi sq. calculated)	0.10	0.17	0.22	0.07	<0.001	0.17	0.28	0.01

emp: Empirial selection method
trt: Trait based selection method
Source: Annual Report ACIAR, project number CS1/97/114 "More efficient breeding of drought resistant peanuts in India and Australia"

Table 8. Performance of best five genotypes for specific leaf area (SLA), transpiration efficiency (TE) and harvest index (HI)

SLA		TE		HI	
High	Low	High	Low	High	Low
ICG 476	TMV_2NLM	TMV_2NLM	ICG 8048	TAG 24	ICG 4230
ICG 221	ICGS 76	ICGS 76	ICGV 86742	TG 26	ICG 5263
JL 220	ICGV 86031	ICGV 86031	ICG 3518	ICG 476	ICG 4747
TG 3	DRG 102	DRG 102	ICGV 86635	TG 17	DRG 103
ICG 3098	ICG 4886	ICG 4446	ICG 3098	GG 2	ICG 3845
ICG 3143	ICG 3109	Somnath	TG 3	ICGV 86031	ICG 221

Table 9. Performance of advanced breeding material at research station (NRCG, Junagadh) during rainy (2004 & 2005) and summer (2006) seasons.

Genotype	SLA ($cm\ g^{-1}$)	Shelling Outturn (%)	100 seed weight (g)	Harvest index	Pod yield ($kg\ ha^{-1}$)
Kharif					
ICR 3	164	66	48	0.29	1950
ICR 4	164	65	44	0.36	1620
JUG 16	158	60	39	0.30	2690
TIR 17	133	67	35	0.42	1270
JAL 42	159	66	41	0.45	2000
GG 2 (local check)	155	67	42	0.20	920
Summer					
ICR 3	150	72	51	0.31	2560
ICR 4	143	68	47	0.36	3500
JUG 16	160	71	43	0.31	3160
TIR 17	132	71	36	0.42	3360
JAL 42	156	59	41	0.44	3120
GG 2 (local check)	151	69	34	0.22	2010

Table 10. Yield attributes and farmer's preference of the genotype evaluated in 5 different locations (villages) in Saurastra during two rainy seasons.

Genotype	Number of pod plant^{-1}	Yield plant^{-1} (g)	Yield Kg ha^{-1} (Based on yield /m^2)	Yield Kg ha^{-1} (Based on yield /plot)	HI	Shelling (%)	100 Seed Wt. (g)	Ranking based on Yield	farmers preference
ICR 3	13	11.9	2647	1958	0.36	71	47	1	1
JBDR 64	11	9.2	2038	1427	0.31	68	40	5	5
ICGV 87846	11	9.9	2194	1584	0.30	68	39	4	3
JAL 42	11	10.5	2322	1676	0.36	70	45	3	2
JBDR 7	10	8.5	1888	1298	0.31	69	40	10	8
ICR 4	12	9.4	2086	1394	0.35	69	39	6	9
ICGV 02266	9	7.4	1636	1053	0.28	67	38	11	11
TIR 17	11	9.0	2003	1325	0.31	67	41	8	7
JUG 45	12	9.0	1995	1388	0.33	68	37	7	10
JUG 16	14	11.2	2497	1688	0.37	72	40	2	4
GG 20 (L)	10	8.7	1925	1303	0.31	69	46	9	6

Table 11. Days to flowering and yield attributes of 10 genotypes grown at experimental site during rainy season

Genotype	Days to first flowering	Days to 50% flowering	SLA (cm^2 g^{-1})	HI (%)	Pod Yield (kg ha^{-1})	Shelling (%)	Test weight (g)
ICR 3	25	30	177	0.41	1082	72	36
ICR 4	23	25	169	0.32	978	67	31
JBDR 7	20	23	209	0.16	761	60	29
JBDR 64	20	25	209	0.29	938	64	28
JAL 42	20	23	214	0.40	1471	62	29
ICGV 87846	23	30	187	0.26	1241	65	32
TIR 17	22	26	179	0.25	916	72	35
JUG 16	21	30	174	0.36	1077	69	29
JUG 40	21	23	244	0.46	944	59	24
GG 20	22	27	194	0.32	794	70	38
Mean	21.7	26.2	195.7	0.32	1020.1	65.9	31.1
CD	2.2	1.9	11.9	0.10	412.5	3.9	3.7
CV %	6.0	4.2	3.5	17.21	23.5	3.4	7.0
Maximum	25	30	244	0.46	1471	72	38
Minimum	20	23	169	0.16	761	59	24

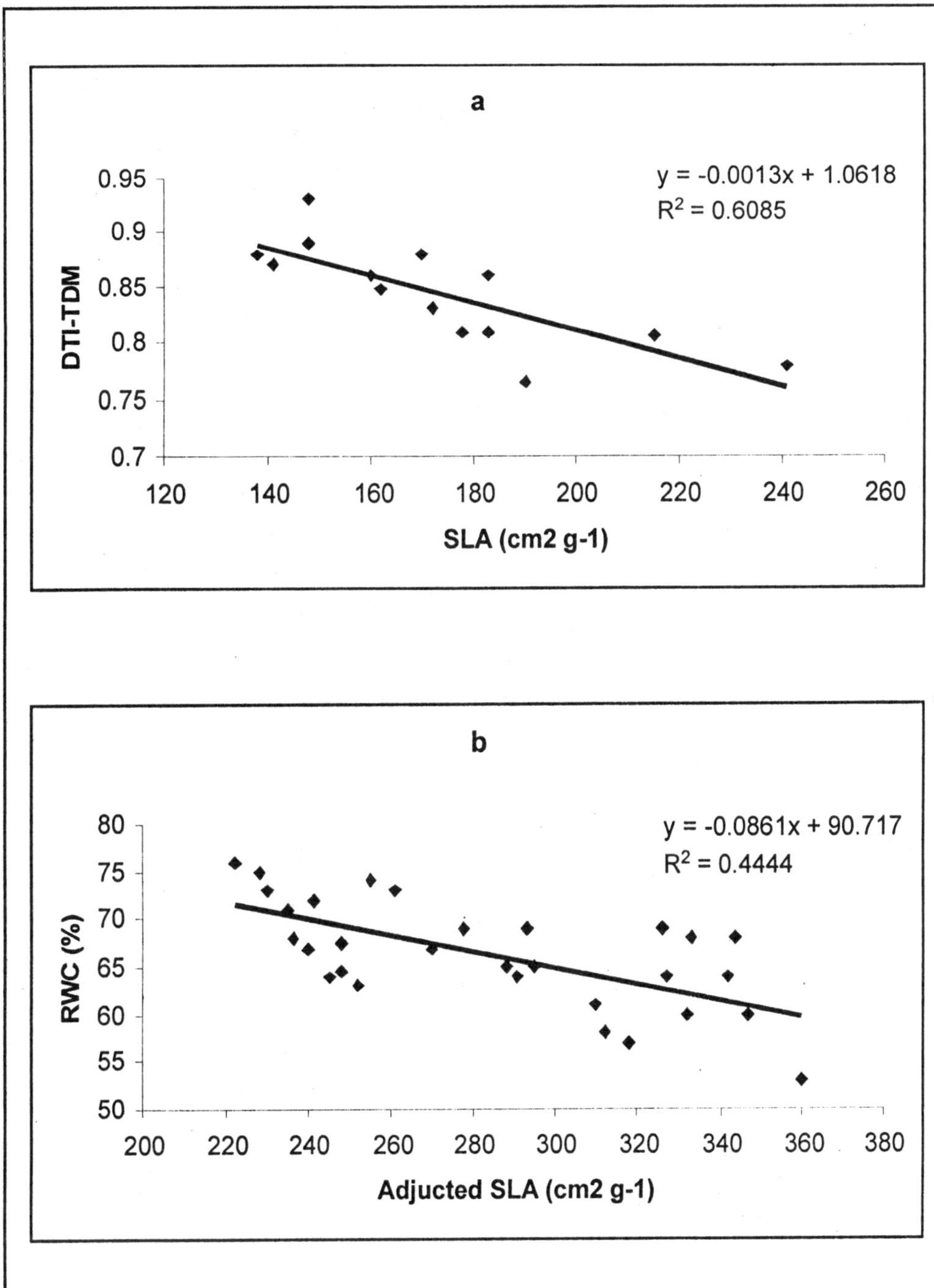

Figure 1: Relationship between drought tolerance index based on biomass (DTI-BIO) and SLA in groundnut cultivars grown in simulated drought conditions under rain-out-shelters (ROS).

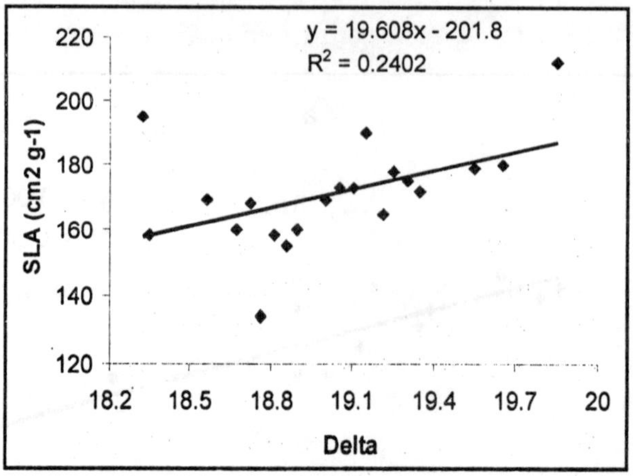

Figure 2: Relationship between SLA and carbon isotope discrimination in groundnut genotypes.

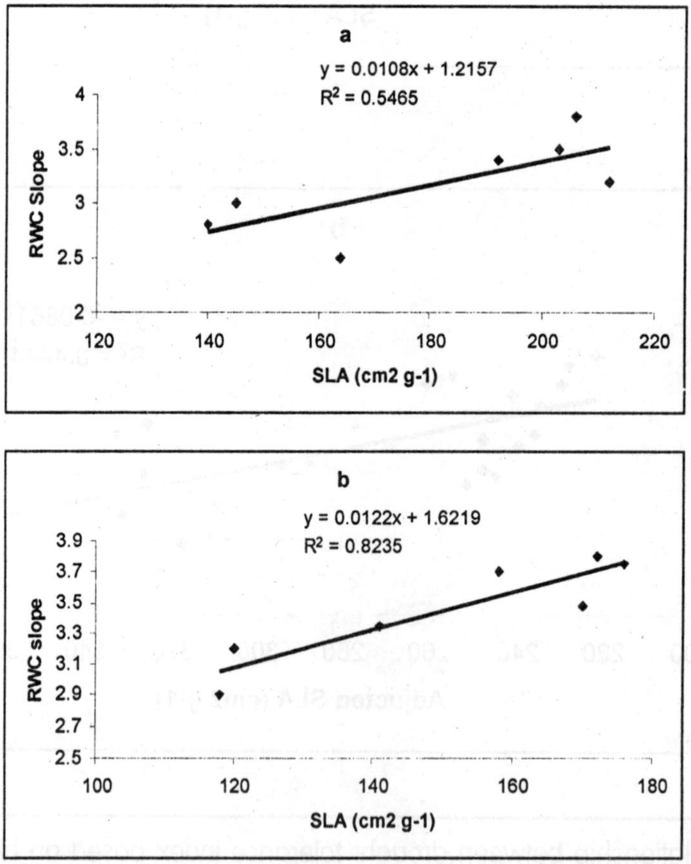

Figure 3. Relationship between the slopes of RWC, calculated with soil moisture during progressive stress and SLA (a=rainy season; b= summer season).

Discussion

It is suggested that in crop improvement there is an evidence of increasing the marginal returns from conventional breeding and it is high time to seek more efficient breeding methods (Fehr 1994, Hammer *et al.*, 1998). Therefore, this study was focused on improving WUE or drought tolerance in groundnut. It is now a proven fact that WUE could be enhanced by trait based breeding by performing selections in segregating population for WUE based on SLA (thicker leaves). It is also possible to combine the WUE and HI together to increase productivity. In this study, initially large numbers of germplasm lines were screened for WUE and HI. The selected high WUE and high HI lines were used in trait based as well as in conventional breeding programme. The groundnut genotypes developed for high WUE were tested in on farm trials for drought tolerance in Gujarat and found promising for the cultivation in rain-dependent system. The approach followed was to work out the relationship between Ä and SLA to use SLA as surrogate of Ä, a convenient and easy method for selection for WUE and the inverse relationship between SLA and Ä allowed us to use SLA as a surrogate for WUE. Therefore, the segregating populations were screened for WUE taking low SLA as trait for drought tolerance or high WUE. The relationship between WUE and Ä was initially established by Hubick *et al.* (1986) and later on Nageswara Rao and Wright (1993) used SLA as surrogate to Ä and suggested its use as selection tool. The WUE and Ä has also been studied by Wright *et al.* (1994) and suggested that WUE as measured following mini-lysimeters allowed accurate measurement. In addition, they found a strong negative relationship between WUE and SLA and between Ä and SLA. The genetic variation in Ä was also attributed to Rubisco content per unit leaf area (Nageswara Rao *et al.* 1995). Thus the cultivar variations in biomass production, pod yield, HI and DTI under the same soil moisture-deficit suggested there is intrinsic variation in water uptake and its utilization in production of biomass and partitioning to reproductive parts.

Further, a significant relationship between SLA and DTI-BIO ($r=-0.71$, $P=<0.01$) showed the importance of WUE in water-deficit conditions. Also the relationship between SLA in control and stressed plants ($r= 0.90$, $P= 0.01$), indicated a low G x E interaction for this trait. The low G x E interaction is also reported in the literature (Nageswara Rao and Wright, 1993). The correlations between SLA and leaf water relations and gas exchange parameters showed that SLA has an important role to play under water-deficit conditions. The relationship between Ä and WUE for those plants which continue to grow under adequate moisture from those, which were subjected to progressive water-deficit, differed. This is also evident from the values of correlation coefficient that under adequate moisture the SLA of cultivar does not have any relationship with their RWC, photosynthetic rate and conductance. However, relationship between SLA and RWC acquires significance if the plants are subjected to water stress. Thus a low SLA type may not acquire any benefit under adequate moisture but is likely to perform better than high SLA types under water-deficit. Results on the field experiment showed significant cultivar variation in photosynthesis per unit area and conductance, indicating that the low SLA types i.e. ICGV 86031 and TMV_2LMN maintain higher photosynthetic rate at comparable rates of conductance under stress. Similar cultivar and seasonal variation for photosynthetic rate has been reported in groundnut (Bhagsari and Brown, 1976; Pallas, 1982, Ravindra *et al.* 1995; Nautiyal *et al.* 1999). It is presumed that among the low and high SLA types, variation in leaf water retention capacity and photosynthetic rate may be due to the differences in thickness of palisade and spongy parenchyma layers, and the capacity of water storage cells (Katring *et al.* 1982 and Nautiyal, unpublished data). Under water deficit, reduced leaflet area with corresponding increase in the stomatal frequency and number of stomata on the abaxial surface showed relationship with high WUE in cultivar TAG 24 and Somnath (Badigannavar *et al*, 1999). Earlier studies have also shown a correlation with low SLA with higher carboxylation capacity (Bowes *et al*, 1972). Thus the higher photosynthetic capacity is achieved by comparable or slightly higher stomatal conductance (Wong *et al.*, 1979). This seems to be characteristics of low SLA types. The highest stomatal density as recorded in ICR 3 may vindicate the above mentioned point and possess high WUE or drought tolerance, which has also been proved in the on farm trials and preferred by the farmers for the cultivation in rain-dependent situations. The higher stomatal density on both abaxial and adaxial surfaces could have allowed the maintenance of lower leaf temperature at high irradiance resulting in a lower vapour pressure deficit and lower transpiration ratio (Table 12). Since, the photosynthetic rate had been shown to depend on thickness of leaves and the size of the mesophyll cells (Wilson and Cooper, 1967), it is also possible that lower SLA types might result in higher C_i (internal CO_2 concentration) which leads to higher WUE or drought tolerance. A lower C_i arises from a greater fractional contribution of stomatal resistance to total resistance in photosynthesis i.e. greater decrease in transpiration rate than photosynthesis (Noble, 1991). The overall relationship obtained in the field and pot culture experiments between adjusted SLA and RWC confirmed the consistency in the above relationship when SLA was corrected for the prevailing VPD effects. Under water-deficit, the ratio of stress and control observations (S/C) has been suggested as the best indicator of the capacity of a cultivar to maintain hydration under drying soil conditions (Katring, 1985). The rate of reduction in RWC per unit reduction in soil moisture percent (RWC slope) indicates the ability of a cultivar to maintain leaf-water status to support metabolic activities as well as to maintain favorable leaf temperatures. Drought tolerance in groundnut cultivars is characterized by the maintenance of relatively higher RWC under water-deficit conditions (Joshi *et al.*, 1988; Ravindra *et al.*, 1989; Nautiyal *et al.*, 1995). The regression between SLA and RWC slopes provides conclusive evidence for the role of SLA in maintaining leaf-water status during stress as well as significant cultivar variation for this trait. The physiological basis of the relationship between drought tolerance and SLA is still not very clear and needs further studies, especially in relation to the superiority of the low SLA type in maintaining higher RWC under soil water-deficit. Traits like low SLA (high DTI) and high HI may be combined by making crosses between cultivars with high HI such as TAG 24 and high DTI such as ICGV 86031, and the selection for both the traits resulted in cultivars with improved performance under rain-dependent conditions such as ICR 3. Though, a negative association between low SLA or WUE and HI has been reported in groundnut (Wright *et al.*, 1991 and 1996), but it seems to be possible to combine the high WUE and high HI in one genotype following trait based approach. The population thus developed provided good opportunity in the breeding cycle by the direct and indirect selection approaches followed by in the multilocation trials. So far, most of the breeding programmes for improved yield of grain legume rely on empirical selection for superior yield and quality across the wide target environment. Thus crop physiology can contribute to the science by assisting in quantification and intimate exploitation of yield determining process (Whan *et al.*, 1991, 1993; Turner and Takeda, 1993). There has been a considerable amount of literature on particular physiological attribute that could be used as direct selection criteria to improve the efficiency of selection of superior type in large scale breeding programmes (Ludlow and Muchow, 1990). For example, it has been well documented that in cereal crops the genetic improvement in the grain yield has been through increase in harvest index (HI). In the present investigation where indirect traits derived from the physiological model were used in breeding programmes after ascertaining the information on physiological parameters related with drought tolerance. However, the heritability, the extent of genetic variation for yield and components of traits and genetic correlation between component traits and yield are required to be studied for further improvement in drought tolerance breeding. The analysis of heritability of component trait i.e. T, WE and HI showed in broad sense that the traits based selection is relatively better as compared to the yield alone. Gene action for surrogate traits of WUE and HI in groundnut was analysed and material effects have been reported for SLA and Ä indicating a crucial role of female parent in improvement of these traits. Variation in WUE can occur due to both environment and genetic factors. From the genetic point of view, it has been shown that decreases in *pi/pa* (the inter-cellular and atmospheric vapour pressures for CO_2) at the leaf and canopy level will increase WUE at a given vapour pressure deficit (VPD) between the leaf and air. Significant genetic variation among groundnut genotype in *pi/pa,* and hence TE, has been reported in the literature (Hubick *et al.*, 1986; 1988). In groundnut, it appears that increase in leaf assimilation rate (A) relative to g_s, which cause *pi* to fall and TE to increase, are responsible for the observed genotypic variation (Hubick *et al.*, 1988; Wright *et al.*, 1988; 1994).

Thus it seems that WUE can strongly influenced by environment change in VPD between the air and leaf. For instance growing crop in semi-arid condition in high VPD will substantially reduce WUE. Thus the same genotype grown in the environment in the contrasting VPD will have substantially different WUE values (Wright and Nageswara Rao, personal communication). Tanner and Sinclair (1983) introduced a simple concept to enable comparison of WUE among species and cultivars, independently of VPD. They stated that WUE was inversely proportional to the average VPD, with k being the constant of proportionality, i.e. WUE = k/VPD. Thus, using this simple analysis, it is possible to use k to compare WUE, independently of VPD, and hence identifying genetic variation without confounding the effects due to environment. The second important trait i.e. HI can be easily estimated in the conventional breeding programs by the additional measurement of total dry matter (TDM) at maturity. These estimates are reliable when leaf material is retained until maturity, however there can be problems in groundnut, in case there is severe drought stress or foliar disease causing the leaf fall before the maturity of the crop. Unfortunately, T is currently much more difficult to measure/estimate than either TE or HI. While total evapotranspiration (ET) can be quantified using gravimetric techniques under both glasshouse and field conditions, and by other technologies such as neutron probe and Time Domain Reflectrometry (TDR) in the field, there are still many problems in accurately apportioning water loss due to crop water use (i.e. T) and soil evaporation (Turner, 1986). There are practical problems in screening large numbers of genotypes in a breeding program using these techniques; therefore measuring the WUE by the SLA is economic and easily measurable technique, which could be followed in large number of germplasm accession or the segregating breeding populations.

Table 12. Stomatal density on the abaxial and adaxial surfaces of the leaves of ten groundnut genotypes.

Genotypes	Abaxial (stomata/mm^2)	Adaxil (stomata/mm^2)	Mean
ICR 3	569	457	513
JBDR 64	313	351	332
ICGV 87846	260	324	292
JAL 42	313	303	308
JBDR 7	276	324	300
ICR 4	244	281	263
TIR 17	415	335	375
JUG 45	452	346	399
JUG 16	362	271	317
GG 20	441	394	418

Further research is required to investigate the extent and nature of this association to accelerate the rate of progress in yield enhancement in breeding programmes. Variation in genetic differences in physiological traits measured showed that there is substantial difficulty in measuring them in large number of genotypes and segregating populations (Nautiyal *et al.*, 2002). Therefore, a measure out come of this study has been the development of cheap and rapid, easily measured surrogate measures for each of these traits and allowing their potential qualification in large number of breeding populations. Similarly, a number of germplasm line which posses both high WUE and high HI (hence higher pod yield), indicated that the previous measured association between these traits were probably genetically linked and could be easily broken by breeding.

Conclusion & Perspectives

Groundnut yield can be increased by combining WUE and HI. It is clear that the conventional breeding programme followed so far has narrowed down the genetic base. Thus there is urgent need to widen the genetic base and incorporate drought tolerant traits in cultivated groundnut. Our current research thrust is to evaluate germplasm for morphological, physiological and molecular characters associated with drought and drought-induced high temperature stresses. The major objectives of this programme are: i.) to study genetic diversity in root architecture and its association with drought tolerance; ii.) to study canopy function and its interaction drought both in terms of physiological and biochemical metabolisms such as leaf water relations, osmotic adjustments, efficiency of photosystem (PS II), partitioning of photosynthates *vis-a-vis* trait-trait interaction; iii.) role of heat-shock proteins (sHSP) in drought-induced heat stress, and iv.) identification of potential molecular markers (SSR and ESTs) related to drought and heat stresses. Thus identification of suitable genotype for use in subsequent breeding programme and to initiate crosses between identified susceptible and tolerant genotypes to make a platform for studies on QTLs associated with drought and drought-induced heat stresses. In addition, the genetic diversity of major traits may allow conventional and marker-assisted breeding approaches to be refined. Identification of combination of key traits in pre-breeding lines could be a useful tool in molecular markers assisted selection (MAS) in trait based breeding programmes.

Acknowledgements

This work includes the research conducted in the collaborative project "ACIAR-ICAR-ICRISAT" WUE (PN 1296) i.e. screening of groundnut germplasm for WUE and in the second phase breeding for high WUE was initiated under the project "PN CS1/97/114". The breeding material developed for high WUE was tested in on farm trial in the project "Farmers-participatory groundnut improvement in rainfed cropping systems" sponsored by the Ministry of Agriculture, Govt. of India under ISOPOM scheme. Therefore, financial assistance and technical know-how rendered by these agencies such as ACIAR, QDPI (Australia), ICRISAT (India) and ISOPOM (India) is acknowledged.

REFERENCES

Badigannavar, A. M., Kale, D. M., Bhagwat, S. G. and Murthy, G.S.S. (1999). Genotypic and seasonal variation in stomatal characteristics in Trombay groundnut varieties. *Trop. Agric. Res. Ext.* 2: 10-12.

Barrs, H.D. and Weatherley, P.E. (1962). A re-examination of the relative turgidity technique for estimating water deficit in leaves. *Aust. J. Biol. Sci.* 15:413-428.

Bhagsari, A. S. and Brown, R. H. (1976). Photosynthesis of peanut (*Arachis*) genotypes. *Peanut Sci.* 3: 1-9.

Bowes, G., Orgen, W. L. and Hageman, R. H. (1972). Light saturation, photosynthesis rate, RUBP carboxylase activity and specific leaf weight in soybeans grown under different light intensities. *Crop Sci.* 12, 77-79.

Farquhar, G. D., O'Leary, M. H. and Berry, J.A. (1982). On the relationship between isotope discrimination and the intercellular carbon dioxide concentration in leaves. *Aust. J. Plant Physiol.* 9:21-137.

Fehr, W. R. (1994). In: Genetic Contributions to yield grains of five major crop plants (ed). CSSA special publication No. 7. Madison, Wisconsin, USA.

Gutschik, V. P. (1988). Optimization of specific leaf mass, internal CO_2 concentration and chlorophyll content in crop canopies. Plant Physiol. *Biochem.* 26: 525-537.

Hammer, G. L., Butler, D. G., Muchow, R. C. and Meinke, H. (1998). Integrating physiological understanding and plant breeding via crop modeling and optimization. In Plant Adaptation and Crop Improvement (eds. M. Cooper and G. L. Hammer), *pp* 419-440.

Hubick, K.T., Farquhar, G.D. and Shorter, R. (1986). Correlation between water-use efficiency and corbon isotope discrimination in diverse peanut (*Arachis*) germplasm. *Aust J. Plant Physiol.* 13: 803-816.

Hubick, K.T., Shorter, R. and Farquhar, G.D. (1988). Heritability and genotype x environment interaction of carbon isotope discrimination and transpiration efficiency in peanut (*Arachis hupogaea* L.). *Aust J. Plant Physiol.* 15: 799-813.

Joshi, Y. C., Nautiyal, P. C., Ravindra, V. and Dwivedi, R. S. (1988). Water relations in two cultivars of groundnut (*Arachis hypogaea* L.) under soil water deficit. *Trop. Afric.* (Trinidad) 65:182-184.

Joshi, Y.C., Nautiyal, P.C. and Ravindra, V. (1993). Drought research at National Research Centre for Groundnut and drought patterns in Junagadh region. In: Selection for water-use efficiency in grain legumes (eds. G.C. Wright and R.C. Nageswara Rao). ACIAR technical report 27. Australian Centre for International Research, Canberra, *pp*. 34-35.

Katring, D. L., Brown, R. H., Sullivan, G.A. and Johnson, B. B. (1982). Growth physiology. In: Paatee, H.E., Young, C. T. (Eds.), Peanut Science and Technology. American Peanut Research and Education Society, Yaokum, TX, USA, *pp*. 411-457.

Katring, D. L. (1985). Physiology response of groundnut to temperature and water deficit-breeding implications. In: Sivakumar, M.V.K., Virmani, S.M. (Eds.), Proceedings of the International Symposium on Agrometeorology of Groundnut, ICRISAT Sahelian Centre, Niamy, Niger, *pp*. 135-143.

Ludlow, M. M. and Muchow, R. C. (1990). A critical evaluation of traits for improving crop yields in water limited environments. *Adv. Agron.* 43:107-153.

Nagaswara Rao, R.C. and Wright G.C. (1993). Stability of the relationship between specific leaf area and carbon isotope discrimination across environment in peanut. *Crop Sci.* 34: 98-103.

Nagaswara Rao, R.C., Udayakumar, M., Farquhar, G.D., Talwar, H.S. and Prasad, T.G. (1995). Variation in carbon isotope discrimination and its relationship to specific leaf area and ribulose-1, 5-bisphophate carboxylase in groundnut genotypes. *Aust. J. Plant Physiol.* 22:545-551.

Nautiyal, P.C., Ravindra, V. and Joshi, Y.C. (1995). Gas exchange and leaf water relations in two peanut cultivars of different drought tolerance. *Biol. Plant.* 37:371-374.

Nautiyal, P.C., Ravindra, V. and Joshi, Y.C. (1999). Net photosynthetic rate in peanut (*Arachis hypogea* L.): influence of leaf position, time of day, and reproductive sink. *Photosynthetica* 36:129-138.

Nautiyal, P. C., Nageswara Rao Rachaputi and Joshi, Y. C. (2002). Moisture-deficit-induced changes in leaf-water content, leaf carbon exchange rate and biomass production in groundnut cultivars differing on specific leaf area, *Field Crop Research* 74:67-79.

Nautiyal, P.C. and Joshi, Y.C. (2003). Carbon discrimination and exchange rate and partitioning of dry matter under natural and simulated drought conditions in groundnut, (Abstract). 2nd Interational Congress of Plant Physiology, Jan. 8-12, New Delhi, India. *pp*. 164.

Nautiyal, P.C. Rajgopal, K. P.V. Zala, Dipti S. Pujari, Manojit Basu, Bhuwaneshwariba Dhadhal and Bharat M. Nandre (2008). Evaluation of Wild Arachis species for abiotic stress tolerance: I Thermal stress and leaf water relations. *Euphytica*, 159:43-57

Noble, P.S. (1991). Physiocochemical and environmental plant physiology. Accademic Press, New York, *pp*. 425-453.

Pallas Jr, J. E., 1982. Photosynthetic traits of selected peanut genotypes. *Peanut Sci.* 9: 14-17.

Passioura, J. B. (1977). Grain yield, harvest index and water use of wheat. *J. Aust. Inst. Agri. Sci.* 43: 117-120.

Ravindra, V., Nautiyal, P. C. and Joshi, Y. C. (1989). Physiological analysis of drought resistance and yield in groundnut (*Arachis hypogaea* L.). *Trop. Agric.* (Trinidad) 64: 290-296.

Ravindra, V., Nautiyal, P. C and Joshi, Y. C. (1995). Ontogenetic changes in growth and net photosynthetic rate of two peanut (*Arachis hypogaea* L) cultivars. *Biol. Plant.* 37 (2): 225-232.

Tanner, C. B. and Sinclair, T. R. (1983). Efficient water use in crop production: research or research? In: "Limitations to Efficient Water Use in Crop Plants" (H. M. Taylor, ed.), *pp*. 1-27, Crop Science Society of America, Madison. WI.

Turner, N. C. and Takeda, G. (1993). Use of physiological attributes in plant breeding-discussion. In: "International Crop Science I" (D.R. Buxton, R. Shibles, R. A. Forsberg, B. L., Blad, K.H., Asay, G. M. Paulsen, and R. F. Wilson, eds.), *pp*. 619-620. *Crop Science Society of America*, Madison, W.I.

Turner, N .C. (1986). Crop water deficits: a decade of progress. *Advances in Agronomy* 39:1-51.

Vivekanandan, A.S. and Gunesena, H.P.M. (1976). Lysimetric studies on effect of soil moisture tension on the growth and yield of maize (*Zea mays* L) and groundnut (*Arachis hypogaea* L.) Beitr. *Trop. Landwirtsch Veterianacrmedizin* 14:369-378.

Whan, B.R., Carlton, G.P., Siddique, K. H. M., Regan K. L., Turner, N.C., and Anderson, W.K. (1993). Integration of breeding and physiology: lessons from a water-limited environment. In: "International crop science I. In: D.R. Buxton, R. Shibles, R. A. Forsberg, B. L. Blad, K. H. Asay, G. M. Paulsen and R. F. Wilson, (eds), *pp*.607-614. Crop Science Society of America, Madison, WI.

Whan, B.R., Anderson, W.K., Gilmour, R.F., Regan, K.L., and Turner, N.C. (1991). A role for physiology in breeding for improved wheat yield under drought stress. In: E. Acevedo, A.P. Conesa, P. Monneveux and J. P. Srivastava (eds.), Physiology breeding of winter cereals for stressed Mediterranean environments. No. 55, *pp*. 179-194. INRA, Paris.

Wilson, D., and Cooper, J. P. (1967). Association of lolium in relation to leaf mesophyll. *Nature*. 214: 989-992.

Wong, S. C., Cowan, I. R. and Fraquhar, G.D. (1979). Stomatal conductance correlates with photosynthetic capacity. *Nature*. 282: 424-426.

Wright, G. C., Hubick, K. T. and Farquhar, G. D. (1988). Discrimination in carbon isotopes of leaves correlates with water-use efficiency of field-grown peanut cultivars. *Aust. J. Plant Physiol*. 15: 815-825.

Wright, G.C., Hubic, K.T., and Farquhar, G.D. (1991). Physiological analysis of peanut cultivar response to timing and duration of drought stress. *Aust.J.Agric.Res*.42: 453-470.

Wright, G.C., Nagaswara Rao, R.C. and Farquhar, G.D. (1994). Water use efficiency and carbon isotope discrimination in peanut under water-deficit condition. *Crop Sci*. 34: 92-97.

Wright, G.C., and Redden, R.J. (1995). Variation in water use efficiency in common bean (*Phaseolous vulgaris*) genotype under rainfed condition. ACIAR Food Legume Newslett. 23: 4.

Wright, G. C.., Nageswara Rao, R. C. and Basu, M. S. (1996). A physiological approach to the understanding of genotype by environment interactions. A case study on improvement of drought adaptation in groundnut. In: Cooper. M., Hammer, G. L. (eds.), Plant Adaptation and Crop Improvement. CAB International in association with the International Rice Research Institute and the International Crop Research Institute for the Semi-arid Tropics, *pp*. 365-381.

□□□

Chapter 31

Field Screening of Greengram (*Vigna radiata* (l.) Wilczek) Genotypes to Drought Tolerance

VINCENT, S.,* H. VIJAYARAGAVAN*, AND K. BALAKRISHNAN**
Department of Crop Physiology*, Tamilnadu Agricultural University, Coimbatore- 641003
**Professor of Crop Physiology, Agricultural College & Research Institute, Madurai-625104

Introduction

Greengram (*Vigna radiata* (l.) Wilczek) is extensively grown on all types of soil under varying climatic conditions. It is mainly grown under rainfed conditions, experiences erratic and rapidly diminishing rainfall and, residual soil moisture is the major limiting factor to increase in yield (Singh and Yadav, 1978). Under such situations, the crop encounters varying degrees of water stress, which affects the various physiological processes associated with growth and development (Sionit and Kramer, 1977). Seasonal variability in available moisture appears to be the major constraint to production under rainfed agriculture. The low and erratic rainfall along with high temperature in these areas induces periods of water stress during the crop growth. Thus, the ability of the crop to grow and yield in such an environment depends on the performance of cultivars adapted to drought. During the course of growth under natural field conditions, the pulse crops are exposed to a number of different abiotic stresses. Water stress constitutes the major abiotic stress that limits the yield of pulses to about 25 per cent of their potential yield. As regards greengram, the impact of water stress on its physiology is not always equal. At different stages during its life cycle, the greengram crop will be differentially sensitive to water deficit. Moisture stress that occurs late in the development of the crop will have little to no impact on yield, but the same level of water stress that occurs early in crop development may reduce yield severely. So, it is imperative to identify the specific growth phase, when plants are particularly sensitive to the influence of water stress (Boyer and Mcpherson, 1975). Hence, the present investigation on field screening of greengram genotypes was carried out to find out the tolerant genotypes to moisture stress during various growth stages.

Materials and Methods

Sixty genotypes of greengram were collected from the Department of pulses, School of Genetics, Tamil Nadu Agricultural University, Coimbatore and National Pulses Research Centre, Vamban. The investigations, which were undertaken both under laboratory and field environments during the years, 2003 and 2004 consisted of four experiments.

Screening of greengram genotypes under field conditions were carried out in the research fields of the Department of Plant Breeding and Genetics, Agricultural College and Research Institute, Madurai which is located at 9^0 54′ N, 78^0 54′ ^{0}E longitude and an altitude of 147 metres above MSL. The soil physical and chemical properties of the experimental field were studied before the start of the experiment. The soil of the experimental field is sandy clay loam in texture with low available nitrogen, medium phosphorus and high potassium.

Water stress was imposed in the experimental field during vegetative, flowering and pod filling stages to identify the genotypes tolerant to water stress. The control plots were irrigated throughout the cropping period.

Results and Discussion

Field evaluation is the most widely adopted method for mass screening of genotypes for drought tolerance (Maibangsa, 1998). In the present investigation conducted under field conditions, water stress was imposed on the plants during all the growth stages of the crop. The crops possessed different adaptive mechanisms for survival under water stress. There existed wider adaptability among greengram genotypes to withstand different degrees of water stress (Kanaka Durga *et al.*, 2003). The genotypes were studied with respect to their morpho-physiological, biochemical and yielding behaviour in the natural drought stress environment under field condition. Observations on traits pertaining to the above processes were recorded and, the results obtained are discussed hereunder.

Morphological Traits as Affected by Water Stress

The plant height of greengram varied with genotypes irrespective of duration. The plant height got significantly reduced due to water stress in plants belonging to different duration groups. However, minimum reduction in plant height was noticed in Pusa Bold (short duration) followed by VBN 2 (medium duration) and CO 4 (long duration). Plant height is generally amenable to be altered by environmental factors and especially, moisture stress will have a greater impact on this trait. Maibangsa (1998) also indicated that water stress at any stage reduced height of rice genotypes. Singh *et al.* (1998) observed that mung bean showed a marked reduction in the weekly height increments as compared to unstressed. The findings on plant height reduction in this study are in conformity with those of Surya Kumari *et al.* (2003) in blackgram and Kanaka Durga *et al.* (2003) in mung bean and urd bean.

Root length is another important trait in a crop like greengram with an indeterminate flowering habit. The depth of root system is considered to be a valuable stress index particularly under low available soil moisture regimes. Root length got reduced due to the impact of water stress. Moisture stress at all stages of crop growth invariably reduced the root growth (Anderson Amalan Kumar, 1998). In the present study, although all the genotypes suffered reductions in root length due to water stress, genotypes such as Pusa Bold under short duration and VBN 2 in medium and CO 4 under long duration were able to maintain longer root system in their respective duration categories. Similar results were observed in blackgram by Murugeswari (1994), in chickpea by Singh (2000) and in mung bean and urd bean by Kanaka Durga *et al.* (2003).

Several screening studies conducted in pulse crops demonstrated that the morphological indices were related to the growth and productivity of plants under condition of water stress, where the tolerant cultivars showed higher metabolic activity (Singh, 2000). So, the above mentioned morphological characters could be regarded as useful tools in the selection of greengram genotypes adapted to moisture-limited condition.

Dry matter accumulation is the most integrative parameter indicative of the magnitude of stress effect. The total biomass production as indicated by

DMA was reduced due to water stress, but the amount of reduction varied with genotypes and duration of greengram. However, the minimum reduction in respect of DMA under stressful environment was observed in the genotypes of Pusa Bold, VBN 2 and CO 4 and, this was suggestive of their greater ability for drought tolerance than the other genotypes. Similar studies by Pankaj Kumar *et al.* (2001) showed that the reduction in biomass was higher in susceptible cultivars of chickpea.

Leaf area index, an important morphological attribute, reflects the assimilatory capacity of the crop. The leaf growth, being most sensitive to water deficit, is greatly affected under limited water availability. In the present investigation aimed at screening of genotypes for drought tolerance, the leaf area and specific leaf weight (SLW) underwent marked reductions under moisture stress in all the cultivars. The genotypes with drought tolerance characteristics withstood in a better way the impact of water stress as indicated by the least reductions in leaf area. Among the different genotypes, Pusa Bold, VBN 2 and CO 4 maintained relatively a greater leaf area throughout the growth period. As suggested by Singh (2004), presence of sufficient foliage at the time of drought might increase the efficiency of photosynthate production and hence, genotypes with higher LAI and SLW could be preferred for adoption under dry land conditions. Similar view was also expressed by Surya Kumari *et al.* (2003) and Kanaka Durga *et al.* (2003), who studied the drought tolerant attributes of mung bean and urd bean and considered LAI and SLW selection indices for screening for drought. Similar views were expressed by Anderson Amalan Kumar (1998) in mung bean, Murugeswari (1994) and Singh (2004) in blackgram and Pankaj Kumar *et al.* (2001) in chick pea.

Physiological and Biochemical Traits as Affected by Water Stress

Relative water content is considered to be a reliable index for identifying genotypes with ability for drought tolerance. Although all the greengram genotypes had lower tissue water content under the impact of stress, the tolerant genotypes maintained higher RWC as compared to the susceptible types. This might be as a result of low transpirational water loss as evident from their lower transpiration rates. The tolerant genotypes like Pusa Bold, VBN 2 and CO 4 maintained higher RWC because of lower transpiration. The degree of reduction in RWC under stress was used as a measure for screening crop varieties for drought tolerance. The results of this study regarding RWC corroborated the findings of Seetharani (1990) in blackgram, and Nautiyal *et al.* (2002) in groundnut cultivars.

Leaf water potential (LWP), another important physiological attribute associated with turgor, constitutes a selection index used for screening drought tolerant genotypes. Although a decrease in available soil moisture reduced the LWP in all the greengram genotypes, minimum reduction was observed in Pusa Bold, VBN 2 and CO 4.

Biochemical characters like chlorophyll stability index (CSI), soluble protein and nitrate reductase activity (NRA) were also drastically decreased by moisture stress in all the greengram genotypes. These three parameters are closely associated with the metabolism of the plant and as such, drought stress has got a greater impact on these attributes. Plant types with ability to ward off major damage to these characters under stress are thought to be in a better position to cope with the onslaught of drought stress. In the present study, Pusa Bold, VBN 2 and CO 4 were identified as better genotypes in their respective duration groups, having the capacity to avoid major reductions in the afore-said biochemical traits. In his work on screening rice genotypes for drought tolerance, Maibangsa (1998) found these three characters as valuable indices for drought screening. Similar findings on CSI were reported in blackgram (Murugeswari, 1994), soluble protein in mung bean (Anderson Amalan Kumar, 1998) and NRA in blackgram (Murugeswari, 1994) and rice (Maibangsa, 1998).

Yield and Yield Characters as Affected by Water Stress

Number of flowers, number of pods and 100 seed weight are the major determinants of seed yield and water stress at any time of the growth inflicted irreversible damage to these traits in greengram. The magnitude of reduction in flower production, pod number, seed weight and seed yield varied with the genotypes. The response among the cultivars to water deficit in respect of flower and pod production varied in both greengram and blackgram (Chandrababu, 1990; Sadasivam, 1996). Chiang and Hubbell (1978) cited the reduction in number of pods per plant as the main factor responsible for reduced yield under water stress. According to Muhammed Saeed *et al.* (1998), water stress significantly decreased seed weight and increased the number of sterile plants in maize.

The resistance to drought refers to the capacity of a genotype for survival, reproduction and economic yield under limited availability of water. It appears that the mechanisms responsible for the survival of one species are quite different from those, which give a maximum economic yield under stress (Pankaj Kumar *et al.*, 2001).

Table 1. Effect of water stress on morphological and physiological parameters of greengram genotypes at vegetative stage

Genotypes	Vegetative Stage								
	Plant height (cm)	Root length (cm)	Relative Water Content (per cent)	Leaf water potential (-MPa)	Leaf area (cm^2 plant^{-1})	Specific Leaf Weight (mg cm^{-2})	Chlorophyll Stability Index (per cent)	Reductase activity (μ moles NO_2 $h^{-1}g^{-1}$)	Dry Matter Accumulation (g plant^{-1})
Short duration									
COGG 910	20.08	4.93	80.65	0.61	147.75	3.52	69.15	0.82	1.27
COGG 912	20.20	4.75	79.20	0.58	145.75	3.54	71.85	0.83	1.22
COGG 913	19.58	5.00	81.05	0.57	143.25	3.48	70.13	0.83	1.26
COGG915	19.15	4.80	80.83	0.55	151.00	3.56	71.15	0.85	1.27
COGG917	20.05	5.18	81.80	0.58	158.75	3.57	71.90	0.83	1.26
Pusa Bold	20.35	6.05	82.73	0.54	182.25	4.31	74.15	1.06	1.41
VGG112	17.53	4.45	81.23	0.61	154.50	4.09	70.85	0.81	1.29
VGG114	17.25	4.43	80.65	0.59	149.50	4.18	70.35	0.83	1.28
VGG117	16.20	4.33	80.78	0.57	158.75	4.14	70.33	0.83	1.26
VGG124	17.38	4.40	81.08	0.56	168.25	4.13	72.00	0.84	1.26
SM-03-05	16.40	4.55	80.88	0.55	146.50	4.08	71.30	0.84	1.32
SM-03-08	16.85	4.50	81.58	0.57	149.75	4.19	71.28	0.90	1.29
SM-03-15	17.60	4.53	81.15	0.57	162.50	4.21	70.03	0.79	1.29
SM-03-18	17.40	4.43	81.10	0.56	165.25	4.13	71.23	0.83	1.30
SM-03-24	17.13	4.40	81.68	0.58	153.75	4.07	70.83	0.88	1.28
2KM-3-3	16.48	4.45	81.68	0.57	145.75	3.96	70.68	0.90	1.29
2KM-3-7	16.70	4.30	81.00	0.57	144.25	4.09	70.78	0.88	1.27
M-702	16.75	4.45	81.10	0.57	152.25	4.20	72.85	0.91	1.28
M-703	16.93	4.30	81.23	0.57	149.00	4.15	71.38	0.87	1.29
M-704	16.08	4.48	80.60	0.56	143.25	4.17	70.15	0.84	1.30
M-705	16.45	4.18	80.25	0.58	143.75	4.16	70.78	0.82	1.22
M-706	15.80	4.40	80.30	0.57	152.75	4.15	71.35	0.83	1.25
M-707	16.33	4.30	80.85	0.57	154.25	4.14	70.40	0.88	1.26
Medium duration									
COGG 918	14.85	5.08	81.40	0.58	151.25	3.53	69.88	0.83	1.22
COGG 919	18.95	5.00	79.25	0.58	141.50	3.41	67.95	0.85	1.22
COGG 920	19.10	4.95	79.08	0.57	154.25	3.45	71.10	0.87	1.16
COGG 921	19.53	4.63	79.70	0.56	146.75	3.37	70.20	0.85	1.27
COGG 922	19.53	4.88	80.20	0.55	147.00	3.47	71.40	0.84	1.26
COGG 923	19.43	4.48	80.48	0.55	150.50	3.53	67.58	0.85	1.18
COGG 924	19.85	4.58	78.80	0.61	151.50	3.43	69.38	0.78	1.23
COGG 925	19.30	4.85	81.35	0.58	160.00	3.57	71.63	0.90	1.25
COGG 926	20.35	5.08	80.30	0.56	158.50	3.49	71.23	0.86	1.18
COGG 927	20.48	4.70	81.33	0.58	147.00	3.46	70.75	0.85	1.16
COGG 928	20.20	4.63	80.15	0.56	146.50	3.38	71.88	0.88	1.18
COGG 930	20.53	4.75	79.23	0.58	155.75	3.49	71.00	0.88	1.25
COGG 123 A	20.60	5.30	79.85	0.56	155.00	3.51	68.68	0.86	1.23
PLS 269(A)	19.68	4.70	79.60	0.55	138.00	3.47	71.50	0.88	1.14
LM 253	20.25	4.65	79.40	0.55	151.75	3.43	71.13	0.85	1.18
AC 246	20.00	4.50	80.40	0.57	138.25	3.41	70.73	0.78	1.21
AKM 9243	19.88	4.63	80.00	0.60	143.00	3.51	71.10	0.90	1.19
VGG 119	17.48	4.40	81.00	0.59	147.00	4.17	68.30	0.87	1.32
VGG 126	16.73	4.53	80.45	0.57	160.50	4.13	71.15	0.84	1.33
VGG 128	16.70	4.50	81.15	0.54	166.00	4.20	65.73	0.85	1.31
2KM-3- 7	16.40	4.40	80.90	0.59	149.50	4.24	71.55	0.85	1.29

Contd...

2KM-3 43	16.33	4.33	80.48	0.55	151.50	4.08	71.00	0.91	1.27
2KM-3-36	16.75	4.30	81.98	0.58	153.25	4.11	72.30	0.89	1.23
K1	17.28	5.00	82.78	0.53	189.75	4.25	73.58	0.99	1.36
K 851	16.98	4.68	82.48	0.53	191.50	4.21	72.18	1.00	1.37
KM 2	17.58	5.00	82.30	0.54	178.75	4.26	72.75	0.99	1.36
KM 17	18.00	4.98	81.73	0.54	173.75	4.28	72.93	1.00	1.35
M 709	16.50	4.25	80.20	0.57	151.50	4.02	71.15	0.89	1.28
VBN 2	20.25	5.05	84.18	0.52	200.75	4.41	77.40	1.06	1.43
VRM 1	17.05	4.73	82.28	0.53	165.50	4.23	72.78	0.98	1.37
Long duration									
COGG 929	20.85	4.88	80.90	0.55	150.75	3.40	72.03	0.87	1.16
CO 3	20.38	5.40	82.33	0.54	171.00	3.54	73.00	0.99	1.33
CO 4	22.38	5.59	83.90	0.53	200.00	4.49	77.53	1.06	1.38
CO 5	20.83	5.13	82.30	0.55	170.75	3.50	73.18	0.98	1.35
CO6	20.90	5.03	81.98	0.54	167.00	3.53	73.08	0.94	1.37
MDU 2013/1-2	20.38	4.75	80.23	0.58	124.00	3.38	72.20	0.85	1.15
PLS 269(D)	19.90	4.53	79.48	0.59	146.25	3.46	70.60	0.87	1.16
Mean	**18.45**	**4.71**	**80.95**	**0.56**	**156.14**	**3.85**	**71.27**	**0.88**	**1.27**
SED	0.5646	0.3647	0.2629	0.0088	6.912	0.0365	1.2973	0.82	1.27
CD (0.05)	1.1180	0.7221	0.5205	0.0175	13.686	0.0722	2.5688	0.83	1.22

Table 2. Effect of water stress on morphological and physiological parameters of greengram genotypes at flowering stage

Genotypes	Flowering Stage								
	Plant height (cm)	Root length (cm)	Relative water content (per cent)	Leaf water potential (-Mpa)	Leaf area (cm² plant⁻¹)	Specific leaf weight (mg cm⁻²)	Chlorophyll stability index (per cent)	Reductase activity (μ moles No_2 $h^{-1}g^{-1}$)	Dry matter accumulation (g $plant^{-1}$)
Short duration									
COGG 910	38.38	8.40	78.28	0.60	352.50	3.46	75.98	1.09	5.27
COGG 912	39.33	8.05	77.28	0.57	347.25	3.53	77.73	1.11	5.34
COGG 913	37.63	8.43	77.70	0.58	341.00	3.46	76.95	1.10	5.28
COGG915	38.35	8.25	78.88	0.61	345.50	3.49	76.50	1.13	5.24
COGG917	38.00	8.10	78.50	0.57	339.00	3.51	77.35	1.12	5.38
Pusa Bold	39.26	8.40	82.35	0.55	381.00	4.04	77.68	1.42	5.90
VGG112	36.65	7.55	78.73	0.59	353.75	4.10	76.83	1.10	5.20
VGG114	33.83	7.98	78.83	0.60	344.25	4.16	75.30	1.12	5.51
VGG117	33.38	7.78	78.63	0.57	338.00	4.13	74.45	1.12	4.98
VGG124	38.18	7.43	78.83	0.61	341.25	3.87	77.40	1.09	5.00
SM-03-05	37.13	7.25	78.20	0.59	339.25	3.94	76.58	1.13	5.13
SM-03-08	36.15	7.33	78.90	0.64	345.50	3.97	76.40	1.11	5.26
SM-03-15	37.85	7.95	78.70	0.66	353.00	4.14	77.45	1.13	5.01
SM-03-18	39.70	7.70	78.65	0.64	350.25	3.93	77.58	1.22	5.32
SM-03-24	37.90	7.95	78.80	0.57	359.75	4.18	76.80	1.19	5.11
2KM-3-3	36.83	8.33	79.05	0.61	350.75	3.95	75.80	1.11	5.32
2KM-3-7	35.05	7.85	79.40	0.62	349.75	4.18	78.18	1.10	4.97
M-702	35.30	8.18	79.48	0.59	361.25	4.25	76.93	1.13	5.18

Contd...

M-703	34.23	7.93	79.15	0.59	353.75	4.24	76.63	1.16	5.35
M-704	33.33	7.23	78.25	0.64	364.00	4.22	76.20	1.14	5.29
M-705	32.63	6.93	77.95	0.64	345.50	4.28	76.93	1.14	5.13
M-706	33.30	6.73	78.85	0.62	353.75	4.27	76.78	1.12	5.35
M-707	33.28	7.18	79.33	0.62	347.50	4.28	76.53	1.10	5.26
Medium duration									
COGG 918	44.48	9.08	77.40	0.60	338.25	3.54	77.10	1.10	5.45
COGG 919	41.93	8.58	78.25	0.62	335.75	3.50	77.98	1.14	5.46
COGG 920	43.68	8.05	77.38	0.64	346.50	3.46	78.08	1.23	5.38
COGG 921	45.33	8.75	78.38	0.65	350.50	3.44	76.40	1.08	5.39
COGG 922	44.63	9.20	78.80	0.64	347.50	3.44	75.10	0.99	5.44
COGG 923	42.18	8.60	78.48	0.62	345.25	3.49	75.65	1.16	4.76
COGG 924	44.43	8.38	78.43	0.57	348.75	3.45	77.88	1.12	5.55
COGG 925	43.65	8.48	78.40	0.56	347.25	3.55	77.35	1.17	5.59
COGG 926	44.25	8.95	78.95	0.60	348.75	3.53	78.10	1.08	4.99
COGG 927	43.60	8.78	78.63	0.62	348.75	3.42	77.20	1.09	5.54
COGG 928	45.58	8.73	78.60	0.66	343.00	3.44	78.05	1.19	5.20
COGG 930	46.88	8.90	78.95	0.62	341.50	3.48	77.88	1.13	5.38
COGG 123 A	46.78	9.50	79.03	0.57	344.25	3.48	77.35	1.16	5.53
PLS 269(A)	43.05	7.80	77.85	0.46	339.00	3.48	77.80	1.17	5.50
LM 253	43.05	7.60	77.38	0.61	339.25	3.50	76.98	1.09	5.58
AC 246	41.60	7.23	78.15	0.66	338.25	3.41	77.05	1.09	5.39
AKM 9243	39.65	7.48	78.65	0.64	335.75	3.45	76.35	1.14	5.40
VGG 119	38.93	8.13	78.60	0.57	345.25	4.13	78.20	1.12	5.08
VGG 126	40.58	7.13	77.25	0.63	342.50	4.00	76.60	1.11	5.03
VGG 128	36.23	7.48	78.45	0.63	344.75	3.98	77.70	1.18	5.08
2KM-3- 7	34.30	7.50	79.05	0.60	342.50	4.22	77.08	1.09	5.20
2KM-3 43	35.00	7.05	79.08	0.58	342.50	4.01	77.85	1.19	5.11
2KM-3-36	38.53	7.30	78.88	0.61	356.50	4.23	80.30	1.17	5.20
K1	40.20	9.73	79.63	0.55	360.50	4.32	79.65	1.28	5.60
K 851	39.78	10.35	79.70	0.57	368.75	4.34	78.20	1.24	5.52
KM 2	40.75	9.63	79.73	0.56	368.25	4.36	78.88	1.19	5.54
KM 17	39.73	10.08	79.58	0.57	372.00	4.06	79.25	1.20	5.58
M 709	32.53	7.20	78.95	0.63	340.25	4.22	77.95	1.14	5.38
VBN 2	50.91	13.15	82.28	0.53	406.50	4.50	81.35	1.40	5.85
VRM 1	37.90	8.48	79.48	0.57	366.50	4.27	78.75	1.28	5.58
Long duration									
COGG 929	44.78	8.45	78.83	0.64	343.75	3.42	78.25	1.18	5.30
CO 3	47.13	11.78	78.88	0.57	361.75	3.56	78.95	1.32	5.64
CO 4	51.71	13.68	82.78	0.55	413.75	4.25	81.65	1.41	5.98
CO 5	45.85	10.80	79.18	0.56	354.50	3.55	78.20	1.29	5.66
CO6	46.15	10.78	79.00	0.57	356.75	3.54	75.70	1.26	5.65
MDU 2013/1-2	42.73	7.40	78.73	0.57	343.25	3.42	77.48	1.10	5.37
PLS 269(D)	40.05	7.35	78.35	0.61	340.50	3.41	75.80	1.16	5.50
Mean	**40.07**	**8.44**	**78.85**	**0.60**	**351.11**	**3.84**	**77.42**	**1.16**	**5.35**
SED	0.5251	0.5534	0.1539	0.60	6.512	0.1919	0.1717	0.0309	0.1855
CD (0.05)	1.0397	1.0957	0.3048	0.57	12.895	0.3800	0.3401	0.0611	0.3673

Table 3. Effect of water stress on morphological and physiological parameters of greengram genotypes at pod filling stage

Genotypes	Pod filling Stage								
	Plant height (cm)	Root length (cm)	Relative water content (per cent)	Leaf water potential (-Mpa)	Leaf area (cm^2 $plant^{-1}$)	Specific leaf weight (mg cm^{-2})	Chlorophyll stability index (per cent)	Reductase activity (μ moles $No_2\ h^{-1}g^{-1}$)	Dry matter accumulation (g $plant^{-1}$)
Short duration									
COGG 910	48.15	11.75	78.08	0.71	391.00	3.30	51.35	0.78	10.68
COGG 912	50.30	12.43	77.65	0.69	415.25	3.30	51.85	0.80	10.47
COGG 913	49.33	12.05	76.80	0.73	414.75	3.30	51.73	0.81	10.38
COGG915	48.40	12.15	76.98	0.73	406.75	3.33	52.05	0.83	10.43
COGG917	51.08	12.13	77.25	0.71	384.25	3.30	52.10	0.83	10.86
Pusa Bold	45.93	12.95	80.55	0.63	466.00	4.14	52.10	1.00	10.99
VGG112	44.98	11.80	78.28	0.73	398.25	4.00	51.75	0.79	10.71
VGG114	42.45	11.50	78.05	0.71	391.25	3.95	52.00	0.80	10.54
VGG117	41.25	11.48	78.68	0.68	387.75	4.04	52.03	0.80	10.51
VGG124	44.75	11.90	78.70	0.78	383.50	3.99	51.93	0.84	10.66
SM-03-05	44.85	12.38	77.43	0.72	387.00	4.04	52.25	0.92	10.55
SM-03-08	44.55	11.90	77.15	0.73	389.75	4.10	52.03	0.81	10.32
SM-03-15	45.13	12.38	77.50	0.72	388.50	3.98	52.73	0.80	10.51
SM-03-18	46.33	11.10	77.75	0.76	377.00	3.97	52.90	0.87	10.74
SM-03-24	45.13	11.40	78.13	0.74	394.25	4.08	52.33	0.87	10.75
2KM-3-3	44.10	10.85	78.73	0.70	404.50	4.03	52.10	0.88	10.43
2KM-3-7	41.68	11.00	78.05	0.68	384.50	4.01	52.43	0.79	10.62
M-702	43.10	11.75	78.35	0.67	408.50	4.08	52.00	0.84	10.60
M-703	41.18	11.50	78.23	0.68	393.50	4.04	52.03	0.86	10.48
M-704	42.65	11.80	77.65	0.71	401.50	4.02	52.18	0.79	10.47
M-705	40.80	11.93	76.95	0.69	396.75	4.08	52.20	0.79	10.54
M-706	42.38	12.05	77.60	0.75	408.75	4.07	52.40	0.84	10.37
M-707	42.88	11.45	78.20	0.65	407.75	4.01	51.48	0.79	10.67
Medium duration									
COGG 918	55.83	12.20	77.50	0.76	419.25	3.28	51.88	0.82	10.52
COGG 919	53.23	13.10	76.70	0.71	425.25	3.31	51.98	0.87	10.97
COGG 920	53.95	13.23	77.60	0.69	427.75	3.33	52.55	0.92	10.88
COGG 921	54.88	11.93	77.28	0.67	399.50	3.34	52.25	0.80	10.34
COGG 922	58.25	11.93	77.05	0.67	413.50	3.26	52.45	0.80	10.18
COGG 923	55.18	12.45	77.30	0.68	428.75	3.29	52.28	0.91	10.98
COGG 924	52.93	12.43	77.88	0.71	431.00	3.34	52.18	0.86	10.70
COGG 925	54.78	11.70	77.98	0.69	439.50	3.30	51.48	0.89	10.14
COGG 926	55.68	12.80	77.48	0.67	412.50	3.31	51.90	0.79	10.37
COGG 927	57.00	11.73	77.48	0.70	410.75	3.32	52.10	0.79	10.40
COGG 928	54.55	12.50	76.95	0.72	413.50	3.38	52.13	0.94	10.69
COGG 930	59.08	11.70	76.93	0.67	413.50	3.27	51.40	0.85	10.92
COGG 123 A	56.75	12.50	77.23	0.66	430.50	3.26	51.90	0.91	10.75
PLS 269(A)	56.18	11.35	76.98	0.69	395.25	3.31	51.60	0.87	10.60
LM 253	57.15	12.15	77.40	0.71	379.25	3.34	52.20	0.78	10.22
AC 246	50.73	11.45	77.75	0.66	408.50	3.32	52.18	0.78	10.85
AKM 9243	52.65	11.83	77.95	0.66	396.00	3.34	51.65	0.87	10.65
VGG 119	43.23	11.20	78.03	0.70	389.00	3.89	52.70	0.83	10.30
VGG 126	47.23	10.70	78.38	0.75	412.50	4.08	52.48	0.82	10.80
VGG 128	40.90	11.38	77.85	0.80	395.75	3.98	52.50	0.88	10.54

Contd...

2KM-3- 7	42.33	11.23	77.73	0.72	405.75	4.07	51.95	0.80	10.53
2KM-3 43	41.35	11.28	77.70	0.72	420.75	4.02	51.45	0.89	10.79
2KM-3-36	45.80	11.70	78.33	0.66	403.75	4.06	51.65	0.81	10.66
K1	49.03	13.15	78.50	0.88	462.25	4.17	53.73	1.00	11.81
K 851	50.18	14.03	78.93	0.65	433.75	4.16	53.20	0.96	11.67
KM 2	49.98	13.95	78.98	0.63	427.75	4.18	52.98	0.95	11.60
KM 17	50.65	11.98	79.05	0.64	426.25	4.17	52.65	0.94	11.48
M 709	42.35	11.60	77.93	0.69	401.00	3.98	52.18	0.81	10.54
VBN 2	45.98	16.80	81.18	0.62	485.75	4.34	56.15	1.03	12.50
VRM 1	43.25	14.10	79.20	0.67	468.75	4.19	52.62	0.98	11.61
Long duration									
COGG 929	58.38	12.45	77.15	0.75	404.00	3.29	52.23	0.92	10.30
CO 3	60.95	14.93	78.35	0.65	484.25	3.39	53.38	0.98	11.74
CO 4	62.33	16.19	80.65	0.63	519.50	4.49	55.40	1.05	12.48
CO 5	60.08	14.80	78.58	0.64	481.25	3.37	52.63	0.99	11.76
CO6	59.90	14.35	78.28	0.64	486.50	3.39	52.73	0.98	11.51
MDU 2013/1-2	51.75	11.90	77.33	0.70	415.50	3.32	51.88	0.80	10.56
PLS 269(D)	54.10	11.60	77.40	0.67	403.50	3.32	51.73	0.82	10.31
Mean	**49.50**	**12.30**	**77.96**	**0.69**	**415.88**	**3.72**	**52.30**	**0.86**	**10.78**
SED	0.9619	0.2476	0.1790	0.0677	24.482	0.0267	0.2587	0.0090	0.1460
CD (0.05)	1.9047	[illegible]	0.3543	0.1340	48.477	0.0528	0.5123	0.0178	0.2891

Table 4. Effect of water stress on flower, pod numbers,100 seed weight and grain yield per plant of greengram genotypes

Genotypes	Yield characters			
	Flowers/ Plant	Pods/ Plant	100 seed weight	Grain yield/ Plant
Short duration				
COGG 910	22.45	12.03	2.55	3.40
COGG 912	21.25	13.33	2.67	3.29
COGG 913	22.43	13.18	2.66	3.32
COGG915	22.03	12.43	2.61	3.12
COGG917	22.83	12.55	2.54	3.12
Pusa Bold	27.20	13.30	4.72	4.56
VGG112	22.68	13.45	2.57	3.73
VGG114	22.88	13.23	2.48	3.81
VGG117	24.75	13.00	2.54	3.63
VGG124	23.33	13.48	2.64	3.36
SM-03-05	23.73	15.00	2.68	4.24
SM-03-08	23.30	13.15	2.82	3.81
SM-03-15	23.83	13.88	2.60	4.19
SM-03-18	23.43	15.33	2.62	3.43
SM-03-24	23.85	15.20	2.55	3.22
2KM-3-3	25.03	14.15	2.62	3.52

Contd...

2KM-3-7	26.73	14.35	2.64	4.20
M-702	25.80	17.20	2.60	4.73
M-703	23.90	13.93	2.54	4.35
M-704	23.33	13.10	2.48	4.55
M-705	23.00	14.20	2.54	4.30
M-706	22.80	13.85	2.45	3.97
M-707	23.25	13.35	3.05	4.01
Medium duration				
COGG 918	22.35	12.60	2.62	3.74
COGG 919	23.08	12.20	2.48	4.07
COGG 920	22.13	12.75	2.47	3.29
COGG 921	22.20	14.13	2.69	2.71
COGG 922	22.88	12.93	2.62	3.52
COGG 923	22.35	12.75	2.55	3.24
COGG 924	21.95	13.73	2.62	3.21
COGG 925	23.15	13.90	2.56	3.26
COGG 926	22.55	14.33	2.49	3.85
COGG 927	22.05	13.95	2.69	4.07
COGG 928	21.85	13.65	2.59	3.19
COGG 930	22.45	14.08	2.53	3.23
COGG 123 A	22.43	13.30	2.65	3.13
PLS 269(A)	23.18	14.58	2.76	3.52
LM 253	21.68	14.00	2.86	3.62
AC 246	22.53	13.90	2.60	3.52
AKM 9243	21.95	13.45	2.59	3.59
VGG 119	23.30	14.63	2.56	4.44
VGG 126	24.23	14.38	2.55	3.72
VGG 128	22.98	12.95	2.59	4.18
2KM-3- 7	26.75	14.80	2.51	4.58
2KM-3 43	25.63	15.15	2.47	3.87
2KM-3-36	25.38	15.25	2.62	3.70
K1	30.35	20.18	3.05	5.05
K 851	30.20	20.45	2.83	4.92
KM 2	30.98	20.00	2.93	4.86
KM 17	30.43	19.08	2.92	4.84
M 709	23.93	14.25	2.96	3.69
VBN 2	37.03	21.28	3.31	5.55
VRM 1	30.00	17.58	2.97	4.99
Long duration				
COGG 929	21.43	14.23	2.59	3.10
CO 3	26.15	17.60	3.31	4.65
CO 4	32.38	26.07	3.49	5.48
CO 5	27.03	17.20	3.15	4.72
CO6	27.78	17.18	3.13	4.62
MDU 2013/1-2	21.83	14.15	2.50	3.62
PLS 269(D)	22.35	13.13	2.63	3.60
Mean	**24.41**	**14.77**	**2.73**	**3.91**
SED	0.1987	0.3248	0.0762	0.1137
CD (0.05)	0.3934	0.6431	0.1509	0.2252

The current study provided comprehensive information on the existence of varying degrees of adaptation in greengram genotypes to restricted moisture availability. The variability in yield was primarily due to the variation in yield contributing characters under the impact of moisture stress. The screening experiments thus made it possible to obtain cultivars with wide adaptation and tolerance to drought (Queiroz *et al.*, 1997). Among the genotypes, the three varieties, namely, Pusa Bold, VBN 2 and CO 4 exhibited higher yield potential. The major yield contributing attributes as exhibited in these genotypes were higher flower production, pod number per plant and 100 seed weight. Surya Kumari *et al.* (2003) demonstrated that genotypes having higher number of pods, number of seeds and seed weight under water deficit condition expressed better genotypic efficiency in translocating photosynthates to the reproductive parts. In the case of pigeon pea also, Kamaldeep Singh Virdi *et al.* (2004) used the afore-said yield characters for identifying genotypes with better drought tolerance ability.

In the light of the explanations drawn from the results of the screening study, it could be concluded that yielding ability of greengram genotypes under water stress was modified by yield attributing parameters *viz.*, flower number, pod number, 100-seed weight and grain yield per plant. Since the final grain yield of a crop is the outcome of morpho-physiological and biochemical reactions, the genotypes possessing these traits at a higher level even under drought stress, exhibited higher physiological efficiency and, hence greater ability to perform better in dry land agriculture under rainfed situation. The above parameters may thus serve as potential selection indices for higher productivity of greengram genotypes.

Based on the above conclusions, the genotypes, namely, Pusa Bold under short duration, VBN 2 under medium duration and CO 4 under long duration were selected for further study to be carried out under imposed water stress conditions during chosen stages of crop growth.

REFERENCES

Anderson Amalan Kumar, A. (1998). Physiological studies on water stress and its amelioration in greengram (*Vigna radiata* (L.) Wilezek) Genotypes. Ph.D. Thesis, Tamil Nadu Agrl. Univ., Coimbatore, India.

Boyer, J.S. and McPherson, H.G. (1975). Physiology of water deficits in cereal crops. *Adv. Agron.*, 27 : 1-23.

Chiang, M. Y. and Hubbell, J.N. (1978). Effect of irrigation on mung bean yield. pp.93-96. In : Robert Cowell (ed.). First Int. Mungbean Symp. Proc. AVRDC. Shanhua, Taiwan.

Chandrababu, R. (1990). Physiological basis for yield improvement in greengram (*Vigna radiata* L. Wilczek) cultivars. Ph.D. thesis, Tamil Nadu Agrl. Univ., Coimbatore, India.

Kamaldeep Singh Virdi., Sidhu, P.S. and Sarvjeet Singh. (2004). Relationship of morpho-physiological traits with yield and its components for identifying efficient plant type in pigeonpea. *J. Res. Punjab Agric. Univ.*, 4 (2): 175-182.

Kanaka Durga, K., Murali Rao, M., Reddy, M.V. and Koteswara Rao, Y. (2003). Effect of early drought stress on morpho-physiological characters and yield of mungbean and urd bean cultivars. *Indian J. Pulses Res.*, 16 (2): 133-135.

Maibangsa, S. (1998). Morpho-physiological, biochemical and molecular mechanism of water stress tolerance in rice (*Oryza sativa* L.) Ph.D. Thesis, Tamil Nadu Agrl. Univ., Coimbatore, India.

Muhammed Saeed., Saifi,M.Y., Akhtar, M. and Mohsan, S. (1998). Differential genotypes response to drought stress in maize. *Sarhad J. Agri.*, 14 (1): 49-55.

Murugeswari, R. (1994). Physiological studies on management of water stress for improvement of productivity in blackgram (*Vigna mungo* (L.) Hepper). M.Sc. (Ag.) Thesis, submitted to Tamil Nadu Agrl. Univ., Coimbatore, India.

Nautiyal, P.C., Rachaputi, N.R. and Joshi, Y.C. (2002). Moisture-deficit-induced changes in leaf-water content, leaf carbon exchange rate and biomass production in groundnut cultivars differing in specific leaf area. Fld. *Crops Research*, 74 (1): 67-79.

Pankaj kumar., Hiremath, S.M., Deshmukh, P.S. and Kushwaha, S.R. (2001). Studies on foliar application of growth regulators on biomass production, harvest index and yield in soybean (*Glycine max* (L.) Merrill). *Ann. Agric. Res., New Series.* 22 (2) : 221-224.

Queiroz, M. F. D., Fernandes, P.D., Almedia, F.D.A.C. and Queiroga, V.D.P. (1997). Tolerance of seed germination of bean cultivars to water stress induced by Polyethylene glycol. IRRIGA, 2 (1): 115-123.

Sadasivam, R. (1996). Seasonal influence on growth and production physiology of blackgram (*Vigna mungo* (L.) Hepper) cultivars. Ph.D. Thesis, submitted to Tamil Nadu Agrl. Univ., Coimbatore, India.

Seetharani, M. (1990). Studies on the effect of different ameliorants under water stress conditions in blackgram (*Vigna mungo* (L.) Hepper). Ph.D. Thesis, Tamil Nadu Agrl. Univ., Coimbatore, India.

Singh, D. N. (2000). Physiological constraints limiting the productivity of chickpea in India-Breeding strategies to improve yield: In National Seminar on Physiological panadigm for fostering Agro and Biotech and Augmenting Environmental productivity in millennium 2000. Nov. 7-9, Lucknow, India.

□□□

Chapter 32

Screening of Guar Genotypes for Drought Tolerance Traits Under Simulated Moisture Stress Conditions

CHAKRABORTY, A[1]., P. SUDHAKAR[2], A. SIVA SANKAR[3], P. LATHA[4] AND B.G. SINGH[5]
1, 3, 5-College of Agriculture, ANGRAU, Rajendranagar, Hyderabad, A.P
2, 4- Regional Agricultural Research Station, ANGRAU, Tirupati, A.P

Summary

Guar or clusterbean is an arid legume, belonging to the family Papilionaceae having 19-23 per cent galactomannan gum having high export value. The crop is warm loving and adopts well in arid conditions of India. However identification of genotypes conferring tolerance to drought, which is a potent threat in arid conditions is research priority. Ten guar genotypes were screened for drought tolerance under field conditions by imposing moisture stress from 35 DAS to 70 DAS (flowering to pod filling). Different drought tolerance traits viz., specific leaf area(SLA), SPAD Chlorophyll meter readings (SCMR), Relative water content(RWC) and Leaf Peroxidase contents were recorded 30 days after imposing moisture stress (65 DAS) along with irrigated treatments. The genotypes WSP – 50, GG-1, RGM-111 recorded lower SLA, higher SCMR, RWC values with higher peroxidase activity under moisture stress conditions, which establish drought tolerance nature of these genotypes and can be used as good donar source for developing drought tolerant genotypes.

Key words : Guar genotypes, drought tolerance traits

Introduction

Drought affects almost all functions in plants. Plants respond to drought by initiating a number of developmental, physiological, biochemical and molecular changes. Guar or cluster bean (*Cyanopsis tetragonoloba* L.), is an arid legume and its endosporium contains 19-43% of galactomannan gum having higher export value. This crop confers drought tolerance and sensitive to low temperatures.

In the recent past, various morphological (Chaves *et al.* 2003), physiological (Nageswara Rao and Wright, 1994) and drought tolerant traits (Sudhakar *et al.*, 2006) contributing to drought tolerance were reported and substantial genetic variability among genotypes were observed for these traits. Such variability in guar genotypes for these traits are yet to be explored.

This crop is introduced into new areas like Andhra Pradesh, Karnataka as an alternate crop under dry land conditions. Hence this study is taken up to screen guar genotypes which possess higher tolerance traits, so as to recommend to these newly adopted areas.

Materials and Methods

The present investigation was conducted in the Semi Arid conditions of Hyderabad in sandy clay loam (College of Agriculture, Rajendranagar) during late rabi season, 2006-07. There were two main treatments i.e irrigated and simulated mid season moisture stress (35 DAS to 70 DAS). The sub treatments consisted of ten guar genotypes viz. WSP-50, Vikas -35, RGC-1038, GG-1, HGS-884, GUAG-9808, GUAG-003, GUAG-9708, RGC 1078 and RGM-111. The treatments were arranged in split plot design and replicated thrice. The seeds were sown on flat beds of 3x2 m size plot with 30 cm x 10 cm spacing. Prophylactic measures were taken for protecting the crop from diseases and pests. The total amount of rainfall received was 17.8 mm during experimental period and incidentally during period of imposing moisture stress treatment no rainfall was recorded. The average maximum and minimum temperatures during crop period was 34.16^0 C and 18.90^0 C respectively (Table 1). Soil moisture data shows that after stress imposition, mean soil moisture gradually reduced from 8.02% to 1.76% in stress imposed plots (Table 2) confirming the depletion of soil moisture.

Drought tolerance traits viz. specific leaf area (SLA), SPAD (Soil plant analytical development) Chlorophyll Meter Reading (SCMR), Relative water content(RWC) and peroxidase activity were measured during imposition of moisture stress and seed yields were recorded for harvest. SLA and SCMR were recorded on third leaf from the apex. SLA was calculated from the measured values of leaf area (using leaf area meter LI-COR Model-3100) and leaf dry weight. SCMR was measured using Minolta SPAD-502 chlorophyll meter. The relative water content (RWC) of leaf samples was calculated based on fresh weight, turgid weight and dry weight of leaf samples using the following formula

RWC= (Fresh weight - Dry weight) / (Turgid weight- Dry weight) x100

The peroxidase activity was measured using guaiacol as substrate (Sadasivam and Manickam, 1997) and expressed as enzyme activity units per liter.

Table 1. Meteorological data during crop growth period (January, 07 to April, 07)

Standard week	Period	Temperature (°C) Max.	Min.	RH (%) I	II	Rainfall (mm)	Rainy days	Mean Temp. (°C)
4	22-28 Jan	30.7	15.5	85	34	0.0	0	23.1
5	29Jan-04 Feb	31.4	15.5	84	26	0.0	0	23.5
6	05-11	31.9	16.5	83	31	0.0	0	24.2
7	12-18	30.6	16.1	82	34	0.0	0	23.4
8	19-25	30.5	14.3	82	26	0.0	0	22.4
9	20 Feb-04 Mar	33.9	16.6	67	22	0.0	0	25.2
10	05-11	34.6	21.4	80	29	0.0	0	28.0
11	12-16	34.4	19.3	70	25	0.0	0	26.8
12	19-25	37.6	22.4	61	19	0.0	0	30.0
13	26 Mar-01 Apr	37.4	20.6	70	19	0.0	0	29.0
14	02-08	38.8	22.0	75	20	0.0	0	30.4
15	09-15	37.6	23.4	74	27	2.6	0	30.5
16	16-22	34.8	22.2	76	36	15.2	2	28.5
	Mean	**34.16**	**18.90**	**76.07**	**26.77**	**1.37**		

Results and Discussion

The mean relative water content of leaves were lowered by 35 per cent in moisture stress imposed treatments compared to irrigated treatments. GG-1, WSP - 50, GUAG-9708 and RGM-111 maintained higher RWC compared to other genotypes conferring higher drought tolerance (table 2). Maintenance of higher RWC under moisture stress conditions, maintain leaf temperature, leaf diffusive resistance and net photosynthetic rate through stomatal regulation and mesophyll resistance (Sandhu and Horton, 1978). Higher SPAD chlorophyll meter reading (SCMR) in moisture stress condition is indication of genotypes capability for higher carbon assimilation. WSP-50 recorded significantly high SCMR followed by GG-1 and RGM-111 compared to other genotypes under simulated moisture stress conditions.

Table 2: SPAD chlorophyll meter readings (SCMR), Specific leaf area (SLA), Relative water content (RWC) and peroxidase activity of Guar genotypes at 30 days after imposing moisture stress (DAIS)

GENOTYPES	SCMR			SLA(cm^2/g)			RWC (%)			PEROXIDASE ACTIVITY (UNITS)		
	Irrigated	Stress	Mean	Irrigated	Stress	Mean	Irrigated	Stress	Mean	Irrigated	Stress	Mean
WSP	74.13	65.17	69.95	67.64	93.6	80.66	90.79	62.11	76.45	245	301	273
Vikas	62.34	53.55	57.95	99.4	94.92	97.18	90.78	58.84	74.81	227	281	254
RGC-1038	66.27	50.70	58.48	122.4	100.6	115.3	88.26	57.80	73.03	222	287	254.5
GG-1	71.73	63.13	67.43	78.6	91.3	84.98	87.14	64.70	75.92	240	298	269
HGS-884	68.50	54.70	61.60	117.0	99.9	108.48	90.21	57.50	73.86	237	286	261.5
GUAG-9808	68.73	43.86	56.30	128.3	114.8	121.58	88.74	56.10	72.42	227	275	251
GUAG-003	66.33	52.57	59.45	128.8	102.8	114.33	91.86	58.40	75.13	232	282	257
GUAG-9708	54.40	50.33	52.36	105.0	91.6	98.33	90.31	60.0	75.16	219	279	249
RGC-1078	66.47	53.27	59.86	113.2	92.12	102.68	89.15	55.06	72.11	221	284	252.5
RGM-111	71.07	62.67	66.87	78.0	97.32	87.68	90.0	59.80	74.90	232	289	260.5
MEAN	66.20	54.29		111.25	96.60		81.37	60.27		228.5	285.8	
	SEM	CD at 5%		SEM	CD at 5%		SEM	CD at 5%		SEM	CD at 5%	
IRRIGATION TREATMENT	0.47	1.35		0.53	1.54		0.49	1.41		0.75	2.13	
GENOTYPE	1.05	3.03		1.20	3.44		1.10	3.17		1.66	4.77	
INTERACTION	1.50	4.3		1.70	4.87		1.56	4.48		2.36	6.75	

In contrast, specific leaf area (SLA) values were significantly lower for these three genotypes viz., WSP-50, GG-1 and RGM-111 compared to other genotypes and SLA recorded a negative correlation with SCMR (r2= 0.3835) (Fig 1). Wright *et al.*, (1994) reported that if SLA is higher, the leaf thickness would be smaller and also observed an inverse relationship between SLA and water use efficiency. Similar results were also reported in groundnut (Latha, 2004) and in blackgram, greengram (Sudhakar *et al.*, 2006).

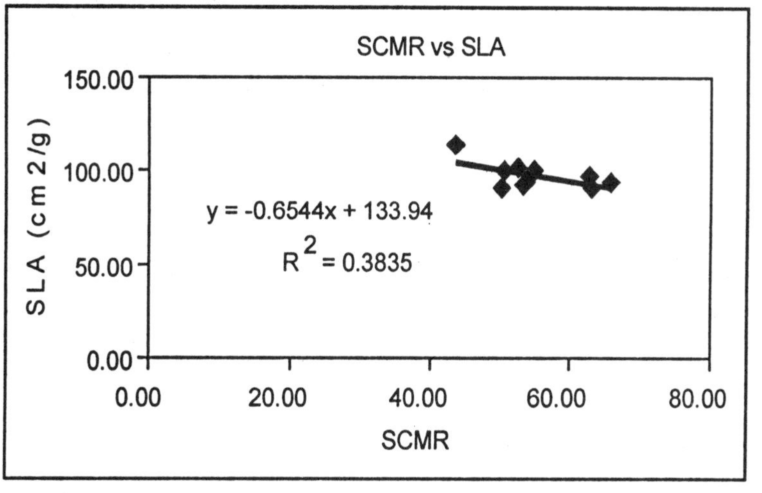

Figure 1: Correlation between SCMR and SLA at 25 days after imposing moisture stress of guar genotypes under moisture stress condition

Leaf peroxidase activity recorded at 60 DAS significantly varied between irrigated and moisture stress treatments. WSP-50, GG-1 and RGM-111 also recorded higher peroxidase activity under moisture stress conditions which further establish drought tolerant nature of these genotypes.

The present study revealed that WSP-50, GG-1 and RGM-111 confers drought tolerance in terms of higher RWC, SCMR and leaf peroxidase activity and lower SLA values compared to other genotypes. As both SLA and SCMR are considered to be surrogate methods to measure water use efficiency, a negative correlation between these two traits was established in the guar also. This study concludes that genotypic differences in water use efficiency traits and the identified genotypes can be used as good donar source for breeding drought tolerant genotypes.

REFERENCES

Chaves, M. M., Pereira, J. S. and Marco, J. (2003). Understanding plant response to drought-from genes to whole plant. *Functional Plant Biotechnology*, 30: 1-26

Latha, P. (2004). Genotypic variation for water use efficiency in advanced breeding lines of groundnut developed through trait and empirical methods under moisture stress. M.Sc. (Ag.) Thesis, ANGRAU, Hyderabad.

Sadasivam, S. and Manickam, A. (1997). Biochemical methods. *New Age International Publishers*:108-171

Sandhu, B. S. and Horton, M. L. (1978). Temporal response of oats to water stress in the field. *Agricultural Meteorology*, 19 : 329-336.

Sudhakar, P., Latha, P. and Babitha, M. (2006). Physiological traits contributing to grain yields under drought in black gram and green gram. *Indian Journal of Plant Physiology*, 11(4): 391-396

Wright, G. C., Rao, R. C. N. and Farquhar, G. D. (1994). Water use efficiency and carbon isotope discrimination in peanut under water deficit conditions. *Crop Science*, 34(1): 92-97.

□□□

Chapter 33

Effect of Bio-Regulators on the Yield of Greengram

ANANTHI K.[1], MALLIKA VANAGAMUDI[1], M. GOMATHY[2], K. KRISHNA SURENDAR[1], S. SRIMATHI[1] AND T. SIVASHANKARIDEVI[2]
[1]Department of Crop Physiology, TNAU, Coimbatore
[2]Department of Microbiology,TNAU, Coimbatore

Introduction

Pulses occupy a strategic position in intensive as well as subsistence agriculture, as they are excellent source of dietary protein for millions of people, nutritious feed for livestock and a mini nitrogen plant having profound ameliorative effect on the soil. Greengram is extensively grown in all types of soil under varying climatic conditions. Greengram is considered as the most whole some among pulses, free from heaviness and flatulence. The green pods are used as vegetable and haulms are used as fodder. On an average, the biomass has 1.5 % N on fresh weight basis, serving as a good source of animal feed. Flower and pod shedding is a common problem in greengram. There is also a possibility to overcome these constraints by foliar application of bio-regulators at the pre-flowering stage, which is one of the latest trends in agriculture. These bioregulators include humic acid (HA), Naphthalene Acetic Acid (NAA), Benzyl Adenine (BA), Salicylic Acid (SA), Brassinosteroid (BR) etc. These bioregulators in general, help to increase the number of flowers on the plant when applied at the time of flowering. The flower and pod drop may be reduced to some extent by spraying various growth regulators on foliage (Ramesh and Thirumurugan, 2001). The foliar application of PGRs and urea significantly increased seed yield per plant (Patil *et al.,* 2005). The HA increased the yield of plants more especially, with insufficient nutrient supply than with optimum nutrient supply (Pagel, 1960). The NAA is one of the important synthetic, stable and also widely used form of auxins. Auxins had been very effective in promoting growth and development in pulse crops. Upadhyay (1994) reported that foliar spray of NAA at bud initiation and pod formation stage of chick pea increased plant height, number of branches, number of flower bud, number of flowers and yield. The main role of NAA rests with the efficient transport of sugars from the photosynthesizing parts of the plant (source) to the developing grains (sinks) and also facilitating nitrogen accumulation that probably resulted in higher total dry matter production. Foliar spray of 25 ppm NAA recorded significantly higher seed yield by 21 to 22 per cent than control through increased flower production, clusters per plant, pod setting percentage and pods per plant in mungbean (Patil *et al.,* 2005). Cytokinins are adenine derivatives characterized by the ability to induce cell division in the presence of auxins. Cytokinin improves crop growth and yield through redirecting the metabolic balance of growth, delayed leaf senescence, increased photosynthetic CO_2 uptake and partitioning of assimilates (Brault and Maldiney, 1999). Ramesh and Thirumurugan (2001) reported that foliar spray of 25 ppm BA registered a higher DMA in soybean. Salicylic Acid belongs to the group of plant phenolics. A concentration of 100 and 500 ppm of SA affected both seed germination and seedling growth in two blackgram cultivars (*viz.,* CO 5 and T 9), but induced more flowering at 10 and 100 ppm levels respectively (Anandhi and Ramanujam, 1997). Brassinosteroid is a novel plant growth promoting steroidal lactone which is the most biologically active compound and acting as bioregulators in crop plants. Bindhu Joseph (2000) recorded an increased yield in groundnut by the foliar application of BR. Senthil *et al.* (2003) reported that foliar spray of 0.5 ppm BR increased the chlorophyll content in soybean.

Materials and Methods

A field experiment was conducted at Millet Breeding Station, TNAU, Coimbatore during Kharif (July, 2006 - October, 2006) in green gram (CO 6) in randomized block design with nine treatments and replicated thrice. Duration of the crop was 75 days. Sowing was done on 20.07.2006 with a spacing of 30 x 10 cm between rows and between plants. The soil was low in available nitrogen and phosphorus content and high in potassium content. The cultural management and plant protection measures were undertaken as and when needed. The treatment details were given below.

Treatment Details

T_1 - Control

T_2 - HA 0.1 % alone

T_3 - HA 0.1 % + 10 ppm Benzyl Adenine (BA)

T_4 - HA 0.1 % + 100 ppm Salicylic acid (SA)

T_5 - HA 0.1 % + 0.1 ppm Brassinosteroid (BR)

T_6 - HA 0.1 % + micronutrient mixture

T_7 - HA 0.1 % + micronutrient mixture + 10 ppm Benzyl Adenine (BA)

T_8 - HA 0.1 % + micronutrient mixture + 100 ppm SA

T_9 - HA 0.1 % + micronutrient mixture + 0.1 ppm BR

Sprays of 2 % DAP and 40 ppm NAA are common for all the above treatments. First spray was taken up at peak vegetative phase and second spray at 10 days thereafter.

The important components contributing to the yield potential of the crop were recorded from the commencement of flowering to the harvest of the crop. Different yield parameters were recorded and analyzed statistically.

Results and Discussion

The data collected on various pod characters as well as the yield components are presented in Tables 1,2 and 3.

Growth regulator spray on crop growth period had significant influence on the number of clusters per plant. Foliar spray of 0.1 % HA + 0.1 ppm BR spray (T_5) produced significantly more number of clusters per plant (7.7), which was closely followed by the foliar spray of 0.1 % HA + 10 ppm BA (T_3) with comparable value (7.6) (Table 1). The treatment of NAA was effective in increasing the number of clusters per plant (Patil *et al.*, 2005) in greengram. A high correlation was also observed between number of clusters per plant and grain yield in pigeonpea (Dani, 1979). Different treatments had significant influence on the number of pods per cluster which is an important yield component. Higher number of pods per plant was recorded with the 0.1 % HA + 0.1 ppm BR spray (T_5) (31.7), while the minimum number was rested with control (21.6) treatment (Table 1). Rajavel (2004) pointed out that the application of BR significantly influenced pods per cluster resulting in higher yield in black gram. Maximum number of seeds per pod was recorded by the treatment of 0.1 % HA + 0.1 ppm BR (9.75), which was followed by T_3 (9.48) and quite comparable. Both the treatments were significantly different from control. Minimum number of seeds per pod was however, produced by the control treatment (6.60) (Table 1). The influence of BA on number of seeds per pod was explained by Upadhyay (1994).

Table 1: Effect of foliar spray of liquid humic acid with growth regulators on yield and yield components in greengram

Treatment	Days to 50% flowering	Days to maturity	No. of clusters plant^{-1}	No. of flowers plant^{-1}	No. of pods clusters^{-1}	No. of pods plant^{-1}
T_1	30.6	67.5	5.7	74.6	3.8	21.6
T_2	29.8	67.1	6.0	77.3	4.0	23.9
T_3	27.1	65.2	7.6	94.5	4.1	31.2
T_4	27.7	65.5	7.3	92.6	4.1	29.9
T_5	26.3	64.7	7.7	95.5	4.1	31.7
T_6	29.6	67.6	6.2	79.5	4.0	24.7
T_7	28.5	66.6	6.9	89.5	4.1	28.2
T_8	28.9	67.1	6.7	86.8	4.0	26.8
T_9	28.1	66.3	7.4	90.5	4.1	26.6
Mean	28.5	66.4	6.8	86.7	4.0	21.65
SEd	0.3	0.8	0.2	1.5	0.1	1.2
CD (P=0.05)	0.5	1.9	0.4	3.1	0.3	2.7

No significant differences were observed in seed - pod ratio due to growth regulator / nutrient sprays. However, treatment T_3 produced higher seed - pod ratio (0.96) which was closely followed by the spray of 0.1 % HA + 0.1 ppm BR (0.93). The least value was recorded by the control (0.84) (Table 2). Ghosh *et al.* (1991) pointed out that the spray of NAA at pre-flowering stage significantly increased the shelling percentage. Similar results were obtained by Gomathi (1996) in mung bean.

The grain yield per plot of green gram was significantly influenced by various treatments tried. Among the treatments, foliar spray of 0.1% HA + BR 10 ppm (T_5) ranked first with higher grain yield per plot of 1047 g plot^{-1}, followed by T_3 (925). This was 23.32 and 8.95 per cent increase over the unsprayed control (849). The grain yield of greengram was significantly influenced by the foliar spray of different growth regulators and nutrients. Significantly higher yield was registered (873) with the treatment of 0.1 % HA + 0.1 ppm BR (T_5) in comparison with the control (708). The increased yield in the best treatment was found to be 23.31 percent over control and the second best treatment was T_3 (771) (Table 3 and Fig 1). The production of higher seed yield due to growth regulators may be attributed to the fact that plants treated with growth regulators remained physiologically more active to build up sufficient food reserves for developing flowers and seeds. Thus, the plant showed early flower initiation and better seed development which ultimately resulted in higher seed yield in the treated plants. Increased grain yield was recorded by BR spray in greengram as reported by Bhatia and Jatinder Kaur (1997). Application of NAA also enhanced the number of flowers, number of pods per plant, seed yield per plant and hundred seed weight in greengram (Patil *et al.*, 2005).

Table 2: Effect of foliar spray of liquid humic acid with growth regulators on fertility co-efficient and pod characters in greengram

Treatment	No. of seeds pod^{-1}	Pod weight plant^{-1} (g)	Seed weight plant^{-1} (g)	Seed / Pod ratio
T_1	6.60	11.24	9.68	0.84
T_2	7.28	11.61	10.58	0.86
T_3	9.48	13.58	12.25	0.96
T_4	8.84	13.26	11.67	0.90
T_5	9.75	13.83	12.55	0.93
T_6	7.58	11.90	10.77	0.87
T_7	8.57	12.51	11.30	0.89
T_8	8.33	12.15	10.94	0.90
T_9	8.74	12.81	11.34	0.90
Mean	8.35	12.54	11.23	0.89
SEd	0.14	0.47	0.48	0.03
CD (P=0.05)	0.30	0.99	1.02	0.06

Table 3: Effect of foliar spray of liquid humic acid with growth regulators on yield and yield components in greengram

Treatment	Pod length (cm)	Hundred seed weight (g)	Grain yield (g plot^{-1})	Grain yield (kg/ha)	Harvest Index (%)
T_1	5.8	3.28	849	708	26.24
T_2	5.7	3.52	875	729	26.88
T_3	7.1	4.13	925	771	30.57
T_4	6.8	4.05	912	760	30.01
T_5	7.3	4.27	1047	873	31.64
T_6	5.9	3.53	892	743	27.71
T_7	6.5	3.85	907	756	28.85
T_8	6.4	3.79	901	751	28.21
T_9	6.6	3.97	912	760	29.97
Mean	6.4	3.94	916	761	29.24
SEd	0.10	0.28	47	57	0.99
CD (P=0.05)	0.19	0.59	100	120	2.13

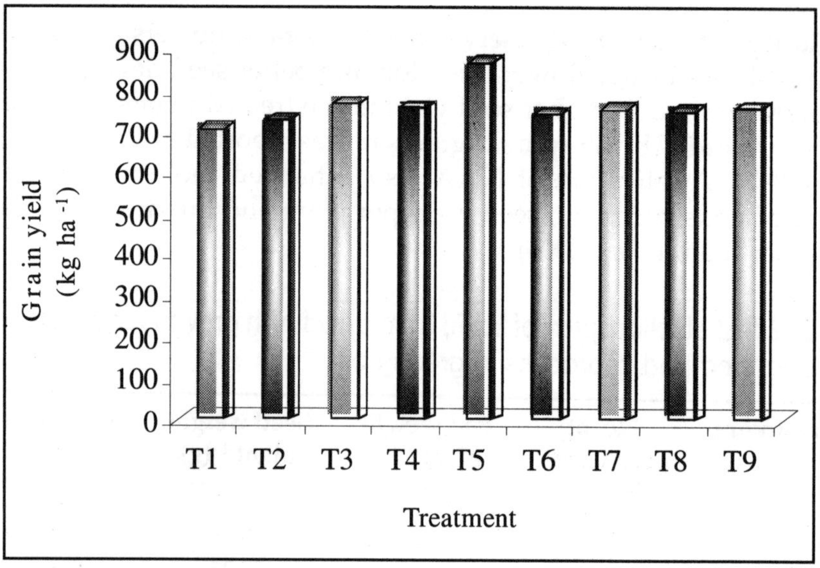

Figure 1. Effect of foliar spray of liquid humic acid with growth regulators on grain yield (kg/ha) in greengram

REFERENCES

Anandhi, S. and Ramanujam, M.P. (1997). Effect of salicylic acid on blackgram cultivars. *Indian J. Plant Physiol.,* 2: 138-141.

Bhatia, D.S. and Jatinder kaur. (1997). Effect of homobrassinolide and humicil on chlorophyll content, hill activity and yield components in mungbean (*Vigna radiata* (L.) Wilczek). *Phytomorphol.,* 47: 421-426.

Bindhu Joseph. (2000). Physiological and biochemical effects of Brassinolide on the productivity of groundnut. M.Sc. (Ag.) Thesis, Tamil Nadu Agricultural University, Coimbatore.

Brault, M. and Maldiney, R. (1999). Mechanism of cytokinin action. *Plant Physiol. Biochem.,* 37: 403-412.

Dani, R.G. (1979). Variability and association between yield and yield components in pigeonpea. *Indian J. Agric.Sci.,* 49: 507-510.

Ghosh, R.K., Bikash Kumar Mandai. and Chatterjee, B.N. (1991). Effect of growth regulators on the productivity of some major oilseed crops. *J. Agron. Crop Sci.,* 167: 221-228.

Gomathi, R. (1996). Chemical manipulation on yield improvement in greengram (*Vigna radiata* (L.) wikzek) M.Sc. (Ag.) Thesis, Tamil Nadu Agricultural University, Coimbatore.

Pagel, H. (1960). Uber den Einfuss Von Humussofen auf dus pfalnzenwachstum. I. Einfluss. Von Humusstoffent auf keimug and Warzerwahstum. *Albrecht. Thaer. Arctiv.,* 4: 450-468.

Patil, S.N., Patil, R.B. and Suyawanshi. Y.B. (2005). Effect of foliar application of plant growth regulators and nutrients on seed yield and quality attributes of mungbean (*Vigna radiata* (L.) Wilezeli. *Seed Res.*, 33: 142-145.

Rajavel, S. (2004). Physiological studies on increasing yield potential in blackgram (*Vigna mungo* (L.) Hepper) Genotypes with plant growth regulators. M.Sc. (Ag.) Thesis, Tamil Nadu Agricultural University, Coimbatore.

Ramesh, K. and Thirumurugan, V. (2001). Effect of seed pelleting and foliar nutrition on growth of soybean. *Madras Agric. J.* 88: 465-468.

Senthil, A., Pathmanaban, G. and Srinivasan, P.S. (2003). Effect of bioregulator on some physiological and biochemical parameters of soybean (*Glycine max* L.). *Leg Res.*, 28: 54-56.

Upadhyay, R.G. (1994). Effect of bioregulators on growth, development, flowering behaviour and yield of chickpea (*Cicer arietinum* L.). *Legume Res.*, 17: 60-62.

□□□

Chapter 34

A Step-wise Protocol for Standardization of *In Vitro* Pollen Germination Medium for Pigeonpea and Its Wild Species

JAYAPRAKASH, P.
Department of Agricultural Botany, Faculty of Agriculture, Annamalai University
Annamalainagar -608002, Tamil Nadu

Introduction

Legume pollen is difficult to germinate in artificial medium. Standardization of *in vitro* pollen germination is a pre-requisite for pollen storage, *in vitro* fertilization etc., In the present study, Brewbaker and Kwack medium was used to germinate the pollen of pigeonpea and two of its wild species *viz.*, *Cajanus volubilis* and *C. platycarpus*. Pigeonpea requires complex medium for germination as compared to its wild species. Addition of poly ethylene glycol and ?-amino caproic acid (EACA) improved the germination over 95 %(Jayaprakash and Sarla, 2001). A complete *in vitro* pollen germination medium was reported for all these species for the first time(Jayaprakash, 1998).

Few hundreds of combinations were tried changing the concentration of constituents of the BK medium. A simplified procedure for standardization of *in vitro* pollen germination of these species has been developed which involved two sets of key media. Pollen of all species were incubated in the key media, the medium giving best germination was selected and modified.

Materals and Methods

Cajanus cajan var. Pusa 33, *C. platycarpus* and *C.volubilis* were used for *in vitro* pollen germination. The seeds were sown in June and most of the *in vitro* pollen germination studies were carried out during September and March of 1997–99 at the Indian Agricultural Research Institute, New Delhi, India.

Pollen Germination Media (PGM)

In this study, Brewbaker and Kwack (BK) medium (Brewbaker and Kwack , 1964) (10% sucrose, 100 mg l^{-1} boric acid, 300 mg l^{-1} calcium nitrate, 200 mg l^{-1} magnesium sulphate, and 100 mg l^{-1} potassium nitrate) was used. Two sets of key media were used initially. Set I consisted of BK medium with 1% agar at different levels of sucrose viz., 10, 20, 30, and 40 percent (A, B, C & D). With addition of 15% PEG 4000 to each of A, B, C & D; the media E, F, G & H were obtained. These were designated as Set II.

Key Media

Set I	Set II
A- 10% sucrose +BK salts+1% agar	E- medium A + 15% PEG
B-20%sucrose +BK salts+1% agar	F- medium B + 15% PEG
C-30%sucrose +BK salts+1% agar	G- medium C + 15% PEG
D-40%sucrose +BK salts+1% agar	H- medium D + 15% PEG

The selected basal medium was then used in subsequent studies and further improved by altering the concentrations of sucrose, boric acid and calcium nitrate one by one, respectively. Since liquid medium gave inconsistent results, 1% agar was added in all the media used and the results were consistent in replications.

Pollen Collection and Culture

Pollen was collected and cultured according to the method described earlier (Singh *et al.*, 1992). Five replicates were used for each medium combination/treatment. All observations on pollen germination and pollen tube growth were taken 3 h after incubation. At least 250 pollen was counted from 5–10 random fields. A pollen grain was considered germinated if pollen tube length was greater than the diameter of the pollen grain.

Results

Pollen is germinated in the key media. The criteria for selection of medium is based on the bursting of pollen, initiation of germination (budding), pollen tube bursting and pollen with smooth pollen tube (Fig.1) and then it is modified. The procedure is depicted schematically (Fig. 2).

Cajanus Cajan

The best germination for pollen of pigeonpea was observed in medium G. Further, increasing sucrose concentration to 37.5% controlled pollen bursting substantially. Then other constituents of this medium were altered to improve pollen germination. A set of five media (G_1-G_5) with different permutations and combinations of boric acid and calcium nitrate showed an improved pollen germination of 53–87% (Table 1). However, pollen and pollen tube bursting was a common feature in all these media used. The maximum mean pollen tube length of 19.63 μm was observed in medium G_3.

Table 1 : *In vitro* germination of pollen and pollen tube growth of *C. cajan*

Medium	Concentration of		Germination%	Pollen tube growth(μm)
	Boric acid(mg/l)	Calcium nitrate(mg/l)		
G	200	300	53.70±10.63	5.14±0.295
G1	250	300	87.44±5.139	10.0±0.501
G2	300	300	74.52±2.558	9.65±0.345
G3	250	100	66.75±2.734	19.63±0.632
G4	250	200	87.25±4.873	12.56±0.508
G5	250	300	53.70±10.63	5.14±0.295

C. Platycarpus

The pollen of *C.platycarpus* showed complete germination (> 99%) in medium B with mean pollen tube length of 172.56 μm (Table 2). The germination was 94.25% in medium C. However, there was irregular pollen tube growth and pollen tube bursting was observed.

Table 2 : *In vitro* germination of pollen and pollen tube growth of *C. platycarpus*

Medium	Germination	Pollen tube growth(μm)
A	61.60 ±0.661	83.88
B	99.15 ±0.315	172.56
C	94.25 ±0.139	33.31

C. Volubilis

When the pollen was germinated, a highest percentage of 20 was observed in medium G of Set II. Then pollen was germinated in medium G with varying concentration of its constituents(G1-G5). It was observed that a medium having 250ppm boric acid and 100 ppm calcium nitrate gave 94.26 % germination(G3) followed by G4 with 83.12% (Table 3).

Table 3 : *In vitro* germination of pollen and pollen tube growth of *C.volubilis*

Medium	Concentration of		Germination%	Pollen tube growth(µm)
	Boric acid(mg/l)	Calcium nitrate(mg/l)		
G	200	300	<20	-
G1	250	300	62.12	-
G2	300	300	31.03	-
G3	250	100	94.26	27.47
G4	250	200	83.12	18.42
G5	250	300	61.23	16.16

Discussion

In the present study, Brewbaker and Kwack's medium (Brewbaker and Kwack, 1964) with some minor modifications was used for *in vitro* pollen germination of pigeonpea and its wild species. Pigeonpea pollen requires a complex medium for germination so the medium (PGM) was also supplemented with PEG 4000. In these media, 53–87% germination was observed with profuse pollen and pollen tube bursting. Addition of PEG improved the percentage germination in pigeonpea. It has also been found useful in another legume, chickpea (Shivanna *et al.*, 1997). Use of a liquid medium consisting of 40% sucrose, 250 mg l^{-1} boric acid and 200 mg l^{-1} calcium nitrate to germinate pigeonpea pollen was reported earlier for *C. cajan*, but only 48.7% pollen germinated (James *et al.*, 1987). Later, Singh *et al.* (1992) modified the same medium by adding 0.7% agar and achieved a maximum of 43.1% pollen germination at 22 °C. A complete pollen germination medium for pigeonpea was improved by the addition of EACA(Jayaprakash and Sarla, 2001). They also noticed that there are genotypic differences for pollen germination.

Legume pollen in general is difficult to germinate. Few hundreds of media combination has been used for germination of *Cajanus* spp. Based on the observation, two sets of key media were developed. A scheme proposed shows a systematic way of standardizing media constituents for pollen germination.

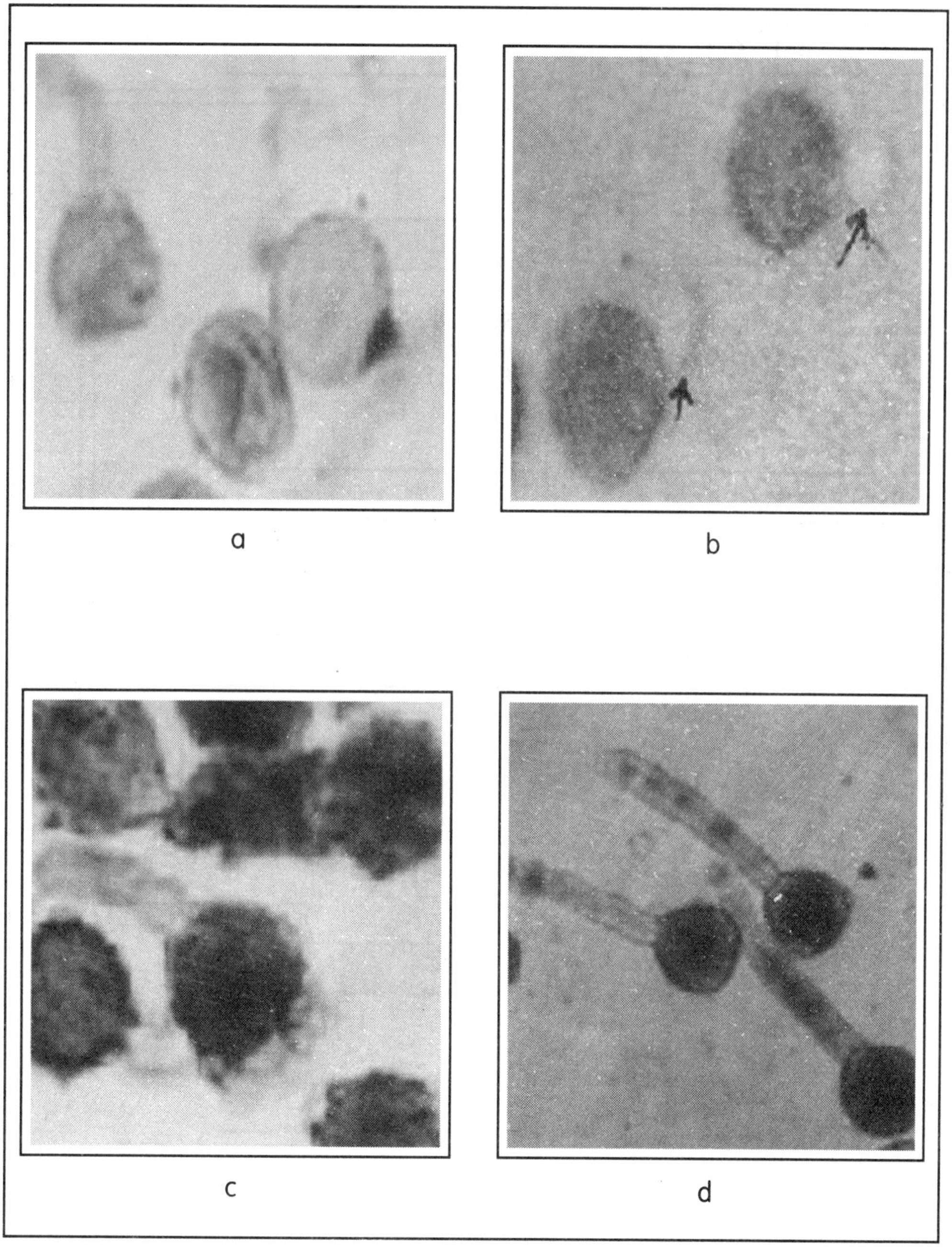

Figure 1: Response of pollen in pollen germination medium a) bursting of pollen, b) budding, c) pollen tube bursting d) pollen with smooth pollen tube

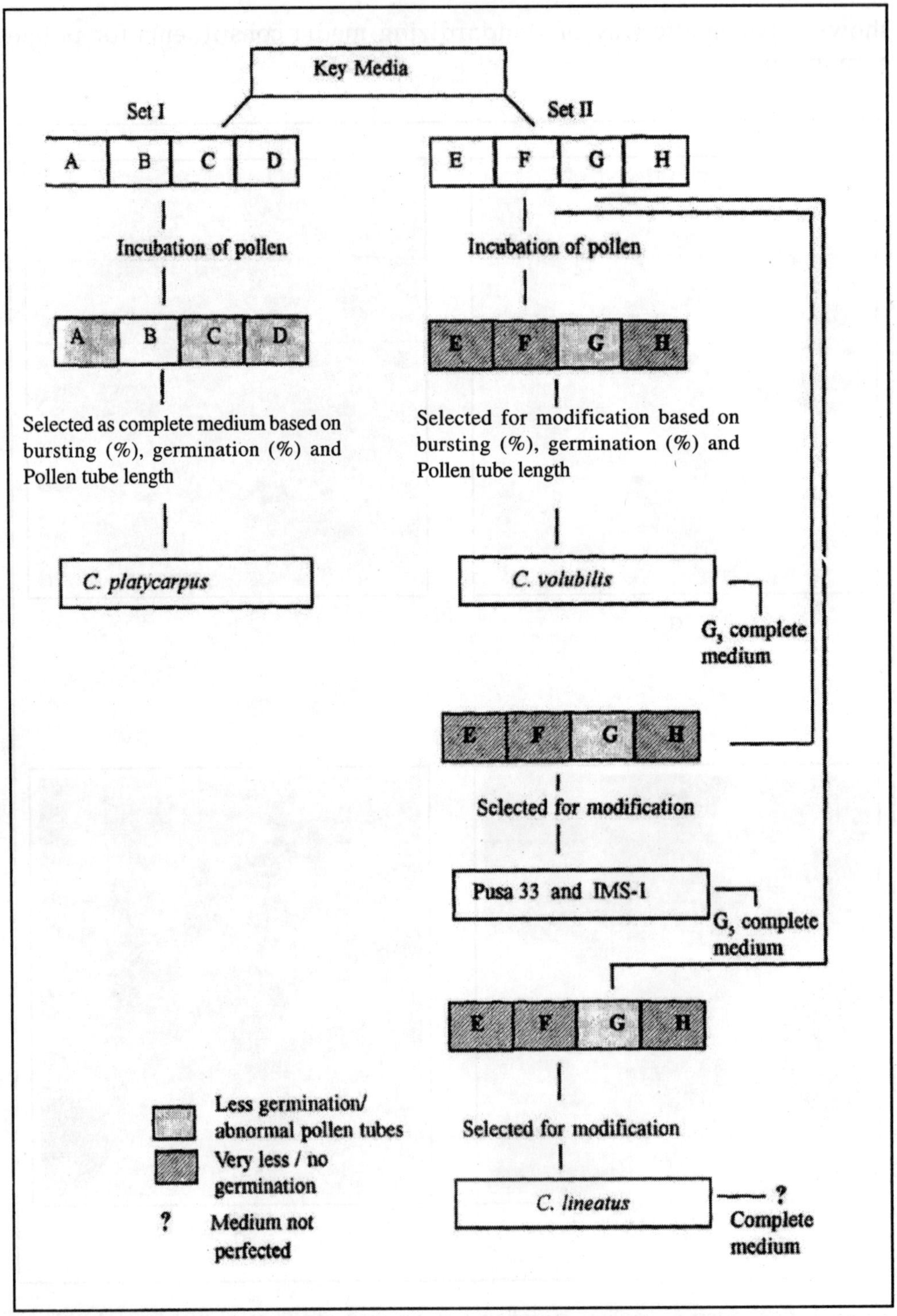

Fig. 2. Procedure for standardizationof *in vitro* pollen germination media.

REFERENCES

Brewbaker, J.L., Kwack, B.H. 1964. The calcium ion and substances influencing pollen growth. In: Linskens HF, ed. Pollen physiology and fertilization. Amsterdam: Elsevier North Holland, 145–151.

James, D., Ariyanayagam, R.P., Dungan, E.J. 1987. Comparative studies of *in vitro* germination of pigeonpea (*Cajanus cajan* (L.) Millsp.) and *Atylosia platycarpa* Benth. Tropical Agriculture 64, 313–346.

Jayaprakash, P. 1998. Studies on crossability of *Cajanus cajan* (L) Millsp. and its wild species. Ph.D. thesis, Department of Genetics, Indian Agricultural Research Institute, India.

Jayaprakash, P. and Sarla. 2001. Development of an improved medium for germination of Cajanus cajan *(L.) Millsp. pollen* in vitro . *Journal of Experimental Botany, 52(357): 851-855.*

Shivanna, K.R., Saxena, N.P., Seetharama, N. 1997. An improvised medium for *in vitro* pollen germination and pollen tube growth of chickpea. *International Chickpea and Pigeonpea Newsletter* 4, 28–29.

Singh, I.S., Bharti, I.S., Nandwal, A.S., Goswami, C.L., Verma, S.K. 1992. Effect of temperature on *in vitro* pollen germination in pigeonpea. *Biologia Plantarum* 34, 461–464.

❑❑❑

SECTION - IV
CROP PROTECTION

Chapter 35

Effect of Age of Plants and Standardisation of Inoculum Level of *Macrophomina phaseolina* Causing Root Rot of Blackgram

RETTINASSABABADY. C. AND C.ISSAC SUNIL
Department of Plant Pathology, Pandit Jawaharlal Nehru College of Agriculture & Research Institute, Karaikal-609 603, U.T. of Puducherry

Introduction

Among the various diseases affecting rice fallow blackgram, root rot incited by *Macrophomina phaseolina* (Tassi.) Goid. was found to reduce the population of the crop in the field leading to considerable reduction in yield (Than *et al.*, 1991). Muthukrishnan *et al* (1995) reported the incidence of root rot in Rice fallow blackgram. The pathogen is both seed and soil inhabitant in nature (Dhingra and Khare, 1973).The soil borne nature of the pathogen is a major hurdle in the successful management of the disease. It causes seedling mortality, root rot and charcoal rot on more than 500 plant species in tropical and sub-tropical countries (Ghaffer, 1988). In soil, the pathogen primarily survives as sclerotia and released into the soil as host tissue decay(Cook *et al.*, 1973). Shokes *et al* (1977) observed that, the ability of *M.phaseolina* to develop at water potentials at which bacteria were inactive and the hosts weakened, explained the higher incidence of rots fairely on dry or alternately wet and dry soil than on wet soil. Hence, an experiment was conducted to study the relationship of age of blackgram plants to root rot development as well as to standardise the inoculum level of *M.phaseolina* towords inciting root rot.

Materials and Methods

Macrophomina phaseolina (Tassi.) Goid was isolated from the naturally infected blackgram plants showing typical dry root rot symptoms during ricefallow season. The auxenic culture of the pathogen was obtained by single hyphal tip method and maintained in PDA slants. About 200 g of sand:maize meal medium(95:5) was prepared in saline bottles of 500 ml size with 35 per cent moisture level. The bottles were sterilised at 15 lbs pressure for two hours on two successive days. *M.phaseolina* was grown on PDA for 7days. Six fungal discs of 9mm size were inoculated into each bottle. The bottles were incubated at room temperature (29+1 ^{c}) for 10 to 12 days. The sand:maize meal culture of *M.phaseolina* was used as inoculum for the experiment. To find out the critical stage of the plant vulnerable to root rot disease development, a pot culture experiment was conducted . Blackgram seeds (ADT 3), surface sterilised with 0.1% mercuric chloride for 2 minutes were sown in pots containing sterilised soils inoculated with *M.phaseolina* (5% inoculum). Five seedlings per pot and five replications were maintained. The observations were recorded at 15 days interval till the crop was harvested. Pot mixture was prepared with field soil and sand in 3:1 ratio. Five kg of this soil mixture were taken in earthen pots of 30 cm diameter and sterilised in an autoclave at 1.4 kg/cm^2 for two hours. The inoculum multiplied in sand:maize medium was added to the pot soil @ 0,2,5 and 10 per cent of soil weight and mixed well. In each pot, ten seeds of blackgram *cv* ADT 3 were sown and irrigated. For each treatment five replications were maintained in completely randomised design. The per cent seed germination was recorded seven days after sowing (DAS). The number of plants showing progressive root rot was recorded on 15,30,45 and 60 DAS and the per cent incidence was calculated.

Results and Discussion

When graded levels of inocula of the pathogen were used, even the lowest level (2%) reduced the seed germination and caused infection (Table-1). Minimum seed germination of 91.70 per cent was observed in 10 per cent inoculum level which was on par with 5 per cent inoculum (94.6%). Samiyappan (1988) also observed that a very low number of propagules of *M.phaseolina* were sufficient to cause infection on blackgram. The root rot incidence was found to increase with increase in days after sowing. The root rot incidence on 15 DAS was found to be higher (10.30) at 10 per cent inoculum level than at 5 per cent inoculum level recording 2.3 per cent. But the disease incidence at 5 per cent inoculum level from 30 DAS was on par with 10 per cent inoculum level of *M.phaseolina*. Sheik and Ghaffer (1979) observed that even 1-2 sclerotia/g of soil caused 12 to 20 per cent infection in urdbean Whereas, Zambolium *et al.* (1983) reported that 38 sclerotia/g of soil caused 50 per cent infection in soyabean. Shankar (1994) observed that 5 per cent *M.phaseolina* inoculum was able to cause 90 per cent root rot incidence coupled with significant reduction in germination of sesame.

Table 1. Effect of different inoculum levels of *M.phaseolina* on seed germination and root rot incidence of blackgram

Inoculum level (%)	Seed germination (%)	Root rot incidence on days after sowing (DAS)			
		15	30	45	60
0	100.00[a]	0.00[aA]	0.00[cA]	0.00[cA]	0.00[cA]
		(0.20)	(0.20)	(0.20)	(0.20)
2	99.90[a]	0.00[cC]	9.60[bB]	20.60[bA]	38.00[bA]
		(0.20)	(17.98)	(26.96)	(29.46)
5	94.60[b]	2.30[bD]	26.60[aC]	43.20[aB]	61.30[aA]
		(8.65)	(31.03)	(41.00)	(51.34)
10	91.70[b]	10.30[aD]	28.00[aC]	45.60[aB]	63.60[aA]
		(18.66)	(31.73)	(42.44)	(52.57)

Figures in parentheses are the angular transformed values
In a column, mean followed by the same letter (lower case) and in a row mean followed by the same letter (upper case) are not different statistically by DMRT.

The experiment conducted to assess the effect of age of blackgram plants on *M.phaseolina* infection, revealed that significant increase in root rot incidence was observed with increase in age of plants (Table-2). Blackgram plants of 45 and 60 days old were severely affected by root rot whereas in soybean, 15 and 30 days old plants had moderate infection as observed by Meyer *et al.* (1974). Shanmugasundaram (1992) reported that sunflower crop was severely affected at later stages by *M.phaseolina*. The susceptibility of blackgram plants in later stages of growth (45 and 60days old) observed in the present investigation might be due to the decline in the level of reducing sugars as reported by Davet and Series (1987) in sunflower.

Table 2. Effect of age of plants on the incidence of root rot

S. No.	Age of Plants (Days)	Per cent root rot incidence
1	15	12.0[a] (20.26)
2	30	21.0[b] (27.27)
3	45	39.5[c] (38.57)
4	60	56.0[d] (48.49)

Figures in parentheses indicate mean angular transformed values.
In a column, mean followed by the same letter are not different statistically by DMRT

REFERENCES

Cook, G.E., Boosalis, M.G., Dunkle, L.D. and Wodvony, G. (1973). Survival of *Macrophomina phseolina* in corn and sorghum stalk residue. *Pl. Dis. Rept.*, 57:873-875.

Davet, P. and Series, H. H. (1987). Relation between the amount of reducing sugars in sunflower tissues and their invasion by *Macrophomina phaseolina* (Tassi.) Goid. *Phytopathology*, 77 :212-226.

Dhingra, O.D. and Khare, M.N. (1973). Biological control of *Rhizoctonia bataticloa* on urdbean. *Phytopathology*, 76: 23-29.

Ghaffer, A. (1988). Soil borne disease research centre, Final research report, 1st jan. 1986-30th june 1988. Karachi, Pakistan, p.110.

Meyer, W.A., Sinclair, J.B. and Khare, M.N.(1974). Factors affecting charcoal rot of soyabean seedlings. *Phytopathology*, 64:845-849.

Muthukrishnan, K., Arjunan, G. and Raguchander, T. (1995), Effect of dates of sowing on the incidence of root rot in ricefallow blackgram. Madras Agric. J.,82:495-496.

Samiyappan, R. (1988). Biological control of blackgram root rot caused by *Macrophomina phaseolina* (Tassi.) Goid. Ph.D.Thesis, TNAU, Coimbatore, India, p.169.

Shankar, P. (1994). Biological control of sesamum root rot caused by Macrophomina phaseolina (Tassi.) Goid. M.Sc., (Ag.) Thesis, TNAU, Coimbatore, India, p.141.

Shanmugasundaram, P. (1992). Studies on charcoal rot of sunflower (*Helianthus annus* L.) caused by *Macrophomina phaseolina* (Tassi.) Goid .M.Sc.,(Ag.) Thesis, TNAU, Coimbatore, India,p. 85.

Sheikh, A.H. and Ghaffar, A. (1979). Relation of Sclerotial inoculum density and soil moisture to infection of field by *Macrophomina phaseolina*. *Pak. J. Bot.*, 11:185-189.

Shokes, F.M., Lyda, S.D. and Jordan, W.R. (1977). Effect of water potential on the growth and survival of *Macrophomina phaseolina*. *Phytopathology*, 67:239- 241.

Than, H., Thein, M.M. and Myint, S.S. (1991). Relationship among *Rhizoctonia bataticola* isolates in rice based cropping system based on colony fusion types, *Int. Chickpea Newsl.*,25:29-31.

Zambolim, L., Schenck, N.C. and Mitchell, D.J. (1983). Inoculum density, pathogenicity and interactions of soyabean root infecting fungi. *Phytopathology*,73:1398-1402.

□□□

Chapter 36

Field Screening of Cowpea Accessions for Resistance Against Selected Insect Pests

SAJAD HUSSAIN MIR AND V. SELVANARAYANAN
Department of Entomology, Faculty of Agriculture
Annamalai University - 608002 Tamil Nadu, India

Introduction

Cowpea (*Vigna unguiculata* (L.) Walp.) is the most important grain legume being cultivated widely in the semi-arid regions as sole or mixed crop. It is highly valued as a major source of protein and considered as "Poor man's meat". In India, cowpea is cultivated both under irrigated and rainfed conditions either as sole/mixed or intercrop.

Many phytophagous insects infest cowpea at various stages of growth (Jackai and Daoust, 1986) and management of these insects by chemical means is discouraged keeping in view of the resistance development in target insects, resurgence of minor pests and residual toxicity to non-target organisms. Hence identifying, exploiting and utilizing host plant resistance available in germplasm of cowpea is a viable alternative to chemicals management of insect pests.

Materials and Methods

Preliminary Field Screening

Fourteen cowpea accessions were gathered and screened for resistance against four insect pests namely leafhoppers, *Empoasca* sp., aphids, *Aphis craccivora*, thrips, *Taeniothrips* sp. and blue butterflies, *Lampides boeticus* Linn. under field conditions. The field trial was carried out at the Experimental Farm, Faculty of Agriculture, Annamalai University, Annamalai Nagar, Tamil Nadu, India.

The seeds of the test accessions were sown along the sides of the ridges by adopting 90 cm between ridges and 45 cm between plants. Regular inter cultivation practices, need based irrigation and organic fertilization were used. Incidence of sap feeding insects such as leafhoppers, thrips, aphids and the pod borer namely blue butterflies was recorded from 30 days after sowing (DAS) at seven days interval till the harvest of the crop.

Three replications were maintained and the observations were recorded from five plants selected at random in each replication at weekly intervals. The incidence of aphids, thrips and leafhoppers were recorded by counting the number of insects on three leaves from each plant *viz.*, one top, middle and bottom leaf each respectively. The population of pod borer larvae was recorded in each plant. The mean values of the weekly observations were computed based on which the accessions were rated for their resistance.

Confirmatory Field Screening

Based on the preliminary field screening of 14 cowpea accessions, ten cowpea accessions were selected and again field evaluated to confirm their resistance potentials by adopting the procedures followed in the preliminary screening.

Resistance Rating of the Selected Accessions Based on Infestation Index

Based on the mean population of various insects, the accessions were grouped according to the classification detailed below.

Insect pest	Population/ plant		Resistance rating
	Preliminary field screening	Confirmatory field screening	
Leafhoppers, thrips and blue butterflies	0.0 - 0.3	0.0 - 0.3	R
	0.4 - 0.7	0.4 - 0.7	MR
	0.8 -1.1	0.8 -1.1	S
	1.2 - 1.5	1.2 - 1.5	HS
Aphids	0.0 - 2.0	0 - 3	R
	2.1 - 4.0	4 - 6	MR
	4.1 - 6.0	7 - 9	S
	6.1 - 10	10 - 12	HS

Results and Discussion

Based on the resistance rating scale developed, in the preliminary screening, among the 14 cowpea accessions screened, the accession P-152 was rated as resistant (R) against leaf hoppers while five accessions namely CO 1, CO 2, CO 4, Vu-1 and V-240 were rated as moderately resistant (MR). Aphid incidence was the least in the accession CO 6 and P-152 which were rated as resistant while five accessions were rated as MR to aphids. Population of thrips was the least in the accession CO 6 and seven accessions were rated as MR to thrips. Based on the incidence of blue butterfly larvae, the accessions P-152, CO 1, CO 2, CO 4 and C0 6 were rated as Resistant.

On consolidating the observations against various insect pests, it is noted that the accession P-152 was rated as resistant (R) against leaf hoppers, aphids and blue butterflies while it was moderately resistant (MR) against thrips. Three accessions namely CO 1 and CO 2 were rated as moderately resistant (MR) against the sap feeders whereas resistant against blue butterflies. Accessions Vu-16, Vu-25, Vu-30 and VBN 1 were found susceptible (S) or highly susceptible (HS) to leafhopper, aphids, thrips and blue butterfly (Table 1).

Table 1. Resistance rating of the cowpea accessions based on preliminary screening under field conditions

S. No.	Accession	Leafhopper Mean	Leafhopper Grade	Aphids Mean	Aphids Grade	Thrips Mean	Thrips Grade	Blue butterfly Mean	Blue butterfly Grade
1.	CO 1	0.7	MR	2.1	MR	0.5	MR	0.3	R
2.	CO 2	0.7	MR	3.0	MR	0.4	MR	0.3	R
3.	CO 4	0.6	MR	4.2	S	0.4	MR	0.3	R
4	CO 6	0.9	S	1.2	R	0.3	R	0.3	R
5.	CoCP 702	0.8	S	3.7	MR	0.6	MR	0.4	MR
6.	Vu-1	0.7	MR	3.2	MR	0.9	S	0.4	MR
7.	Vu-8	0.8	S	4.0	MR	0.5	MR	0.4	MR
8.	Vu-10	0.8	S	5.4	S	0.7	MR	1.0	S
9.	Vu-16	1.4	HS	9.7	HS	0.8	S	1.2	HS
10.	Vu-25	1.1	S	8.4	HS	1.3	HS	0.8	S
11.	Vu-30	1.2	HS	9.6	HS	1.2	HS	1.4	HS
12.	VBN 1	1.2	HS	8.3	HS	1.3	HS	1.4	HS
13.	V-240	0.4	MR	5.7	S	1.2	HS	0.4	MR
14.	P-152	0.3	R	1.2	R	0.4	MR	0.2	R

It is evident from the results obtained in the preliminary field screening tests that out of 14 accessions tested, four accessions namely Vu-16, Vu-30, Vu-25 and VBN-1, were found to be either highly susceptible or susceptible, while the remaining accessions *viz.*, CO 1, CO 2, CO 4, CO 6, COCP-702, Vu-1, Vu-8, Vu-10, V-240 and P-152 were resistant/ moderately resistant or susceptible to various insect pests. Hence, the four susceptible or highly susceptible accessions were discarded and ten accessions that responded as resistant or moderately resistant against most of the insects were selected for confirmatory screening.

In the confirmatory screening, among the accessions evaluated, P-152 was categorized as resistant against all the test insects whereas the accession CO 6 was rated as moderately resistant against sap feeders and resistant against blue butterfly (Table 2). The other accessions recorded varying degrees of resistance against the insect pests as reported earlier against aphids and leafhopper by Jayappa and Lingappa (1988) and Singh and Sangwan (2000) respectively. In general, the incidence of the insect pests was more pronounced during the confirmatory screening than the preliminary screening which led to variation in the resistance or susceptibility of the test accessions (Fig. 1 to 4). Such influence of seasonal factors on the insect population and in turn its influence on insect resistance has been well documented in sesame (Vijai Anandh and Selvanarayanan, 2005).

Table 2. Resistance rating of the selected cowpea accessions based on confirmatory screening under field conditions

S. No.	Accession	Leafhopper		Aphids		Thrips		Blue butterfly	
		Mean	Grade	Mean	Grade	Mean	Grade	Mean	Grade
1.	CO 1	0.32	R	8.97	S	0.65	MR	1.18	S
2.	CO 2	1.10	S	12.27	HS	0.97	S	0.62	MR
3.	CO 4	1.10	S	9.92	S	0.50	MR	0.65	MR
4	CO 6	0.70	MR	4.83	MR	0.42	MR	0.30	R
5.	CoCP 702	1.40	HS	10.41	HS	0.87	S	1.40	HS
6.	Vu-1	0.90	S	4.55	MR	1.20	HS	1.50	HS
7.	Vu-8	0.90	S	5.16	MR	1.57	HS	1.07	S
8.	Vu-10	1.50	HS	10.55	HS	1.42	HS	1.00	S
9.	V-240	1.00	S	5.36	MR	1.70	HS	1.20	HS
10.	P-152	0.32	R	3.81	R	0.32	R	0.25	R

Observations on the various agronomic characters of the selected accessions revealed that the plants of the accession CO 2 were the tallest in contrast to the least in the accession Vu-1. Length and breadth of the pods were the maximum in the accession P-152. Similarly, weight of individual pods was the maximum in the accession P-152 followed by CO 4. The seeds of the accession P-152 were bold and recorded the maximum hundred seed weight. The plants of the accession P-152 had the maximum number of pods per plant. All these characters culminated in recording the maximum seed yield in the case of the accession P-152.

Hence, it is concluded that the accession P-152 being resistant / tolerant to many insect pests, can be used in future breeding programmes.

Table 3. Economic characters of the selected cowpea accessions

S. No.	Accession	Pod length	Pod breadth	Pod weight	No. of pods/ plant	100 seed weight	Yield/ plant
1.	CO 1	14.28	0.63	1.50	6.76	7.82	7.62
2.	CO 2	13.36	0.61	1.39	7.11	7.50	6.10
3.	CO 4	13.30	0.71	1.60	8.67	5.86	8.18
4	CO 6	11.78	0.63	1.58	7.89	8.24	9.88
5.	CoCP 702	13.30	0.62	1.51	6.93	8.40	6.58
6.	Vu-1	10.44	0.57	1.20	7.57	8.36	6.91
7.	Vu-8	12.44	0.65	1.35	8.06	10.70	6.92
8.	Vu-10	12.90	0.61	1.42	8.49	7.52	7.17
9.	V-240	10.36	0.63	1.48	8.57	7.70	7.12
10.	P-152	15.42	0.72	1.70	12.44	13.04	12.44
	S. Ed.	1.04	0.03	0.14	0.93	0.93	1.05
	C. D. (p=0.05)	2.09	0.06	0.27	1.87	1.87	2.12

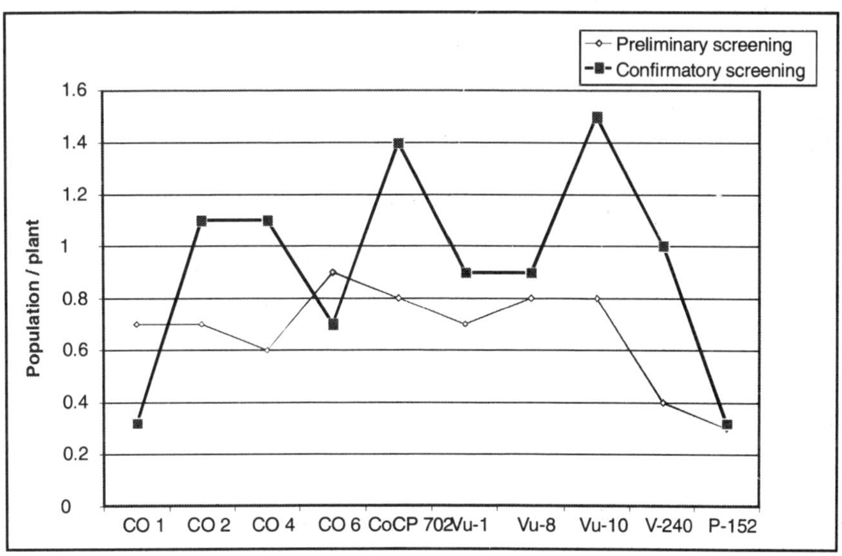

Figure 1. Reaction of cowpea accessions against leafhoppers

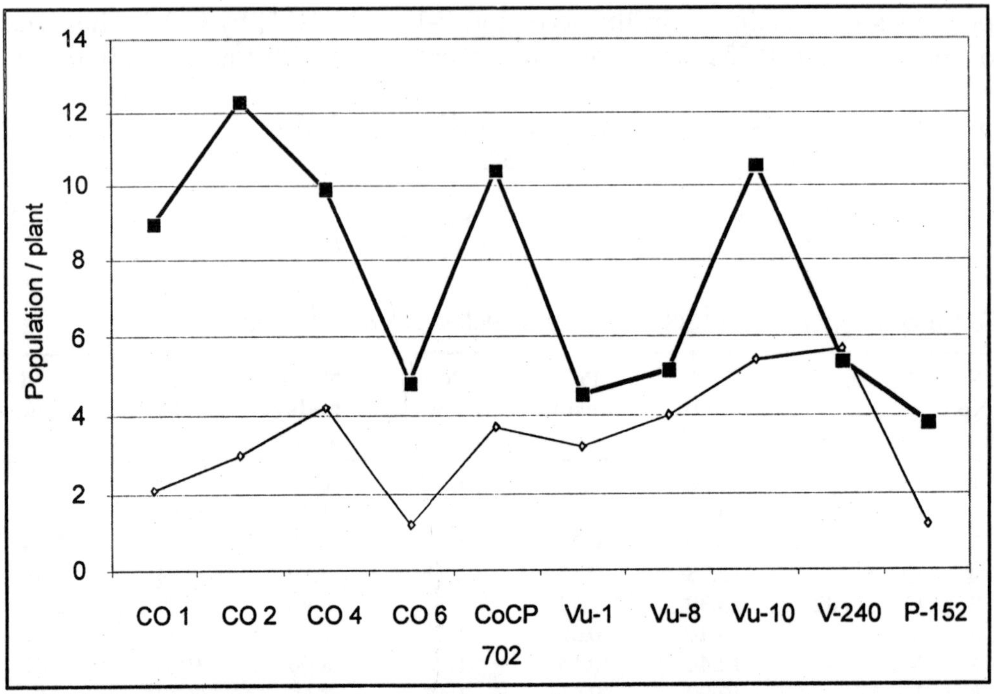

Figure 2. Reaction of cowpea accessions against aphids

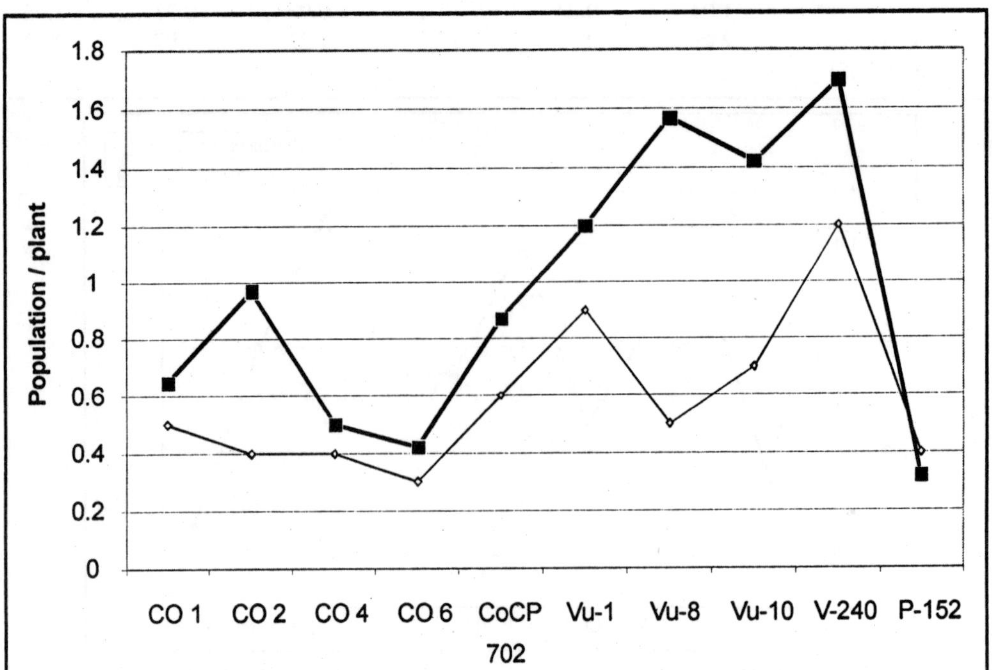

Fig. 3. Reaction of cowpea accessions against thrips

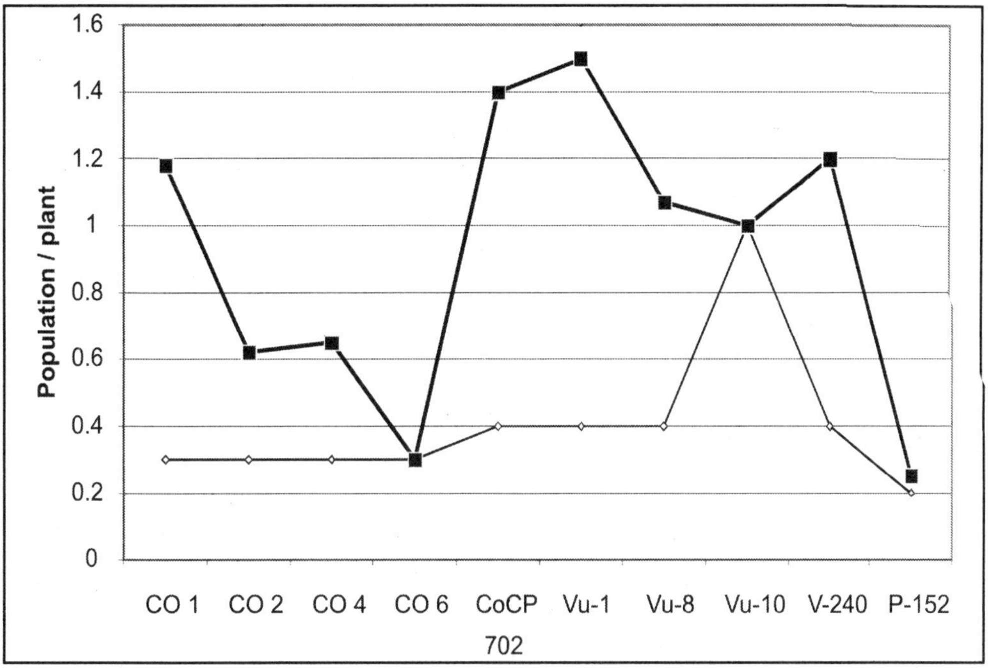

Figure 4. Reaction of cowpea accessions against blue butterfly

REFERENCES

Jackai, L.E.N and Daoust, R.A. (1986). Insect pests of cowpeas. *Ann. Rev. Entomol.*, 31: 95-119.

Jayappa, B.G. and Lingappa, S. (1988). Causes of resistance to *Aphis craccivora* Koch in cowpea germplasm in India. *Trop. Pest. Management*, 34: 59-61.

Singh, S.P. and Sangwan, R.S. (2000). Relative susceptibility of some cowpea genotypes to leafhopper and whitefly. *Haryana J. Horti. Sci.*, 29: 261-263.

Vijai Anandh, G. and Selvanarayanan, V. (2005). Reaction of sesame germplasm against shoot webber, *Antigastra catalaunalis* Duponchel and phyllody. *Indian J. Plant Protec.*, 33(1): 35-38.

□□□

Chapter 37

Study on Efficacy of Neem Based Products Against Pulse Beetle During Storage of Mung Bean (*Vigna radiata* (L.) Wilczek) Seeds

RAMANADANE, T., K. PARAMASIVAM AND C. RETTINASSABABADY
Pandit Jawaharlal Nehru College of Agriculture and Research Institute
Karaikal-609 603, Puducherry (UT)

Introduction

Pulses have a prominent place in daily diet and they are rich sources of vegetable protein, mineral matter and vitamin B. They are of special significance of the people in developing countries like India, who can hardly afford animal protein in adequate quantities. Among the pulses, mungbean (*Vigna radiata* L. Wilczek) is the third important grain legumes grown in India next to chick pea and pigeon pea. Mungbean seeds are damaged to great extent by pulse beetle, *Callosobruchus chinensis* (Linn). The pest being an internal feeder lays eggs on the seed surface which hatch during storage as and when the conditions are favourable (Aslam *et al.*, 2002). Controlling this pest is one of the great tasks for pulses seed growers. The use of synthetic pesticides is hazardous to the human health as well as to the environment. Hence, the present study was undertaken to find out the efficacy of neem products against pulse beetle in stored mungbean seeds with an aim to avoid environmental hazards of the chemical insecticides and to strengthen the use of botanicals.

Materials and Methods

The experiment was conducted at Seed Technology Unit of PAJANCOA & RI, Karaikal during 2005-06 to find out the efficacy of neem products against pulse beetle in mungbean (cv. ADT 3) seed storage. The different neem products *viz.*, Neem Seed Kernel Powder (NSKP) @ 5 g, Neem cake @ 5 g, Neem dry leaf powder @ 5 g, Neem oil @ 5 ml, Neem INDIA @ 2.5 ml , Neem INDIA @ 5 ml, Econeem @ 2.5 ml, Econeem @ 5 ml, Nimbicidine @ 5 ml per kg of mung bean seeds were used and compared with Deltamethrin @ 1ppm and untreated control. After drying in shade, the treated seeds were stored in gunny baglets each of 2 kg capacity in a room under ambient conditions. The observations on percentage seed moisture content and seed germination per cent by adopting ISTA (1999) method and percentage pulse beetle infestation by visual observation were recorded at three months interval upto six months of storage period. The experiment was conducted in three replications under Completely Randomized Design.

Results and Discussion

The results on storage of mungbean seeds treated with neem based products and insecticides revealed that the moisture content did not vary significantly due to treatments during storage (Table 1). However, the different neem based bio-products had significant influence on seed germination and pulse beetle infestation. After six months of storage period, the germination per cent of all the treatments were above Minimum Seed Certification Standard except in Neem cake (52 %), Neem INDIA @ 2.5 ml kg^{-1} (72 %) and Nimbicidine (64 %). Significant reduction in germination might be due to heavy insect infestation in those three treatments. Among the neem products, NSK Powder @ 5.0 g (0.5 %), Neem INDIA @ 5 ml (1.00) and Econeem @ 5 ml kg^{-1} of seed (1.00) were very effective against pulse beetle during storage under ambient condition. The lower concentration of Neem INDIA and Econeem (2.5 ml kg^{-1}) failed to contain pulse beetle infestation in mungbean seeds during storage. The interaction effect of treatment and period of storage also had significant effect on seed germination and insect infestation. Pesticidal efficacy of certain indigenous plant oils (Das, 1986; Yadava, 1985) and neem leaf powder (Pandey and Singh 1995) against pulse beetle (*Callosobruchus chinensis* Linn.) has earlier been emphazised. In the present investigation, NSK Powder @ 5.0 g kg^{-1} of seed, Neem India @ 5 ml kg^{-1} and Econeem @ 5 ml kg^{-1} of seed were found to be the best treatments in improving the storability of mungbean

Table 1. Efficiency of different neem based products for control of pulse beetle infestation and viability of mungbean seed during storage under ambient condition at Karaikal, Puducherry (UT)

Treatments (T)	Seed moisture content (%)			Seed germination (%)			Pulse beetle infestation (%)		
	Initial		Storage (6 months)	Initial		Storage (6 months)	Initial		Storage (6 months)
NSK Powder @ 5.0 g kg^{-1} of seed (T_1)	9.7		10.3	92		90	0.00		0.50
Neem cake @ 5 g kg^{-1} of seed (T_2)	9.7		11.1	90		52	0.00		51.00
Neem dry leaf powder @ 5 g kg^{-1} of seed (T_3)	9.7		10.3	92		82	0.00		4.00
Neem oil @ 5 ml kg^{-1} of seed (T_4)	9.8		10.0	86		90	0.00		1.75
Neem INDIA @ 2.5 ml kg^{-1} of seed (T_5)	10.0		11.3	94		72	0.00		15.0
Neem INDIA @ 5 ml kg^{-1} of seed (T_6)	9.8		10.3	98		90	0.00		1.00
Econeem @ 2.5 ml kg^{-1} of seed (T_7)	10.0		11.0	92		82	0.00		10.0
Econeem @ 5 ml kg^{-1} of seed (T_8)	9.9		10.5	92		92	0.00		1.00
Nimbicidine @ 5 ml kg^{-1} of seed (T_9)	9.9		10.5	96		64	0.00		4.75
Deltamethrin @ 1ppm (40 mg kg^{-1} of seed) (T_{10})	9.9		10.5	94		90	0.00		1.75
Untreated control (T_{11})	9.7		10.3	90		82	0.00		3.50
	T	P	TxP	T	P	TxP	T	P	TxP
SEd	NS	NS	NS	1.84	0.79	2.61	0.71	0.30	1.00
CD (5%)				3.71	1.58	5.25	1.42	0.61	2.01

seeds. Juneja and Patel (2002) also reported that NSK Powder (4%) can be used for recharging the stored green gram at an interval of two months. Similar study was conducted by Bajbai *et al.* (2002) in urdbean and reported that Neemazal was found to be the best in improving the storability of urdbean seeds.

REFERENCES

Aslam, M., Khalid Ali Khan and Bajwa, M.Z.H. (2002). Potency of some spices against *Callosobruchus chinensis* Linn. *Online Journal of Biological Sciences. 2 (7):* 449-452.

Bajpai, V.P., Khan, A.A., Vyas, R.P., Tiwari, L.P. and Tripathi, R.A. (2002). Effect of bio-insecticides on viability of urdbean seed (*Vigna mungo* L.) stored in different containers. *Seed Research. 30(2):*331-333.

Das, G.P. (1986). Pesticidal efficacy of some indigenous plant oils against the pulse beetles (*Callosobruchus chinensis* Linn.). *Bangladesh J. Zoology 14(1):* 15-18.

ISTA (1999). International Rules for Seed Testing, Rules 1999. *Seed Sci. & Technol.,* 27:(Supplement).

Juneja, R.P. and Patel, J.R. (2002). Persistence of botanical materials as protectant of green gram (*Vigna radiata* (L.) Wilczek) against pulse beetle (*Callososbruchus analis* Fabricus). *Seed Research. 30(2):* 294-297.

Pandey, N.K. and Singh, S.C. (1995). Effect of neem leaf powder on survival and mortality of pulse beetle (*Callosobruchus chinensis* Linn.) infesting black gram. *U.P. J. of Zoology 15 (3):*162-164.

Yadava, T.D. (1985). Anti-ovipositional and ovicidal toxicity of neem oil against three species of *Callosobruchus. Neem Newsletter 2(1):* 5-6.

□□□

Chapter 38

Management of Root Rot of *Vigna mungo* (L.) Using Biocontrol Agents Multiplied in Vermicompost

SARAVANAN, T., R. RAMAMOORTHY AND K. SUBBIAH
Agricultural Research Station, Kovilpatti -628 501

Introduction

Blackgram (*Vigna mungo* (L.)) is one of the common pulse crops in tropical countries. The crop is being affected by many fungal, bacterial and viral diseases. Among the diseases, root rot caused by *Macrophomina phaseolina* (Tassi) Goid causes considerable yield loss by reducing plant populations in the field. It is more a devastating disease, right from the establishment phase of the crop (Mohan Babu *et al.*, 2002). The possibility of replacing chemical control of plant root diseases with a biological alternative remains an exciting and challenging objective. Now a day's use of beneficial microorganisms against soil borne disease is a promising tool in plant protection strategy. The micro organisms like *Pseudomonas fluorescens, Trichoderma viride* and *Bacillus subtilis* are widely used in plant disease control. *P. fluorescens* belonging to plant growth promoting rhizobacteria is a best alternative biological approach for the control of many diseases. Several mechanisms viz., production of siderophores, rhizosphere colonizing ability, production of antibiotics and induced systemic resistance have been identified in the disease suppression by various strains of *P. fluorescens* (Thomashow and Weller, 1998). Vermicompost added to chickpea improved dry matter accumulation, grain yield, soil nitrogen and phosphorus and bacterial population (Jat and Ahlawat, 2006). Zaidi and Singh (2004) reported addition of *Trichoderma viride* and *Pseudomonas fluorescens* during vermicomposting increased the plant height and yield. Combined effect of bio control agents and vermicompost against root rot disease in *V. mungo* is to be identified in order to minimize the loss due to the disease.

Materials and Methods

The cultures of *P. fluorescens* (pf1) and *T. viride* were procured from Department of Plant Pathology, Tamil Nadu Agricultural University, Coimbatore, while the vermicompost of earth worms *Eisenia fetida* was obtained from Agricultural Research Station, Kovilpatti. *P. fluorescens* (pf1) was inoculated into King' B medium broth (KMB) and kept in a rotary shaker at 150 rpm min^{-1} for 48h at room temperature (25 ± 2°C). One kg of talc powder was taken in a metal tray under aseptic conditions and its pH was adjusted to 7.0 by adding $CaCO_3$ @ 15 g kg^{-1}. Ten grams of carboxy methyl cellulose (CMC) was added to 1 kg of talc and the mixture was autoclaved for 30 min at 120°C on each of the two days. Then 400 ml of the bacterial suspension containing 9 x 10^8 cfu ml^{-1} mixed with carrier - cellulose mixture under aseptic conditions. After air drying (35% moisture content) overnight under aseptic conditions, the mixture was packed in a polypropylene bag, sealed and stored at room temperature (28±2°C). At the time of application, the population of bacteria in the formulations was 10^8 cfu g^{-1} of talc powder.

The fungal bio control agent was multiplied by transferring the culture of *T. viride* into 250 ml conical flasks containing 70 ml of sterilized yeast molasses medium and was incubated for 10 days. This was used for preparing the fermenter inoculum. The mother culture was added @ 2.5 l / 50 l of yeast molasses medium under aseptic condition and incubated for 15 days and then the content was drained. The fungal biomass was removed and mixed with 100 kg of talc powder at 1: 2 ratio. The mixture was shade dried. The formulation was made by adding CMC @ 5 g / kg of carrier material.

Screen House Experiment

Pits of 3x2x2' were dug out and filled with vermicompost. Talc formulation of *T. viride* or *P. fluorescens* (100 g) was added to each respective pit and mixed thoroughly. Four pits were maintained for each treatment.

Talcum powder alone was added at 100 g in pit as check and mixed well. Inoculated pits were covered with dried sorghum straw and moisture was maintained by regular sprinkling of water. After 30 days of incubation, the vermicompost was used for field experiments.

The seeds of *V. mungo* var CO 5 received from Farm Office, Agricultural Research Station, Kovilpatti wre used in the study. The seeds were surface sterilized using NaoCl for 5 min before sowing. The healthy seeds of CO 5 were sown in a pot size of 75 x 45 x 40 cm. The pot was filled with soil taken from the infected field. The seeds were sown @ 10 seeds/pot. Pot culture soil was mixed with the *T. viride* or *P. fluorescens* or *T. viride* multiplied in vermicompost or *P. fluorescens* multiplied in vermicompost at 5% w/w. Soil without any addition of biocontrol agents served as control. Three pots were kept as a replication and three replications were maintained in each treatment. After 40 days of sowing, the root rot disease incidence was recorded and expressed as per cent of plants infected with root rot.

Field Experiment

The seeds of var CO 5 were sown in Black soil farm at Agricultural Research Station, Kovilpatti during 2004-06. Six treatments including a check were laid out in the field trial and the seeds were sown in a plot size of 4 x 3 mt. Each plot was kept as a replication and three replications were maintained in the experiment. Regular crop production practices were carried out for the crop. The crop was monitored for root rot disease incidence. At 45 days after sowing (DAS) and 60 DAS, root rot incidence was recorded and expressed as per cent of infected plants. At the end of the experiments, number of pods per plant from individual treatments were recorded. The data from the field experiments were analyzed using AGRES statistical analysis and means were compared with other treatments.

Results and Discussion

Screen house experiments were carried out to find out the efficacy of biocontrol agents multiplied in organic amendments against root rot disease in *V. mungo*. As indicated in Table 1, all the biocontrol agents multiplied in organic amendments performed better against root rot disease. Among them, Soil application of *P. fluorescens* has proven as an efficient one in reducing the disease incidence as sole application of biocontrol agents. Soil application of *P. fluorescens* multiplied in vermicompost was the superior in reducing the disease incidence under screen house conditions. Seed treatment of bio control agents also reduced the disease. In the field condition (Table 2), Soil application of *P. fluorescens* multiplied in vermicompost at 2.5 kg/ha have also performed

Table 1. Effect of biocontrol agents multiplied in vermicompost on root rot disease in *V. mungo* under screen house condition

Treatments	*Disease incidence (%)				Germination (%)			
	2004	2005	2006	Mean	2004	2005	2006	Mean
Soil application of Pf multiplied in vermicompost at @ 5 % w/w	19.6 (26.28) h	20.1 (26.64) i	18.3 (25.83) g	19.33 (26.08) h	84.1 (66.50) a	86.2 (68.19) a	83.7 (66.19) a	84.6 (66.89) a
Soil application o f Pf @ 5 % w/w	23.4 (27.63) ef	21.8(28.93) h	21.5 (27.83) g	22.23 (28.13) fg	80.5 (63.79) b	78.8 (62.29) bc	80.1 (63.50) bc	79.8(63.29) bc
Seed treatment with Pf @ 5 % w/w	28.1 (32.01) bcd	29.7 (33.02) e	25.6 (30.39) e	27.8 (31.82) e	78.4 (62.31) bcd	76.1 (60.73) f	78.3 (62.24) bcd	77.6 (61.25) bcd
Soil application of Tv multiplied in vermicompost @ 5 % w/w	22.5 (28.31) fg	23.1 (28.72) fg	20.5 (26.92)	22.03 (27.99)	80.1 (63.51) bc	79.1 (62.79) b	80.8 (64.01) b	79.9 (63.36) b
Seed treatment with Tv @ 5 % w/w	30.2 (33.33) c	38.1 (38.14) c	32.9 (35.00) c	33.7 (35.51) c	75.1 (60.06) efg	77.1 (61.83) bcde	74.4 (59.61) fgh	75.5 (60.33) f
Soil application of Tv @ 5 % w/w	29.2(32.71) bc	34.6 (36.07) d	28.5 (32.26) d	30.77 (33.67) d	76.7 (61.14) def	78.1 (62.52) bcd	77.9 (61.96) cde	77.5 (61.68) de
Soil application of vermicompost @ 5 % w/w	41.2 (39.93) b	44.7 (41.78) b	40.5 (39.52) b	42.13 (40.48) b	78.0 (62.03) bcde	79.1 (62.79) b	76.6 (61.07) ef	77.9 (61.96) bcd
Chemical check (Soil application of ridomil at 5 ml /pot)	24.6 (29.73) e	25.2 (30.13) f	24.0 (29.33) ef	24.6 (29.73) f	72.7 (58.50) i	74.3 (59.54) g	75.6 (60.40) fg	74.2(59.48) fg
Untreated control	62.0 (51.49) a	68.1 (55.61) a	54.5 (47.58) a	61.53 (51.66) a	60.3 (50.94) j	64.2 (53.25) h	63.3 (52.71) i	62.5 (52.24) h

*Mean of disease incidence three replications and each replication contain three pots

Pf - *P.fluorescens*, Tv- *T.viride*

Values in parenthesis are arcsine transformed values. Data are arcsine-transformed before analysis;

Mean followed by same letters are not significantly different.

Table 2. Effect of biocontrol agents multiplied in vermicompost on root rot disease in *V. mungo* in field condition

Treatments	*Disease incidence (%)				Number of pods per plant			
	2004	2005	2006	Mean	2004	2005	2006	Mean
Soil application of Pf multiplied in vermicompost at 2.5 kg/ha	14.3 (22.22) g	12.7(20.88) g	11.1(19.46) g	12.7(20.88) g	12.7 a	10.4 a	11.4 a	11.5 a
Soil application of Pf at 2.5 kg/ha	18.8(25.69) de	17.2(24.51) e	16.9(24.27) de	17.6(24.81) de	8.6 b	9.3 ab	8.1 cd	8.67 b
Soil application of Tv multiplied in vermicompost at 2.5 kg/ha	19.1(25.92) d	19.1(25.92) d	18.3(25.53) d	18.8(25.69) d	8.4 bc	7.9 c	9.5 b	8.6 bc
Soil application of Tv at2.5 kg/ha	25.0(30.00) c	25.3(30.19) c	28.1(32.01) c	26.1(30.65) c	7.0 bcd	6.5 cd	6.4 ef	6.63 e
Soil application of vermicompost at 2.5 kg/ha	37.1(37.53) b	39.1(38.71) b	34.1(35.73) b	36.7(37.28) b	7.0 bcd	5.6 def	6.9 e	6.5 ef
Chemical check (Soil application of ridomil at 5 ml /bed)	17.1(24.43) ef	16.7(24.13) ef	16.0(23.58) f	16.6(24.02) ef	7.0 bcd	8.5 abc	9.2 bc	8.23 bcd
Untreated control	52.6(46.49) a	63.2(52.65) a	49.8(44.88) a	55.2(47.99) a	6.5 cde	5.9 de	5.1 efg	5.83 efg

*Mean of disease incidence in three plots.

Pf - *P.fluorescens*, Tv- *T.viride*

Values in parenthesis are arcsine transformed values. Data are arcsine-transformed before analysis,
Mean followed by same letters are not significantly different.

better in reducing the disease incidence and the same treatment has the higher number of pods per plant.

Use of bio control agents may yield a reasonable control of plant diseases, but may lack broad-spectrum activity responsible for increased germination, plant height and soil health etc. Earthworms are considered as the major invertebrate fauna in soil, which contribute to decomposition of organic matter. Recently Kavitha *et al.* (2005) reported that a significant increase in plant growth was induced by treatment with Plant Growth Promoting Rhizobacteria (PGPR) isolates CBE4 and these PGPR isolates triggered the defense related enzymes involved in phenyl propanoid pathway and phenolics. In crop plants, various mechanisms account for the ability of *Pseudomonas* strains to control plant pathogens, including competition for iron and other nutrients, niche exclusion (Rioux *et al.* 1983) and production of inhibitory compounds or metabolites (Thomashow and Weller 1988). Similarly, *T. viride* expressed parasitism through physical contact, pathogen cell wall lysis and coiling of hyphae (Mukhopadhay, 1994; Patel and Anahosur, 2001). Vermicompost added to various container media significantly inhibited the infection of tomato plants by *Fusarium oxysporum* f.sp. *lycopersici* (Szczech, 1999). Springett and Syers (1979) observed that earth worms altered nutrient availability, altered the plant's ability to take up nutrients or affected the growth mechanisms of the plants. The earth worm casts probably contain an auxin like substance or some substance that modifies the effect of the plant auxins (Lee, 1985). Nielson (1965) proposed presence of plant growth promoting substance in vermicompost of *Aporrectodea caliginosa* and *E. fetida.* Vermicompost acts as a substrate for rapid multiplication of bio control agents and the presence of growth inducing substance may be responsible for increased height in the seedlings. Hence application of bioagents *P. fluorescens* multiplied in vermicompost will give a tremendous control of root rot disease and increase the number of pods per plant.

REFERENCES

Jat, R.S. and Ahlawat, I.P.S. (2006). Direct and residual effect of vermicompost, biofertilizers and phosphorus on soil nutrient dynamics and productivity of chickpea - fodder maize sequence. *Journal of Sustainable Agriculture* 28: 41 - 54.

Kavitha, K, Mathiyazhagan, S., Senthilvel, V., Nakkeeran, S., Chandrasekar, G. (2005). Development of bioformulations of antagonistic bacteria for the management of damping off of Chilli (*Capsicum annuum* L). *Archives of Phytopathology and Plant Protection* 38: 19-30.

Lee, K. E. (1985). Earthworms: their ecology and relationships with land use; Academic press, 411.

Mohan Babu, R., Sajeena, S., Seetharaman, K. (2002). Management of root rot of Blackgram (*Vigna mungo* L.) in problem soils. *Journal of Biological Control*, 17: 145-149.

Mukhopadhay, A.N. (1994). Biocontrol of soil borne fungal plant pathogens, current status: Future prospect and potential limitations. *Indian Phytopathology*, 47: 119-124.

Nielson, R. L. (1965). Presence of plant growth substances in earthworms demonstrated by paper chromatography and the went pea test. *Nature* 208: 1113- 1114.

Patel, S. T and Anahosur, K.H. (2001). Potential antagonism of *Trichoderma harzianum* against *Fusarium* spp., *Macrophomina phaseolina* and *Sclerotium rolfsii*. *Journal of Mycology and Plant Pathology*, 31: 365.

Rioux, C., Jordan, D. C. and Rattray, J. B. M. (1983). Colorimetric determination of catechol siderophores in microbial cultures. *Annals of Biochemistry* 133: 163 - 169.

Springett, J. A and Syers, J. K. (1979). The effect of earth worm casts on rye grass seedlings. In proc. 2nd Australian Conf. Grassl. Invert. Ecol. (Crosby, T.K. and Pottinger, R.P. eds), Wellington: Government Printer, 44 - 47.

Szezech, M.M. (1999). Suppressiveness of vermicompost against *Fusarium* wilt of tomato. *Journal of Phytopathology* 147: 155-159.

Thomashow, L. and Weller, D. M. (1988). Role of phenazine antibiotic from *Pseudomonas fluorescens* in biological control of *Gaummannomyces graminis* var. *tritici*. *Journal of Bacteriology* 170: 3499- 3508.

Zaidi, N. W. and Singh, U. S. (2004). Development of improved technology for mass multiplication and delivery of fungal (*Trichoderma*) and Bacterial (*Pseudomonas*) biocontrol agents. *Journal Mycology Plant Pathology* 34: 732-741.

□□□

SECTION - V
POLICIES AND PROGRAMMES

Chapter 39

Legume Sector in India: An Analysis of Issues, Potentials and Strategies

SURESH KUMAR, D., S. GURUNATHAN AND M. CHANDRASEKARAN
Department of Agricultural Economics, Tamil Nadu Agricultural University, Coimbatore-641003.

Introduction

Although Indian Agriculture has come a long way, there are certain implications that have to be addressed for national food and nutritional security. Our population is further expected to be 1.4 billion by 2020. The increasing population, coupled with growing income will generate increased demand for food grains and non-food grain crops. Therefore, Indian agriculture has to achieve a higher growth rate targeted at 4 per cent per annum on a sustainable basis. Indian agriculture also has to diversify into high-value crops, raise productivity, restore soil health and enhance the application of modern technologies including biotechnology. Focus is needed on potential areas, regionally-differentiated strategies, crop diversification and scientific management of natural resources to achieve 4 per cent growth rate in agriculture.

India is the world's largest pulse producer, consumer and importer, accounting for 27 per cent of the global pulse production. In India, pulses are grown on 22-23 million hectare area with an annual production at of 12-14 million tons and per hectare yield of 600-650 kg. Pulses account for around 19 per cent of gross cropped area and less than 8 per cent of the total food grain production of the country. The production follows the trend displayed by all agricultural crops peaking on a normal monsoon year and dropping when the rains fail. The major pulses grown in India are - Pigeon pea (Arhar) and Tyson chickpea (Gram or Desi Chana). Their share in the total pulses production is about 20 and 33 per cent respectively. The state of Madhya Pradesh is the largest pulse producer, accounting for about 26 per cent of total production from 1996-97 to 1999-2000. Other states with significant output include Uttar Pradesh (18%), Maharashtra (14%), Rajasthan (14%), Karnataka (5%), Andhra Pradesh (5%), and Bihar (5%).

In the eleventh plan period, the food grain production is expected at 2.73 percent and the pulses production is expected to grow at 5.66 percent to meet the requirement of human population growing at 1.45 per cent. This 5.66 per cent could be an up hill task as the growth rate in pulses was highly fluctuating and in fact it was negative during last five years in India. To meet the requirement of an expected 1196.40 million nationals, we have to produce 337 million tones of food grains and 29.76 million tones of pulse in the year 2011-12 while the production of pulses was only 14.11 million tonnes in 2006-07 as per the advance estimates as on 04.04.2007. In 2005-06, it was still worse at 13.39 million tones. There lays a huge gap between the demand and supply of pulses in the country which has to be bridged in near future. Moreover, over a period of five decades, per capita availability of total cereals rapidly increased from 375 grams during pre green revolution period to 429 grams in the current decade mainly because of drastic increase in per capita availability of wheat from 74 grams to 161 grams and of rice from 184 grams to 204 grams during these periods due to significant effect of green revolution technologies (GOI, 2006). Keeping these issues in view, the present paper aimed to address the issues, potentials and policy issues in relation to the important legume crops *viz.*, redgram, gram, and groundnut.

Trend in Area, Production and Productivity

The trend in area, production and yield of pulses shows that there is a declining growth rate in area, production and yield of total pulses over the periods. For instance, the growth rate in area is 1.45 per cent for the pre-green revolution period, which has declined to 0.23 per cent and further to 0.04 per cent in the post liberalisation period. Similar trend is visualized in the case of production and yield of total pulses in the country (Table.1).

Table 1: Growth rate in area, production and yield of total pulses in India

Particulars		Pre Green Revolution Period (1949-50 to 1965-66)	Post Green Revolution Period (1966-67 to 1991-92)	Post Liberalisation period (1991-92 to 2006-07)	1949 to 2006
Area (Mill ha)		22.51	22.85	22.34	22.61
Production (Mill Ton)		10.66	11.55	13.48	11.79
Yield (Kg/ha)		472.35	504.81	593.80	518.31
% coverage under irrigation		8.76	8.80	12.77	9.82
CGR	Area	1.45	0.23	0.04	0.05
	Production	1.89	0.92	-0.15	0.61
	Yield	0.43	0.68	0.01	0.54
	% coverage under irrigation	-0.07	0.28	1.08	0.79

With wide variations in agro-economic conditions and socio economic diversity, the issue on regional differences in production and productivity of legume crops is of crucial interest. Therefore, a proper understanding of dimensions of regional differences in production and productivity of crops is essential for effective regional planning. The regional differences in area, production and productivity of major legume crops are presented in Tables 2-4. About 80 per cent of total area and total production of pulses is comprised in six states like Madhya Pradesh, Uttar Pradesh, Maharashtra, Andhra Pradesh, Karnataka and Rajasthan. Larger states like Gujarat, Bihar, West Bengal, Jharkand etc, though having shown significantly larger yield than the national average, lacking far behind in cultivating pulses crops in larger extend. Same trend was exhibited in other major pulses like gram and tur also. Policy makers need to examine the context of these states and need to explore possibilities of bringing more area under various pulses crops which will have a balancing effect in agricultural import budget of the nation since we heavily depend on import of pulses to a tune of 15 million tonnes.

Pulses Trade

Global pulse trade has expanded rapidly in the last twenty years. However the trade history is some what volatile due to supply and demand variability. Trade patterns have also shifted during this period. The next twenty year period is likely to see these types of changes continue, as Canada an important pulse exporter puts pressure on the supply side. The factors like high daily and monthly price volatility, high dependence on international prices due to large imports, no substantial government control, no trade monopoly, large shelf-life requiring storage and dispersed consumption in all parts of the country make pulses suitable for future trading.

The stagnant production has led to declining per capita consumption over the past 20 years. The per capita availability has progressively declined from 60 gram in 1950-51 to 32 gram at present. The burgeoning demand-supply gap has led the Government of India to ease the norms related to importing of pulses. The Government of India has allowed unrestricted import of pulses with low duties for more than two decades now. The stagnation in production and the fall in per capita availability have been the reasons for this move. However, despite the unrestricted imports and low tariffs, imports relatively accounted for a smaller share of supply and consumption. Peas, chickpeas, dried beans, moong and tur are the major pulses that are being imported into the country.

India's import of pulses has historically been on the increase, especially since 2001-02 when they peaked 2.23 million tones from just 0.35 million tones in the previous year. The value of imports also has increased from Rs 500 crore to Rs 3164 crore. India's import of pulses is governed by the government policy to control domestic inflationary pressures arising from shortages of pulses. The government of India has allowed unrestricted import of pulses with low duties for more than two decades now. The stagnation in production, the importance of pulses as a source of protein in the Indian diets and the fall in per capita availability has been the reasons for this move. However despite the unrestricted imports and low tariffs, imports relatively accounted for a smaller share of supply and consumption in the 1990's.

While the average production of total pulses showing stagnation around 5 million tones in every five years period from 1980-81, the import of pulse is showing an increasing trend from an average of 0.173 million tones during 1980-81 to 1984-85 and it reached an average of 1.38 million tones during the five years of 2000-01 to 2005-06. The import of pulses was merely 3.68 per cent of the production during the period of 1980-81 to 1984-85. It went up to more than nine times to the tune of 27.27 per cent in 2000-01 to 2005-06 (Table.6.).

While the over all production of pulses has been near stagnant at 5 million tones, the growth rate of national production were 0.65 per cent in the post liberalization period which was much better than the global growth of 0.09 per cent during 1990-2003 where as the growth of imports was much higher at 9.25 per cent in this period (Table.7).

Table 2 Area, production and yield of Gram during 2005-06 in respect of major gram producing states along with coverage under irrigation

States	Area	% of total area	Production	% of total production	Yield
Madhya Pradesh	2.56	36.94	2.37	42.32	926
Maharashtra	1.02	14.72	0.71	12.68	691
Uttar Pradesh	0.74	10.68	0.66	11.79	893
Andhra Pradesh	0.39	5.63	0.63	11.25	1591
Rajasthan	1.08	15.58	0.48	8.57	443
Karnataka	0.42	6.06	0.23	4.11	548
Chhattisgarh	0.23	3.32	0.16	2.86	706
Gujarat	0.17	2.45	0.14	2.50	850
Haryana	0.13	1.88	0.07	1.25	554
Bihar	0.06	0.87	0.06	1.07	902
West Bengal	0.04	0.58	0.04	0.71	913
Orissa	0.04	0.58	0.02	0.36	646
Others	0.05	0.72	0.03	0.54	@
All India	6.93	100.00	5.60	100.00	808

@ - Since area/ production is low in individual states, yield rate is not worked out

Note: States have been arranged in descending order of percentage share of production during 2005-06

Agricultural Statistics at a Glance, 2006, Directorate of Economics and Statistics, Ministry of Agriculture, Government of India

Table 3 : Area, production and yield of Tur during and 2005-06 in respect of major Tur producing states along with coverage under irrigation

States	Area	% of total Area	Production	% of total production	Yield
Maharashtra	1.10	30.73	0.79	28.83	720
Karnataka	0.60	16.76	0.44	16.06	727
Uttar Pradesh	0.38	10.61	0.38	13.87	987
Andhra Pradesh	0.49	13.69	0.30	10.95	609
Gujarat	0.25	6.98	0.28	10.22	1102
Madhya Pradesh	0.32	8.94	0.24	8.76	739
Orissa	0.13	3.63	0.10	3.65	742
Jharkhand	0.09	2.51	0.05	1.82	633
Bihar	0.03	0.84	0.04	1.46	1291
Tamil Nadu	0.04	1.12	0.02	0.73	540
Others	0.15	4.19	0.10	3.65	@
All-India	3.58	100.00	2.74	100.00	765

Agricultural Statistics at a Glance, 2006, Directorate of Economics and Statistics, Ministry of Agriculture, Government of India

Table 4 : Area, production and yield of total pulses during 2005-06 in respect of major total pulses producing states along with coverage under irrigation

States	Area	% of total Area	Production	% of total production	Yield
Madhya Pradesh	4.28	19.12	3.23	24.14	754
Uttar Pradesh	2.75	12.28	2.23	16.67	811
Maharashtra	3.43	15.32	2.01	15.02	584
Andhra Pradesh	1.78	7.95	1.38	10.31	772
Karnataka	1.98	8.84	0.96	7.17	487
Rajasthan	3.44	15.36	0.90	6.73	261
Gujarat	0.78	3.48	0.55	4.11	704
Bihar	0.60	2.68	0.45	3.36	749
Chhattisgarh	0.95	4.24	0.45	3.36	477
Orissa	0.81	3.62	0.34	2.54	416
Tamil Nadu	0.53	2.37	0.18	1.35	337
West Bengal	0.22	0.98	0.17	1.27	785
Jharkhand	0.29	1.30	0.17	1.27	592
Haryana	0.19	0.85	0.12	0.90	622
Others	0.36	1.61	0.24	1.79	@
All India	22.39	100.00	13.38	100.00	598

Agricultural Statistics at a Glance, 2006, Directorate of Economics and Statistics, Ministry of Agriculture, Government of India

Table 5. Area, production and yield of groundnut during and 2005-06 in respect of major groundnut producing states along with coverage under irrigation

States	Area	% of total Area	Production	% of total production	Yield
Gujarat	1.95	28.93	3.39	42.43	1734
Andhra Pradesh	1.88	27.89	1.37	17.15	728
Tamil Nadu	0.62	9.20	1.10	13.77	1775
Karnataka	1.04	15.43	0.67	8.39	645
Rajasthan	0.32	4.75	0.49	6.13	1549
Maharashtra	0.43	6.38	0.41	5.13	958
Madhya Pradesh	0.21	3.12	0.23	2.88	1126
Orissa	0.09	1.34	0.11	1.38	1171
Uttar Pradesh	0.11	1.63	0.09	1.13	851
Others	0.09	1.34	0.13	1.63	@
All India	6.74	100.00	7.99	100.00	1187

Agricultural Statistics at a Glance, 2006, Directorate of Economics and Statistics, Ministry of Agriculture, Government of India

Table 6 : Pulses trade in India

Periods	Production ('000 t)	Import ('000 t)	Export ('000 t)	Import export ratio	Import as % of production
1980-81 to 1984-85	4714.00	173.34	2.78	62.31	3.68
1985-86 to 1989-90	4660.00	578.82	7.67	75.45	12.42
1990-91 to 1994-95	5064.00	630.14	33.90	18.59	12.44
1995-96 to 1999-2000	5720.00	592.99	116.67	5.08	10.37
2000-01 to 2005-06	5060.00	1380.04	202.02	6.83	27.27

Table 7. Compound Growth Rates of pulses production and import during various periods

Periods	Compound Growth Rates (CGRs)	
	Production	Import
1980-81 to 2005-06	0.68	7.53
Pre liberalisation period (1980-81 to 1990-91)	0.05	24.53
Post liberalisation period (1991-92 to 2005-06)	0.65	9.25

Regional Differences in Profitability of Legume Crops

Market driven agriculture necessitates the policy makers to promote high value agricultural crops and value addition in agricultural produce. This transition faces serious challenges amidst of diverse agro-climatic, socio-economic and political environments of the country. Both intra and inter regional differences in farm profitability have been observed by researchers (Singh, 2006). The efficiency of farms and farm profitability significantly vary with farmers' access to technology, credit, market infrastructure, and price supports (Haque, 2006). Similarly, the productivity and production of crops affect the farm profitability. It was found that the managerial factors (Selvarajan *et al.*, 1997) and farm size (Sen and Bhatia, 2004; Haque, 1996) significantly influence the productivity of various crops.

The profitability in this paper is defined in terms of net income1 and the ratio of gross value of output to the cost of cultivation. The net return per hectare is considered as an indicator of farm profitability. As the cost of cultivation and farm harvest prices are available only at current prices, profitability per hectare of agricultural products was worked out at constant prices (1993-94) before arriving the triennium ending. The triennium average was worked out for the years 2001-02 to 2003-04. The profitability in pulses production was analysed and discussed.

It is seen that variation is observed in total cost of cultivation per hectare (Cost C2) and farmers paid out cost (Cost A2 +FL) across states[1]. Since, the total cost of cultivation (Cost C2) includes both the operational and fixed cost, the farmers paid out cost (Cost A2 +FL) are considered for the discussion.

Being an important pulse crop, the tur cultivation, exhibits considerable differences in terms of yield, cost of cultivation and profits per hectare (Table.8). The net income per hectare ranged from Rs. 1806 in Orissa to Rs. 13073 in Uttar Pradesh. The gross income per hectare is worked out at the least in Madhya Pradesh and many of the tur growing states exhibited far less than the national average gross income and the net income in Madhya Pradesh worked out to be negative. Adequate price support, varietal improvement, and technological support including inputs will help maximize farm profits and reduce inter-regional differences in profitability. It is evidenced that there is significant variation in yield, cost of cultivation and profits as the coefficient of variation (CV) were found to be higher across various states. This is worked out to 27.53 per cent, 22.85 per cent and 122.17 per cent respectively for the yield, cost of cultivation (Cost A2+FL) and profits. High CV for the profits is mainly due to differences in price of pulses across states and regions.

Gram, another important pulse crop shows comparatively better profitability among important gram producing states. Except in Haryana and Rajasthan, all other major states exhibited higher net income than the national average with 42.14 per cent of CV (Table.9). Also variation in net income, cost of production etc are much lower than other pulse crops.

---

1 The profit (net income) is defined as the difference between the gross value of agricultural produce and the cost of cultivation.

1 $\text{Constant prices} = \dfrac{\text{Value at current prices X Wholesale Price index in 1993 - 94}}{\text{Wholesale Price index (current period)}}$

1 Note: Cost A1 : All actual expenses in cash and kind incurred in production by the farmers

Cost A2 : Cost A1 + rent paid for leased-in land

Cost A2 + FL : Cost A2 + imputed value of family labour

Cost B1 : Cost A1 + interest on value of owned capital assets (excluding land)

Cost B2 : Cost B1 + rental value of owned land + rent paid for leased in land

Cost C1 : Cost B1 + imputed value of family labour

Cost C2 : Cost B2 + imputed value of family labour

GVCR is the ratio of gross value of output to the cost of cultivation (A2+FL)

Table 8. Cost Estimates and profitability of Tur across major producing states (Rs)

States	Cost A2+FL	Cost C2	Yield (Qtl/ha)	Gross value of output	Net income over C2	Net income over A2+FL	GVCR	Cost of production
Andhra Pradesh	8240	14073	8.76	12834	-1239	4594	1.56	1606.51
Gujarat	8282	12385	8.27	15496	3111	7214	1.87	1497.58
Karnataka	6500	9176	4.90	9522	346	3022	1.46	1872.65
Madhya Pradesh	6431	9912	6.83	1095	-8817	-5336	0.17	1451.24
Maharashtra	7777	11878	9.62	17859	5981	10082	2.30	1234.72
Orissa	4107	6455	4.40	5913	-542	1806	1.44	1467.05
Uttar Pradesh	5598	11622	8.58	18671	7049	13073	3.34	1354.55
Mean	6705	10785.86	7.34	11627	841.29	4922.14	1.73	1497.76
SD	1532.13	2495.53	2.02	6492.12	5320.92	6013.47	0.96	202.14
CV (%)	22.85	23.14	27.53	55.84	632.47	122.17	55.49	13.50

Table 9. Cost Estimates and profitability of gram across major producing states (Rs)

States	Cost A2+FL	Cost C2	Yield (Qtl/ha)	Gross value of output	Net income over C2	Net income over A2+FL	GVCR	Cost of production
Haryana	5849	10617	6.23	9809	-808	3960	1.68	1704.17
Madhya Pradesh	6474	11362	10.87	15401	4039	8927	2.38	1045.26
Maharashtra	9083	14201	10.16	17426	3225	8343	1.92	1397.74
Uttar Pradesh	6768	12039	9.84	15439	3400	8671	2.28	1223.48
Rajasthan	5319	7368	5.05	8535	1167	3216	1.60	1459.01
Mean	6698.60	11117.40	8.43	13322	2205	6623	1.97	1365.93
SD	1446.18	2486.27	2.61	3902.03	1999.22	2791.07	0.35	248.60
CV (%)	21.59	22.36	30.93	29.29	90.68	42.14	17.71	18.20

Groundnut is an important legume crop and grown widely in most of the southern states and parts of western region covering Gujarat and Maharashtra. It could be seen that the net income over the paid out cost is ranged from Rs. -1391 in Tamil Nadu to Rs.1917.31 in Maharashtra Table 10. The profit earned per hectare of groundnut cultivation exhibits not only the inter-regional differences but also within the region. Differences in prices of output make considerable variation within the region and inter-regions. The output price per quintal is varied across states and it is Rs.1582 in Gujarat and Rs.1285 in Maharashtra in 2002-03 (current prices).

Table 10. Cost Estimates and profitability of groundnut across major producing states (Rs)

States	Cost A2+FL	Cost C2	Yield (Qtl/ha)	Gross value of output	Net income over C2	Net income over A2+FL	GVC R	Cost of production
Andhra Pradesh	6481	8930	8.62	7108	-1822	626	1.10	1945
Gujarat	4855	6492	6.26	4848	-1644	-7.21	1.00	1827
Karnataka	11095	15594	14.70	12099	-3495	1004	1.09	2109
Maharashtra	6803	9037	10.04	8721	-316	1917	1.28	2195
Tamil Nadu	10153	12606	11.11	87621	-3844	-1391	0.86	1863
Mean	7878	10532	10.15	8308	-2224	429.94	1.07	1987.87
SD	2634.72	3572.46	3.13	2653.24	1447.25	1233.58	0.15	158.76
CV (%)	33.45	33.92	30.82	31.94	-65.07	286.92	14.41	7.99

Like all other crops, the cost of cultivation is significantly varied across states and regions, which push the farmers in relatively unfavourable situation in realizing returns. Though, the differences in yield may be attributed to access to irrigation, it is not the only factor that determines yield. For instance, in Gujarat, the area under irrigation is only 7 per cent, but the yield is more or less equal to the yield of Maharashtra where the irrigated area under groundnut is 31 per cent. Thus, the other factors like variety, level of use of various other inputs, prices of input and output and market infrastructure do matter in determining the yield and hence the profits. It could be seen that the variation in profitability as reflected by net income is found to be higher than the variation in yield and cost of cultivation. Hence, the price of groundnut is the major factor responsible for differences in profitability in groundnut.

Are Legumes the Slow Growth Crops (SGC) ?

Persisting low productivity of pulses is one of the major bottlenecks in stimulating positive response in pulses economy. Aggregate productivity of pulses in India at 587 kg/ha is about 29 per cent lower than the world average of 824 kg/ha. The Kharif pulses productivity (456 kg/ha) is further lower by 22 per cent from total pulses productivity in the country and by 45 per cent from the world average productivity of pulses. Economically, the partial productivity (production per unit of area), a parameter of growth in production does not truly reflect whether it is because of more use of inputs or improvement in the efficiency of their use. It assumes that the supply of inputs used in the given level of production changed in fixed proportion, which does not hold well especially in growing economies like India. Thus partial productivity neither takes into consideration change of all inputs nor reflects completely sustainability of growth. The best measure is the one that compares out put with the combined use of all resources (Fabricant, 1959). There are quite a large number of factors both internal and external, influencing pulses production sustainability.

The partial productivity is an inadequate measure of sustainability since it ignores time, secondary products like byproducts, inputs use other than land and externalities, of all which should be included in a sustainability measurement (Barnett *et al.*, 1995). This paper works on measuring sustainability of pulses production using Total Factor Productivity (TFP) approach. For any system under consistent management, the TFP index will consist of an underlying trend plus natural year to year fluctuations. It is said to be sustainable if it is growing or at least kept constant over time. In this paper, the growth in TFP during the past decade (eight years in fact) for two major pulse crops like Gram and Tur and the legume oil seed crop, ground nut has been discussed for major producing states of these crops.

Crop Wise and State Wise TFP Sustainability

The total factor productivity attempts to measure the amount of increase in total output, which is not accounted for by increase in total inputs. There is a large residual, measured by total factor productivity, which is the contribution of improvement of technology/knowledge, infrastructural development, human capital improvement and policy interventions. The total factor productivity index is computed as the ratio of aggregate output index to the aggregate input index.

In the calculation of output index and input index, Tornquist-Theil index was used for data on outputs and inputs of different pulses crops in major pulse growing states of India. The Tornquist-Theil index is a superlative index, which is exact for linear homogeneous translog production function (Diewart, 1976). The general formula used for construction of index is as follows:

$$I_t = I_{t-1} \Pi^{n(m)}_{j=1} (Q_{j,t} / Q_{j,t-1})^{1/2 (S_{j,t} + S_{j,t-1})}$$

where

$$S_{j,t} = Q_{j,t} * P_{j,t} / [\Sigma^{n(m)}_{j=1} Q_{j,t} * P_{j,t}$$

where

I_t is the index value of output/input for current year 't' ;

I_{t-1} is the index value of output/input for the previous year 't-1';

$Q_{j,t}$ is quantity of j th output/input for the year 't' when 'j' varies from 1 to 'n' for outputs and 1 to 'm' for inputs used in the production of a pulse crop;

$Q_{j,t-1}$ is the quantity of j th output value in total value of production or j th input in total cost;

$P_{j,t}$ is the current price of j th output/input.

The additional advantage of this index is that it accounts for change in quality of inputs because current factor prices are used in constructing the weights. The quality improvements in inputs are incorporated to the extent that these are reflected in higher wages and rental value (Capalbo and Vo, 1988). The Tornquist-Theil index provides consistent aggregation of inputs and outputs under the assumption of competitive behaviour, constant returns to scale, Hicks-neutral technical change and input-output separability. If QI_t is the index for output in year 't' and XI_t is the index for inputs for the same year, the total factor productivity (TFP) index is equal to QI_t / XI_t.

TFP indices were calculated for tur, gram and ground nut crops for major pulse growing states for which data on outputs produced including the byproducts and inputs used were available in the reports on Cost of Cultivation of Important Crops in India (Directorate Economics and Statistics, various issues). Both the main product and byproduct yields are included in output index of an individual crop. The shares in total revenue estimated using farm harvest prices were applied as the weights to aggregate the outputs. Inputs included in input index were, seed, fertilizers, manures, plant protection chemicals, human labour, animal labour, machine labour, capital (working and fixed) and their interests and irrigation. Inputs were aggregated using their shares in total cost of cultivation as weights. The annual growth rates of output index and input index were calculated assuming exponential trend. Since, $\text{Ln } TFP_t = \text{Ln } QI_t - \text{Ln } XI_t$; its derivation with respect to time gives the growth rates, i.e., $\delta (\text{Ln } TFP_t)/dt = \delta (\text{Ln } QI_t)/ \delta t - \delta (\text{Ln } XI_t)/ \delta t$. Thus growth in TFP index is equal to growth rate in output index *minus* growth rate in input index. The TFP of production of a particular pulse crop was considered sustainable in both the cases when its growth rates were either positive (CGR>0) or non significantly negative.

Tur

The total factor productivity of tur was estimated using data on cost of cultivation from seven major tur producing states- Madhya Pradesh, Maharashtra, Andhra Pradesh, Gujarat, Karnataka, Orissa and Uttar Pradesh. These states accounted more than 90 per cent of total area and more than 93 per cent of total production in the country. The state wise growth rate of output, input and TFP indices were presented in (Table 11). The TFP indices were at a dismal negative growth rate in almost all major tur producing states considered for analysis except in Karnataka which shows a TFP growth rate of 3.24 per cent. The input indices were much higher than the input indices in all states.

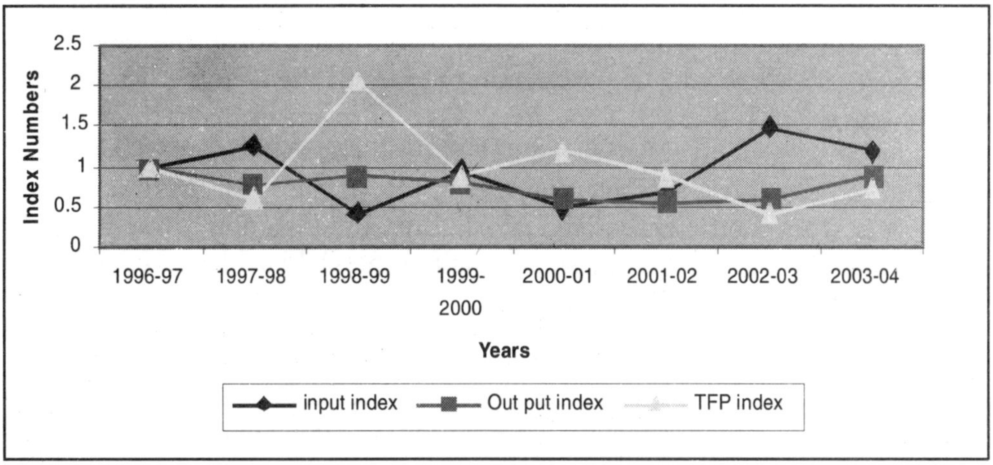

Figure 1. Fluctuations in output, input and TFP indices of tur in Madhya Pradesh

Table 11. State wise growth rates of TF of tur in major tur producing states

States	Period of analysis	Output index	Input index	TFP index
Madhya Pradesh	1996-97 to 2003-04	-7.06	3.06	-10.11
Maharashtra	1996-97 to 2003-04	6.39	10.76	-4.37
Andhra Pradesh	1999-2000 to 2003-04	15.07	13.08	-2.00
Gujarat	1996-97 to 2003-04	2.79	6.41	-3.62
Karnataka	1999-2000 to 2003-04	0.69	-2.55	3.24
Orissa	1996-97 to 2003-04	1.55	21.45	-19.89
Uttar Pradesh	1996-97 to 2003-04	-1.78	16.45	-18.23

It is found that except in Karnataka, all other tur producing states the TFP indices were always below the output and input indices and the decrease in TFP indices were apparent. But in the case of Karnataka the TFP indices were increasing apparently in all these years and the input index was in 0.38 the lowest in 2000-01 showing a negative growth rate of -2.55 per cent during these period. There are wider variations in TFP growth rates as the northern states like MP, Orissa and UP shows much more decline of which Orissa registered the highest of -18.89 per cent as there existed increased input use inefficiencies and lesser out put indices in these states.

Gram

The TFP in gram was estimated using data on cost of cultivation from three important gram producing states like Madhya Pradesh, Rajasthan and Uttar Pradesh for nine years from 1996-97. These states constitute more than 63 per cent of area as well as production of gram in the country. The results are presented in (Table 12). The TFP indices of gram in these three states shows negative growth rates as the out put indices were much lower than the input indices and it was even negative in Rajasthan. The result indicates that inputs were inefficiently used in Madhya Pradesh and Uttar Pradesh and in Rajasthan the yield was less than half of the yield achieved by MP and UP.

Table 12. State wise growth rates of TF of gram in major gram producing states

States	Period of analysis	Output index	Input index	TFP index
Madhya Pradesh	1996-97 to 2004-05	2.65	3.61	-0.96
Rajasthan	1996-97 to 2004-05	-0.70	0.59	-1.29
Uttar Pradesh	1996-97 to 2004-05	0.76	4.52	-3.76

The TFP indices were higher than the output and input indices in all these three states only in 1998-99. Especially after 2000, the gap between input and TFP indices was much higher in the states of Rajasthan and Uttar Pradesh as apparently visible. As the result the TFP growth rates showed higher negative values.

Ground Nut

The TFP in groundnut an important oil seeds legume was estimated using cost of cultivation data for Maharashtra, Andhra Pradesh, Gujarat, Karnataka and Tamil Nadu as these are important groundnut growing states contributing more than 85 per cent of the area and production of ground nut in the country.

The results were presented in (Table 13). In this crop, also the TFP growth rates are pitted at negative values over a period of 8 years from 1996-97. Andhra Pradesh is showing comparatively good result as the decrease in TFP growth was not much prominent than other states. The decline in TFP indices in Maharashtra and Tamil Nadu was more prominent. Input indices grow higher than the output indices at 5.79 and 3.89 per cent in these states since the labour expenditure incurred in these two states were higher than other states. The gross revenue from groundnut in Tamil Nadu was not prominent as the ground nut growers are not getting remunerative implicit price for their produces. Thus, the increase in input use inefficiency in Maharashtra and Tamil Nadu and the decline in out put indices in Gujarat and Karnataka resulted in prominent decline in TFP growth rates in all major groundnut producing states.

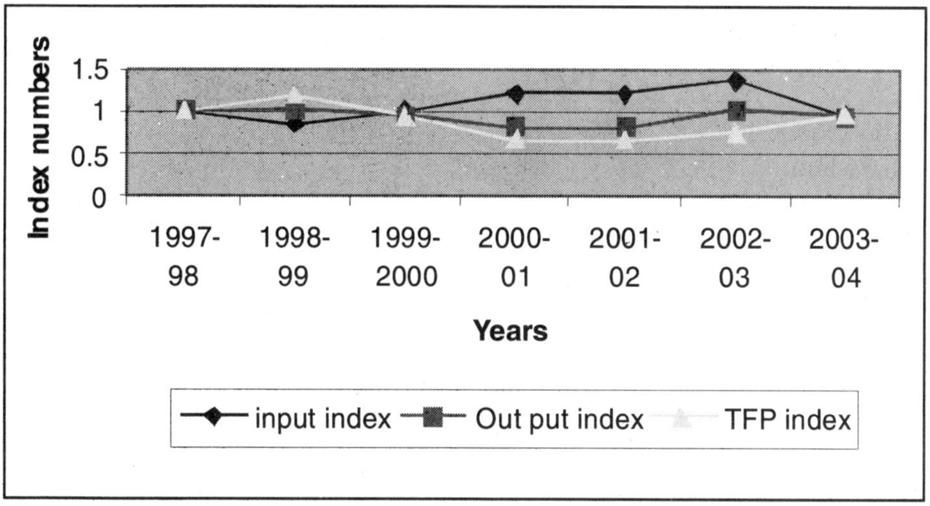

Figure 2. Fluctuations in output, input and TFP indices of groundnut in Tamil Nadu

Table 13. State wise growth rates of TF of ground nut in major ground nut producing states

States	Period of analysis	Output index	Input index	TFP index
Maharashtra	1996-97 to 2003-04	2.07	5.79	-3.72
Andhra Pradesh	1996-97 to 2003-04	-0.60	-0.17	-0.43
Gujarat	1996-97 to 2003-04	-0.48	1.19	-1.67
Karnataka	1998-99 to 2003-04	-8.01	-0.42	-7.59
Tamil Nadu	1997-98 to 2003-04	-0.98	3.89	-4.87

Cross Matrix (State and Crop) of TFP Growth Rates

The states which contribute major share in respect of area and production of important pulses and legumes were taken for analysis and discussion for the estimation of TFP. Only those crops for which cost of cultivation data are available continuously over a period of more than six years were taken for TFP analysis. Thus based on share in area and production and based on availability of cost of cultivation data, states such as Madhya Pradesh, Maharashtra, Andhra Pradesh, Gujarat, Karnataka, Orissa, Uttar Pradesh, Rajasthan and Tamil Nadu and crops like tur, gram and groundnut were only used for TFP analysis. The growth rates of TFP in these states for these crops were presented in (Table 14) as cross matrix.

Table 14. Cross matrix of TFP growth rates

States	TFP indices		
	Tur	Gram	Ground nut
Madhya Pradesh	-10.11	-0.96	
Maharashtra	-4.37		-3.72
Andhra Pradesh	-2.00		-0.43
Gujarat	-3.62		-1.67
Karnataka	3.24		-7.59
Orissa	-19.89		
Uttar Pradesh	-18.23	-3.76	
Rajasthan		-1.29	
Tamil Nadu			-4.87

Among the states, Karnataka is the only state registered a positive TFP growth of tur crop. All other states performed very poor registering negative growth rates in other crops also. For gram and ground nut, the TFP was increasing or stable in no one states. The TFP for gram in Madhya Pradesh and ground nut in Andhra Pradesh shows little better. The fall in productivity of gram in Madhya Pradesh and the rise in input use inefficiency in Andhra Pradesh during 2000-01 and later signals sounds of alarm. Higher input expenditure, labour expenditure (human, animal and machine) in particular and the lower implicit price of the produces played havoc in sustainability of the TFP index for ground nut in Tamil Nadu even though the state registered comparatively higher yield in ground nut.

CONCLUSIONS

Increasing the production and enhancing the profitability of pulses crops is the most needed work to stop the soaring import bill of pulses (22.32 million tones for Rs 3164 crores in 2001-02 and an expected 14.11 million tones in 2006-07). Increasing the production and enhancing the profitability hinges on the sustainability of TFP of pulses in the country and the major pulses growing states especially. The results of Tornquist-Theil index of TFP analysed in terms of growth rates and trends in TFP show continuous decline even after the 1990's.The rate of increase in input indices was much higher than the output indices in all major pulses growing states. Technology being a major contributing factor for TFP growth, greater R&D emphasis on pulses crops is needed. Other factors, which could make a set back on the present unsustainable behaviour of TFP, are the price parity of pulses with competing food crops, market support and irrigation, with pulses being rainfed crops.

There is significant variation in yield, cost of cultivation and profits of major pulse crops as the coefficient of variation (CV) for these factors found to be higher across regions. Thus, our policy focus may be tilted towards development of legume crops by way of developing irrigation infrastructure, access to institutional credit, supply of inputs, adoption of technologies particularly use of high yielding variety seeds, development of market infrastructure and adequate price supports. This would help in a big way to reduce the regional differences in farm profitability.

REFERENCES

Agricultural Statistics at a Glance, (2006). Directorate of Economics and Statistics, Ministry of Agriculture, Government of India, New Delhi.

Barnett, Vic, Roger Payne and Roy Steiner. (1995). Agricultural Sustainability: Economic, Environmental and Statistical Considerations, John Wiley & Sons, Baffins lane, England.

Capalbo, S.M, and Vo, T.T. (1988), A review of the Evidence on Agricultural Productivity and Aggregate Technology. In: Susan M Capalbo and John M antle (Eds.). Agricultural Productivity: Measurement and Explanation, Resources for the Future, Washington, D.C., USA.

Chandel, B.S. (2007). How Sustainable is the Total Factor Productivity of Oilseeds in India?, *Indian Journal of Agricultural Economics*, 62 (2) : 244-258.

Diewert, W.E. (1976). Exact and Superlative Index Numbers. *Journal of Econometrics.* 4(2):115-145.

Fabricant, S. (1959). Basic Facts on Productivity Change, Occasional Paper No. 63, National Bureau of Economic Research, New Delhi.

Fertilizer Statistics, 1999-2000. The Fertilizer Association of India, New Delhi.

Haque, T. (1996). Sustainability of Small holder Agriculture, Concept Publishing Company, New Delhi.

Haque, T. (2006). Resource Use Efficiency in Indian Agriculture. *Indian Journal of Agricultural Economics*, 61 (1) : 65-76.

Nineth Five Year Plan, 1997-2002, Planning Commission, Govt. of India.

Reports of the Commission for Agricultural Costs and Prices. Various issues. Department of Agriculture and Cooperation, Ministry of Agriculture, Government of India, New Delhi

Selvarajan, S., Agarwal, P.L., Pandey, S., Lansigan, F.P. and Bandapadhaya, S.K. (1997), System Approach for Analysing Tradeoff between Income, Risk, and Water Use in Rice- Wheat production in Northern India. *Field Crops Research*, 51 : 147-161.

Sen, Abhijit and Bhatia, M.S. (2004). Cost of Cultivation and Farm Income : State of the Indian Farmer, A Millennium Study, Vol.14, Department of Agriculture and Co-operation, Government of India, New Delhi.

Singh, J.P. (2006). Changing Agrarian Relationships in Rural India. *Indian Journal of Agricultural Economics*, 61 (1): 36-64.

□□□

Chapter 40

Recent Trends in Pulses in Tamil Nadu – An Overview

SUBRAMANIAN, A AND S. MURUGAN
Department of Agricultural Botany, Faculty of Agriculture
Annamalai University, Annamalainagar-608 002

Introduction

Pulses, the food legumes, have been grown by farmers since millennia, and these have contributed in providing nutritionally balanced food to the people of India. While pigeon pea, black gram, green gram, lablab bean, moth bean, and horse gram have definitely originated and domesticated in the Indian subcontinent, there is a probability that chickpea and lentil (Indian type) were also domesticated in the Indian subcontinent. Pea, grass pea, and cowpea were introduced in India millennia ago. Only faba bean was introduced in medieval times. This paper briefly describes about the recent trends in pulses in India and Tamil Nadu in particular.

Yields and Markets

Though substantial progress has been made in evolving techniques to obtain high yields of pulses, their production per hectare has remained the same for the last two centuries. Unfortunately, ancient documents give very little information on yields per unit area. This is due to many reasons, such as

(i) The measurement of land varied in different periods and in different regions;

(ii) Weight measurements were rarely done;

(iii) Volumetric measurements, such as number of bags per unit area, were common, but lacked uniformity; and

(iv) Commercial commodity farming was rarely the goal; the produce was consumed by the farming families and fed to the domestic animals, and only the surplus was marketed.

In addition to these reasons, a millennium of political turmoil in India, after 900 AD, discouraged farmers from making any effort to increase the productivity of land. It is only after the British gained political control of the Indian subcontinent, and aimed at exploiting the resources of the region to strengthen the empire, that we begin to see definite data on yields. Watt (1889) recorded the following information:

1. Uttar Pradesh
   a. Non-irrigated, 460–750 kg ha^{-1} (sole), 550–830 kg ha^{-1} (mixed);
   b. Irrigated, 1100 kg ha^{-1} (sole), 1300 kg ha-1 (mixed)
2. Central India: 1000 kg ha^{-1}
3. Gujarat: 1120 kg ha^{-1}
4. Maharashtra: 740 kg ha^{-1}

It is interesting to note that in Uttar Pradesh, yields of chickpea from a mixed crop were higher than the sole crop yields. One reason could be that chickpea used to be mixed usually with a cereal crop such as wheat that had received manure, which in turn could have benefited chickpea. On the other hand, yields of the sole crop were low because these were rarely manured... The Ain-i-Akbari, which was written around 1590 AD, gives interesting and useful information on market prices of chickpea (Blochmann, 1873). The kabuli chickpea cost twice as much as the desi, and it was 33% more costly than wheat. The kabuli chickpea and the green gram dhal were sold at the same price, thereby showing a high demand for the green gram dhal. Chickpea flour, because of value addition, was sold at par with wheat. The actual prices have not been given, since the units of both weight and money were very different in those times.

Storage

The pulses were stored in large pots, their borders and inner walls were smeared with oil, and ash was spread all around the pots (Risala-Dar-Falahat

c. 1400 AD;Majumdar, 1984). Apparently, ash and oil were commonly used by the Romans (Orlob, 1973), and the technique must have spread to India through West Asia, because the use of oil and ash to protect stored grain has not been made in any of the ancient Indian texts. During the Sultanic period (1206–1555 AD), grain was stored by mixing with pounded bones of elephants and also by placing leaves of pomegranate and Lactuca sp. with the grain, in a ratio of 1 part leaves to 100 parts grain (Naqvi, 1984). Again, there is no mention about the practice of placing leaves (e.g.,neem) so far in any of the ancient Indian texts. Pulses were stored in large pots with their borders smeared with oil, and ash "applied" on all sides (Majumdar, 1984).

Scenario in Tamil Nadu

Major Pulse Growing Districts in Tamil Nadu

This gives a fair idea for the policy makers while making any strategy to improve productivity and production in Tamil Nadu. Since Tamil Nadu is an idly and dosa state the importance of black gram need not be stressed. Such is the requirement of black gram that the prices in different markets in the state vary (from one market to another) much on a given day. The prices in the major producing districts and districts nearer to Andhra Pradesh state are lower than other districts. The contribution of delta districts in the production of pulses especially black gram and green gram is immense since it occupies a huge area in the delta districts (Thanjavur, Tiruvarur, Nagappattinam, Cuddalore, Tiruchi and Pudukkottai). It is to be noted that although black gram and green gram are grown throughout the year excepting in rainy months Tamil Nadu is a net importer of these from neighbouring states all through the year because these two pulses are grown only during January to May in the delta districts as rice fallow. The list of pulses and the major districts in which it is grown is given under.

Pulses	Major districts
Bengal Gram	Coimbatore, Dindigul
Red Gram	Vellore, Krishnagiri, Erode, Theni
Black Gram	Tirunelveli, Thoothukudi, Tiruvarur, Nagappattinam, Cuddalore
Green Gram	Tirunelveli, Thoothukudi, Tiruvarur, Nagappattinam, Tiruvallur, Namakkal
Horse Gram	Dharmapuri, Krishnagiri, Erode, Coimbatore
Cowpea	Coimbatore, Dindigul, Erode, Tirunelveli

Major Varieties Grown in Tamil Nadu

By going through the table given below,it will be interesting to note that the local varieties are found as a major variety in almost all pulse crops. This is due to the fact that the many released varieties of pulses in Tamil Nadu have originated from the local varieties. The common belief in official circles in Tamil Nadu is that varieties released in the past ten years are cultivated in huge area .However it is either the local varieties OR the yester year varieties (older varieties) occupy major area.

Pulses	Major varieties
Bengal Gram	CO 2, CO 3,Local
Red Gram	BSR1,SA1,CO6,Local
Black Gram	ADT 5,ADT 3, KM2, CO5, Vamban 2, Vamban 3, T9,TMV1,local
Green Gram	ADT 2,ADT 3, KM2, CO6, local
Horse Gram	CO 1 Local
Cowpea	CO 6, P152,Local

The PRICE Factor

The sky rocketing price of pulses during the second half of 2006 and frequent revisit of prices to a higher level after cooling down has made pulse crop economically a very viable preposition both for consumption and as well for remunerative purpose for the farmers of TN. The table below indicates the price pattern in the pulses for two years. In fact during last two years the price of black gram seed for sowing purpose during rice fallow season went up to Rs100/kg as against Rs.25 to Rs.30 per kg sold normally during that period. The area under pulses also showed a sudden spurt in all the districts in the past two years because of the price.

Month	Bengal Gram	Red Gram	Black Gram	Green Gram
July '05	2299	2154	2295	2457
Nov '05	2541	2227	2914	2887
Feb '06	2476	2050	2901	2870
Jun '06	2758	2177	3786	3579
2005-06 Average	2493	2164	2945	2881
2004-05 Average	2166	2064	2050	2114
% Var. over previous year	15.06	4.84	43.62	36.3

The Consumer Factor

If you find the tastes of idly, dosa and vada in hotels of Tamil Nadu varying much to your liking it may not have anything to do with their chef masters. It is mainly due to the imported black gram from Burma. The quality of black gram is very important in the tastes of idly, vada and dosa. Black gram has also a lot do with the taste in traditional tamil savouries like murukku, seedai, thattai etc. However, it has been found off late in the trade circles that the imported black gram does not suit to the traditional tastes in these items.

Seed Production and Marketing Issues

- Non availability of pure seeds for multiplication hinders the process of quality seed multiplication.
- Urgency for seeds immediately after rains by the farmers makes it difficult for the seed producers to meet the required quantity and supply in time.
- Yellow Mosaic Virus disease not only affects the yield in commercial plots but also the seed production plot yields .
- Seed Cycle of certain pulse varieties makes it difficult to keep in storage for a very long period. The pulse varieties grown in delta districts are sown only in January to March. These pulses are again sown only during the succeeding year. Since the varieties are very specific to delta region these could neither be taken up in other area also.
- The policy to import grains to meet the domestic consumption requirement (especially black gram) by the government some times find way to the sowing fields as seeds .They have poor seed standards and the field standards are miserable (bold type seed, vegetative ,no reproductive phase,poor yields etc).They also enter into the seed cycle by admixtures with other varieties.
- Inadequate price for quality seeds is another sensitive issue for a seed producer to take up pulses in their portfolio. Farmers normally have a tendency to buy seed from the mundis/market .They buy grains for seed purpose from mundis at a 25% premium to grain price which will be equal to 50% of the cost of seed from the seed dealer for a kg of seed. Despite after 40 years of rigorous extension service farmers still languish in mundis instead of seed dealers whenever there is seed requirement.
- Photo sensitivity nature of some of the varieties add trouble to the seed producers and often varieties tend to exhibit different expressions in

different locations for their vegetative and reproductive phases in the same season (e-g black soil vs red soil, southern districts *Vs* delta districts).

- In the absence of effective storage pests, the ability of the seed producer to keep it for longer period becomes a herculean task.

The FACT File

Very often one finds on the wrong foot while dealing with some of the talks (common beliefs) that go around in the world. Only after sometimes it will be construed as myths. In pulses also similar myths are found .Down under are some of the common beliefs (myths) and the corresponding facts are enlisted below

MYTHS	FACTS
• Black gram consumed in Tamil Nadu is produced in Tamil Nadu and other states	• Most of the black gram produced in Tamil Nadu is exported to NRI living countries at premium prices and what is consumed is (partly) imported from Burma
• Certified seeds are planted in 15% of total acreage	• Certified seeds occupy just 2-5% of the total planted area
• Seeds are supplied from within TN	• AP plays key role in seed supply especially T9 black gram and some private green gram verities
• Higher Purity levels in the field for certified seeds	• Purity levels are 60-70% only.
• Dealers, agri depots are major seed sellers.	• Mundis are the major selling points
• Latest varieties occupy major area	• Older varieties like T9, TMV 1 (Black gram), SA1 (Red gram), KM2 (Green gram) and local varieties occupy lion's share.
• Higher Yield estimates in the official records	• Actual yields are lower than estimated yields.
• Adoption of latest technologies is high	• Adoption of latest technologies is dismally low
• Cost of cultivation of pulses is less	• The major labour force in industrial places in Tamil Nadu (Chennai, Coimbatore, Tirupur, Erode and Karur) are from southern and delta districts where incidentally pulses are grown in large acreage. This has resulted in increasedcost of cultivation of pulses also.
• Yield levels of pulses have increased after green revolution	• Yield levels of pulses reported since the 18th century have not been differentfrom the yield levels being obtained today

Hopeful Solutions

The main aim of this paper is to review and bring out all aspects related to pulses .Starting from sowing seed to commercial production, seed production, storage, marketing ,consumer taste ,varieties ,role of environment and so on without clinging on to one or two aspects.We try to give some suggestions which may be of use to the policy makers.

- Pulses have been and will continue to be an important ingredient in the daily food and nutrition of the people of the Indian subcontinent.
- The research on better storage has not been impressive on practical terms in the past three decades. Effective and efficient methods OR developing newer chemicals / organic materials would be the most important aspect in pulses.
- The "Green" Revolution in cereals and some other crops, characterized by a quantum jump resulting from high levels of management with purchased inputs and irrigation, is not required for pulses. This is because pulses will always remain a 'secondary' food item. Export opportunities being limited, an excess production could result in a market glut and loss to farmers. It is much more sensible to talk about increasing the average yields of pulses to 1275 kg ha^{-1} to meet the needs of the Indian population by 2050 (Nene, 2000).
- Seed coating technology of both commercial pulses and sowing seed for longer storage can mitigate many of the problems
- Development of location / season specific photo insensitive varieties will be key to success.
- Varieties suitable for mechanized farming may be order of the day with prohibitive labour costs in taking up pulses.
- Market Linkage of the farmer with end user will make the farmers reassured of their net returns. In case of falling market prices distress selling could be avoided.
- Biotechnology tools could be the answer to long-standing problems such as Ascochyta blight, pod borers, and sensitivity to salinity. Pulses Improvement in management with regionally adapted and preferred varieties is the key. The focus should be primarily on rainfed pulses.
- Eleven different pulses are grown and eaten in South Asia. In addition, 10 grain legume species were consumed during the past famines (Nene, 2004). These 10 species should be researched as additional crops.

REFERENCES

Blochmann, H. (1873). The Ain-i-Akbari by Abul Fazl Allami. Book I. The Asiatic Society, Calcutta 700 016.

Majumdar, M. (1984). Risala-Dar-Falahat. *Indian Journal of History of Science* 19:341–360.

Naqvi, H.K. (1984). Cultivation under the Sultans of Delhi c. 1206–1555. Indian Journal of History of Science 19:329–340.

Nene, Y.L. (2000). A new vision of pulse pathology research in India. *Indian Journal of Pulses Research* 13(1):1–10.

Nene, Y.L. (2004). Plant species utilized as food during famines and their relevance today. *Asian Agri-History* 8:267–278.

Orlob, G.B. (1973). Ancient and medieval plant pathology. *Pflanzen-schutz-Nachrichten Bayer* 26:65–294.

Watt, G. (1889). A Dictionary of Economic Products of India. Cosmo Publications, Delhi, India. (Reprinted 1972).

□□□

Chapter 41

Problems Faced by Women Agricultural Labourers in Groundnut Farming Operations

BHARATHIRAJA, P. AND SANTHA GOVIND
Dept. of Agricultural Extension, Annamalai University

Introduction

The role of agricultural labourers in the farm front is highly indispensable for carrying out all the operations related to farming. Indian rural women always play an important role and remain as prominent partners of agricultural sector.

International Labour Organization Statistics (2001), shows that women constitute 47.00 per cent of the agricultural work force in India which gives highest degree of female participation in agriculture. About 84.00 per cent of the economically active women are engaged in agriculture and allied activities which include 48.00 per cent as farm women and 46.00 per cent in agricultural labour force.

Chiranjib (1990) stated that in Tamilnadu, 81.50 per cent of the total female workers are engaged in agriculture and 60-70 per cent of labour input is provided by women. Only when women's activities are properly recognized, classified and reported, can it be expected that government organizations and development planners will have an awareness and try to bring women into mainstream of agriculture. Women labourers' participation in crop production is not marginal or insignificant, but is crucial for crop production.

Karl (1995) believed that lack of accurate information on womens' role in agricultural production has resulted in the failure of development programmes and projects because they could not address the needs of farm women.

Women agricultural labourers face several problems while performing various farm operations. Hence, there is a strong need to identify the problems confronted by the women agricultural labourers so as to take certain corrective measures to safeguard the women labourers engaged in agricultural activities. Keeping this in mind, the present study was taken up to identify the problems faced by women agricultural labourers in performing groundnut farming operations.

Materials and Methodology

Villupuram district of Tamilnadu was purposively selected for the present study as it ranked second with 2,02,190 women agricultural labourers. Kallakurchi block of kallakurichi Taluk was selected as it had the maximum number of women agricultural labourers engaged in groundnut farming operations. In order to get sufficient respondents, the first seven villages with maximum number of women agriculture labourers were selected.

A sample size of 120 landless women agricultural labourers was considered to be sufficient for the study. The number of respondents to be selected from each village was arrived at on the basis of proportionate random sampling technique.

Findings and Discussion

The women agricultural labourers were asked to mention the problems faced in participating in groundnut farming activities and the results were ranked based on percentage analysis and the results are presented in Table 1.

Table 1 shows the problems faced by women agricultural labourers while performing the field operations in groundnut farming activities. The major problems faced by cent per cent of women agricultural labourers was that manual weeding caused hip and back pain. The other major problems faced by most of the respondents were that harvesting and transporting caused neck and body pain (97.50 per cent) and groundnut stripping caused hand, neck and hip pain (95.83 per cent). Next major problems confronted were poor overall health due to heavy work load (85.00 per cent). This finding is in line with the findings of Padmavathi (2002).

Table 1. Problems faced by women agricultural labourers in groundnut farming activities.

S.No	Problems	Number	Per cent	Rank
1.	Manual weeding caused hip and back pain	120	100.00	I
2.	Harvesting and transporting caused neck and body pain	117	97.50	II
3.	Groundnut stripping caused hand, neck and hip pain	115	95.83	III
4.	Poor overall health due to heavy work load	102	85.00	IV
5.	Sowing of groundnut seeds caused backache	54	45.00	V
6.	Headache and heat stroke due to working under harsh sun	47	39.16	VI
7.	Seasonal nature of employment in farm activities	39	32.50	VII
8	Discrimination of wage structure in farm	36	30.00	VIII
9.	Hand decortications caused boils and burning in hand	27	22.50	IX
10.	Nasal problem due to inhalation of manure dust	27	22.50	X
11.	Hand operated tools caused physical strain	25	20.83	XI
12.	Spreading manure caused backache	17	14.16	XII
13.	Respiratory allergies	15	12.50	XIII

They further expressed that sowing of groundnut seeds caused back ache (45.00 per cent), head ache and heat stroke due to working under harsh sun (39.16 per cent) and discrimination of wage structure in farm (30.00 per cent).

They felt that groundnut hand decortications caused boils and burning in hand (22.50 per cent), nasal problem due to inhalation of manure dust (22.50 per cent) and hand operation of tools caused physical strain (20.83 per cent). A lower percentage of women agricultural labourers were of the opinion that spreading of manure caused back ache (14.46 per cent) and respiratory allergies (12.50 per cent). Seasonal nature of employment in farm activities (32.50per cent) affected their very survival because of restriction of number of working days.

Participatory research may be initiated on the development of appropriate and drudgery reducing farm tools based on the design and specifications of indigenous tools. Extension programmes aimed at popularization of newly developed women friendly and drudgery reducing tools among women agricultural labourers may be initiated at all levels.

REFERENCES

Anonymous. (2001). The World Employment and World Labour Report. Available at http://www.ilo.org.

Chiranjib, (1990). Female Labour Participation – A Micro level Study, Economic and Political Weekly, 24(2): 24.

Karl, M. (1995). Women and Empowerment- Participation and Decision Making. Zed Books Limited, London.

Padmavathi, M. (2002). Participation and Knowledge of Farm Women Labourers in Rice Farming Operations. Unpublished Ph.D. Thesis, University of Agricultural Sciences, Bangalore.

□□□